DOCUMENT DESIGN

From Process to Product
in Professional Communication

Second Edition

Miles A. Kimball and
Derek G. Ross

With
Ann Hawkins

Cover credit: *Composition with Red, Yellow, and Blue*, Piet Mondrian (1927).
The Cleveland Museum of Art, https://www.clevelandart.org/art/1967.215.

Published by State University of New York Press, Albany

EU GPSR Authorised Representative:
Logos Europe, 9 rue Nicolas Poussin, 17000, La Rochelle, France
contact@logoseurope.eu

For information, contact State University of New York Press, Albany, NY
www.sunypress.edu

Library of Congress Cataloging-in-Publication Data

Names: Kimball, Miles A., author | Ross, Derek G., author | Hawkins, Ann R., contributor.
Title: Document design, second edition : from process to product in professional
    communication / Miles A. Kimball, Derek G. Ross, with Ann R. Hawkins.
Description: Albany : State University of New York Press, [2025] | Includes
    bibliographical references and index.
Identifiers: ISBN 9798855801576 (pbk. : alk. paper) | ISBN 9798855801569 (ebook)
Further information is available at the Library of Congress.

CONTENTS

UNIT 2: PROCESSES

UNIT 3: PRACTICES

PREFACE

To become successful professionals today, students must learn to communicate visually just as well as they learn to write. Whether students go on to work as document designers, web designers, interaction designers, information designers, or even user experience designers, many of the paths for technical communicators focus squarely on visual communication and require a working knowledge of digital tools and formats. Given the complexities of both visual and digital work environments, understanding the fundamentals of document design is more important than ever. Even students who plan careers in business, medicine, industry, or government will find themselves communicating visually through documents in their workplaces. *Document Design, Second Edition: From Process to Product in Professional Communication* introduces students to the basic principles and theories of design, combining practical advice about the design process with a foundation in visual rhetoric and usability. In this preface, we lay out a few of the considerations that form our approach to teaching document design and to writing this book.

A BALANCE OF THEORY AND PRACTICE

Most books on document design lean toward either theory or practice. We offer a balanced approach—theoretically informed practice—that introduces a working vocabulary to help students become reflective practitioners, able not only to create effective and usable designs but also to explain why and how they made their design choices.

The book's structure works to reinforce this balance. Unit 1: Principles introduces students to the practical terms, principles, and theories at the heart of good design. These concepts are then integrated throughout the book in discussions of the choices that designers make as well as in analysis of the sample documents. The theoretical lenses—visual perception, visual culture, and visual rhetoric—are also carried throughout unit 2: Processes, in brief introductions to each chapter's topic. Each chapter in unit 2 focuses on a single important aspect of document design, including format, page layout, typography, graphics, and color. A final chapter in unit 2 guides students in designing those indispensable parts of many technical documents: lists, tables, and forms. Unit 3: Practices will help students succeed in some of the more practical aspects of professional document design, including managing projects strategically, from user and needs analysis to invention, prototyping, and usability testing. Unit 3 also includes a chapter on print production, covering the vocabulary students will need to speak effectively with professional printers.

In this book, we hope to give students the foundation they need to make decisions like designers in any rhetorical situation. Students will learn to negotiate between the needs and desires of both users and clients. Not every design situation calls for a beautiful and intricate document design. In some situations, a cheap and fast document will meet the needs of both clients and users more effectively than something more visually sophisticated. In others, an impressive and expensive design is absolutely essential for the document to fulfill its purposes. *Document Design, Second Edition: From Process to Product in Professional Communication* will help students make choices about what level of design investment is appropriate for the situation at hand given the time, money, and resources available, as well as the client's goals and the users' needs.

ADVICE ABOUT DIGITAL TOOLS, PROCESSES, AND FORMATS

The impact of digital technologies on document design is profound. Whether creating websites or more traditional paper documents, document designers today must work through digital technologies. And more often than not, designs must work in multiple media and formats. For example, the designers of the Amtrak schedules in chapter 1 needed to create information that could be delivered simultaneously in print, on the web, and in PDF.

Recognizing that almost all documents today are at least born digitally, or digitally mediated in some way, *Document Design* helps students understand the dynamics and possibilities of both screen and print media. We offer accessible advice about using technology throughout the design process, such as how to choose typefaces for the screen, how to manipulate bitmap and vector graphics, when to use RGB or CMYK color, and how to use color pickers. We discuss both print and screen media because we want to help document designers make practical, theoretically informed choices about what medium or mixture of media to use for the particular design projects they face.

However, this book is not a book on web design, of which there are many excellent examples on the market. We firmly believe that non-electronic documents—often paper, but in a myriad of other mediums as well—will be with us for a long time to come. As technologies, paper and printing form a convenient, usable, and economical medium for conveying technical information in many situations. To give students a range of flexible design skills necessary for them to succeed professionally, *Document Design* discusses how to create designs for print production through digital tools.

Learning design with a focus on physical output has other aspects to recommend it as well: if you learn to design for print, you've learned many skills that translate well into digital realms. There is also the immediacy of physical design—students and teachers can engage with the products of their labor in a way that encourages play and remediation that digital technologies can't always do in the classroom. Documents may mostly be born digital, but screens still don't do complex folds very well.

ACCESSIBILITY IS NOT AN AFTERTHOUGHT

You'll notice that there is not a specific section in any one chapter of this book that talks directly about accessibility (other than this one). That's because we've woven the idea of accessibility in design throughout this entire text. Generally speaking, accessibility is the consideration of how users interact with documents to ensure that issues of (dis)ability or access—defined broadly and including physical, technological, and economic aspects—do not negatively impact the ability to work with a document as needed. Audience, purpose, and context should be a designer's watchwords—know who might need to use what you design, what your design needs to accomplish, and the contexts in which that document might find use. You'll see this addressed from chapter 1, where we introduce the idea of document design as information design, to chapter 11, where we talk about being aware of the file types your printer might need. Along the way, you'll learn about the importance of contrast, of appropriate image use and Alt tagging images for online environments, of choosing typefaces that increase both legibility and readability in certain situations, and more. You'll start to think about contexts of use—if you're designing a document that's going to be used by first responders in potentially dark or wet environments, for example, what does that document need to do? What technological limitations might apply? Accessibility is embedded in the very idea of human-centered design.

ACCESSIBILITY IN ONLINE ENVIRONMENTS

In primarily online environments, accessibility is typically defined as ensuring that those with disabilities can use the web in a complex sense—not just perceiving content but being able to fully interact and contribute as effectively as those with other abilities (Petrie et al., 2015). Doing that sort of design work is specific to online environments as it involves using active digital technologies to mediate information in complex, adaptive ways. For example, a document born digital and living digitally should, to list just a few guidelines,

- Ensure that color is not the only way of distinguishing content

- Be fully marked up (using the editing language of the internet, such as HTML [Hypertext Markup Language]) so that everything from headings and tables to input fields are engageable via multiple technologies (such as screen readers)

- Allow customization (resizing, voice-over, etc.) that preserves the intent of the design (see Zahra, 2019)

You might consider familiarizing yourself with accessibility principles, such as those offered by the World Wide Web Consortium's Web Accessibility Initiative (WC3's WAI). Doing so will only make you a more effective designer, one more capable of thinking through not only practical aspects of design but also legal and ethical issues (see, e.g., Youngblood, 2013).

ACCESSIBILITY IN GENERAL

It should be immediately obvious that "accessibility in online environments" and "accessibility in general" are different things. A lot of this book deals with design from the physical perspective—we'll be working with everything from liquid inks to folded paper. And for that level of physical accessibility, you need to put your audience first, thinking always (again) of audience, purpose and context. As you work—in your dealings with clients, in your prototyping, even in deciding what file types to share with others and what types of paper or inks might be best for a job—consider always who will be using your work, how they will be using your work, and where they will be using your work. Consider your text size and line length (chapter 5) to determine optimal characteristics for the document you are designing. Create a meaningful typographic system (chapter 6). Ensure that any graphics you use are relevant to the task at hand (chapter 7), and that color, if used, conveys the meaning you intend and emphasizes what you think it should emphasize (chapter 8). And so on. Learn as much as possible as you can about these elements before you even pick up a pencil or open a design program (see chapter 10), and you'll be well on your way to creating documents that don't just look pretty but actually do what they need to do for those that need to use them.

A BOOK THAT SHOWS THE PRINCIPLES OF GOOD DESIGN IN ACTION

We are very proud of the careful design and the variety of real-world examples we use throughout the book. Many of the examples were culled from our own surroundings at home, at work, and in our travels. Rather than including the snazziest documents or creating artificial ones, we have tried to find the real examples that best illustrate the principle we are discussing and present them as we found them. We hope this approach helps students recognize the documents that permeate their own lives.

However, we recognize that some concepts need good visualizations for students to understand, so the book also includes line art to help students see important ideas about design. We have designed the line art to be as minimalistic and clear as possible. In a number of cases, line art accompanies the scanned examples and helps explain to students what they are seeing.

We believe this book practices the principles it preaches.

In addition to sample documents and illustrations, the book includes several helpful reference features:

- *Design Tips* throughout the text offer practical advice for students to apply to their own designs.

- *More about . . .* boxes delve deeper into topics related to chapter discussions to give students a broader understanding of the history and practice of document design.

- Full-color images are integrated into chapter 8, "Color," to illustrate color principles and tools, with several sample documents.

ON THIS EDITION

While an immense amount of work has gone into creating this version of the text, it can still be considered a second edition of the original *Document Design*. So much of the original text remains relevant that entire swaths of the document remain as they appeared in the original. As this is meant to be a pragmatic, introductory text, many of the works cited in the original remain relevant, as do the examples. We've updated images and websites, but where production date didn't matter, we've often left the original examples for a number of reasons. For one, when the intent of an image is to show good contrast, or how headings might appear on a page, it doesn't really matter if the example is from 1806, 2006, or 2026. For another, we tried to think of those users who have been using the original text since it came out: first, thank you!, and second, what you'll find here should be familiar, just freshened up.

Though there are significant changes throughout, those changes often reflect increased knowledge and understanding of audiences, purposes, and contexts; of the changing understanding of what it means to be human and do design. But though much has changed, and we've made great progress in technology and theory, a lot is as it ever was—similarity, enclosure, alignment, contrast, order, and proximity still matter. The history of type has not changed, though we now have access to more typefaces than ever, and how we understand that history might have altered. Image production has changed, somewhat, but not for everything. Color is still color, and though production models have changed (in some ways), a lot remains relevant in how we design those documents destined for eventual production. So what you will find here will feel very familiar to those of you who used that first book, and we hope that this edition serves you just as well.

You'll notice that many of the images look *almost* the same as in the first edition. We've redone every image in the book, from line drawings to photographs, and used the original physical documents where possible (and, as discussed, where still relevant). Some of those original documents are more worn than they were the first time around, but we decided that this is a book about documents, and when we're designing physical documents, they wear. They also have dimension. They reflect light. They fold. And so on. Where we can, we preserved the look of physical documents *being* physical documents—designers learning to design, we thought, should be able to see their own future work reflected (foreshadowed?) in examples from the past.

Where physical documents are concerned, you will find examples from the first book that, at first blush, may seem overly outdated. Perforated mailers? Physical order forms? Our thinking is that designers should know what is possible, as what was done in the past might inspire the present. Will one of us ever need to design a mailer with a tear-off response card? Maybe not, but then again, physical documents have a persistence and power that is unmistakable—they also have the decidedly egalitarian value of requiring no power (or internet) for interaction. Paper documents work even when the power is out, making them invaluable in crisis situations, and require no technological infrastructure to *use* (production and delivery aside). Giving someone a physical document ensures interaction, or at least ensures the *possibility* of interaction, in a way that giving someone a URL might not. That said, we spend plenty of time thinking through designs in electronic environments as well—this is, after all, our world.

Our purpose in writing this book, and in continuing to improve it, is to offer a comprehensive introductory text to the world of information design, particularly as it applies to document design. In each chapter, we offer a few suggestions for further reading—those suggestions are but the tip of an exceptionally large iceberg, and we sincerely hope that you find something in this book to become inspired about, making this just the first stop on your design journey.

ACKNOWLEDGMENTS

This edition of the book only exists because of the book that came before, and so we again thank the many thorough reviewers of that text, including Diana Ashe, University of North Carolina Wilmington; Susan Codone, Mercer University; Gordon Scott Gehrs, Illinois Institute of Technology; Craig J. Hansen, Metropolitan State University; Suguru Ishizaki, Carnegie Mellon University; Carolyn Rude, Virginia Polytechnic Institute and State University; Jennifer Sheppard, New Mexico State University; and Ronald Shook, Utah State University.

We deeply thank those reviewers who gave their time to help us through *this* version.

We thank those involved in the original project, including Leasa Burton, Joanna Lee, Emily Berleth, Ryan Sullivan, Nancy Benjamin, Andrew Ensor; Claire Seng-Niemoeller, Anna Palchik, Martha Friedman, Helane Prottas, Kim Cevoli, Cate Kashem, Karita dos Santos, Nikkole Meimbresse, Joan Feinberg, Denise Wydra, and Karen Henry.

This edition of the book would not have been possible without a few wonderful people: Norman Ed Youngblood, the wonderful folks at the Auburn University Library (especially Jaena Alabi and Greg Schmidt), Jen Ross, and Ann Hawkins contributed in so many vital ways that allowed this text to live.

We offer a very, very special thanks to Lynn Strong. Without her unceasing work, a great many of the images you see here would not exist. From photographing hundreds of documents to creating outlines to seeking permissions, feedback on designs, and endless conversations, she has truly been integral to this work.

We thank our production team at SUNY, including Richard Carlin, Timothy Stookesberry, Caitlin Bean, Carly Miller, and everyone else who had a hand in bringing this text to life.

And we'd be remiss without thanking Albus and Nico. Good dogs.

—Derek G. Ross and Miles A. Kimball

REFERENCES

Petrie, H., Savva, A., & Power, C. (2015). Towards a unified definition of web accessibility. *Proceedings of the 12th International Web for All Conference* (pp. 1–13). https://doi.org/10.1145/2745555.2746653

Youngblood, N. E. (2013). Integrating usability and accessibility into the interactive media and communication curriculum. *Global Media Journal, 12*(23), 1.

Zahra, S. A. (2019). Accessibility principles. *Web Accessibility Initiative (WC3).* https://www.w3.org/WAI/fundamentals/accessibility-principles/

UNIT 1: PRINCIPLES

1

WHAT IS DOCUMENT DESIGN?

LEARNING OBJECTIVES

Upon completing this chapter, you will be able to:

- Discuss what a document is

- Discuss the role of a designer in creating documents

- Understand the complex interactions between client, user, and designer

- Explain and discuss the various needs and expectations of document users

- Articulate levels of design and when to employ them

∾

Take a look around and notice the documents in your life. You will probably discover more than you anticipated—in your office, your desk drawer, your car, your wallet. Documents, from the US Constitution to the manuals for your technology, surround you every day, guiding and shaping your interactions with the world. It's also hard to imagine, but you're very likely *wearing* some documents right now (see figure 1.1).

Figure 1.1. A document you wear. Clothing tags are a humble but intimate example of how document design affects people's lives. This document is designed not only for clarity (we know where the clothing was made, its fabric type and content, how to launder the clothing, the vendor's Wool Product Label [WPL] number, and the style number) but for comfort and durability. It needs to work well both while being worn and after being washed! *Source*: Photo by the author.

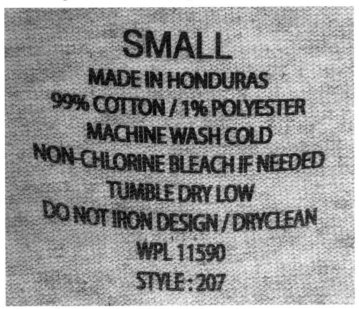

SOCIETY OF THE TEXT

We're not just talking about traditional technical documents in this book, although they have their uses and impact:

- Reports chronicle the recent and distant past.

- Instructions guide us in terms of the present.

- Proposals help us shape the future.

It's no exaggeration to say that documents are the foundation of organized society, a practical mechanism for mediating human action within and among communities, nations, and cultures.

Documents help us interact with technology by giving us clear instructions on how to use computers, equipment, and vehicles. We design not only passive documents, but active ones: moving map displays, for example, help us travel by supplying information about where to go, how to get there, and what to do when we get there.

And even passive documents do a job: they stand ready to speak, to declare themselves when the time comes. And finally, the human body becomes a document itself as we write upon it—not just for decoration, but for information (see figure 1.2)

Ironically, documents are so ubiquitous that at times they seem to be invisible—just a part of the landscape of human society. But if we think about them, it is easy to see how tightly documents are entwined throughout our daily lives. And if you look carefully, you'll find technical documents everywhere around you. Many more technical documents have likely been produced over the centuries than novels, plays, or poems.

Documents like the clothing tag in figure 1.1 do a small but necessary job. They help us as consumers to know whether the clothes will fit, what materials the clothes are made of, where the clothes came from, and how to care for them. Clothing tags also help manufacturers to track clothing models and styles and to advertise their products. Many labels even appear on the outside of the clothes, socially branding them

Figure 1.2. Picture of a no-MRI tattoo, aimed at notifying doctors or emergency response people that the patient has metal within their body and should not be subjected to magnetic resonance imaging. *Source*: Atom Bomb Studio, Las Cruces, NM.

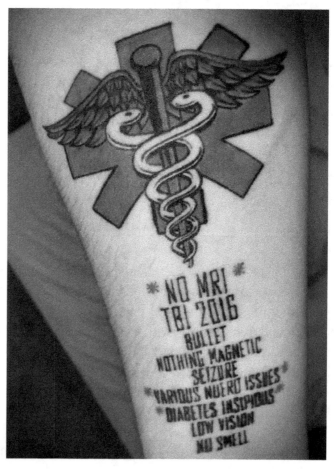

while helping the manufacturer to sell even more clothes. Even just a simplified company logo, absent any sizing or care instructions, does a lot of work: in many cases, the difference in what constitutes desirable clothing for some is not the material that makes up the clothing, but the branding itself. So although they may be a humble example of documents, clothing tags must be usable, readable, understandable, comfortable, and convincing to be effective.

Even designing something as simple as a clothing tag requires dealing with some difficult challenges. How do you display essential information clearly in such a small space? How can you communicate effectively with people who speak many different languages? How do you design a document that is both comfortable to wear and easy to understand? And finally, how do you design a document that can survive the washer and dryer—not just once, but many times?

In this book, you will practice some of the techniques designers use to solve the challenges of successful document design. You will learn essential skills for communicating in the complex, intercultural world of technology and community. And you will learn how to analyze human communication situations so you can create documents that solve human problems. You'll find not only practical techniques and helpful information but also important theories about how to communicate through documents. Having a theoretical awareness as well as practical skills will help you make smart decisions about the documents you design, whatever the situation.

In the rest of this introduction, we'll dig a bit further into the question of exactly what document design is. This will help you understand the practical and theoretical information that comes later.

DOCUMENT DESIGN AS INFORMATION DESIGN

Document design is one part of the broader activity of *information design (ID)*, or *communication design*. These areas of study and practice broadly involve presenting information in the most ideal way for a specific audience, purpose, and context. An information designer or communication designer takes into account who will be using information, what the information needs to do, when users will use it, why they need the information in the first place, and where that information is needed, or where that information needs to be accessed. These are the classic journalistic questions, but reimagined so that "the five *W*'s" can be applied to user-centered design (see Hart, 1996). It wouldn't make any sense, for example, to develop a dive computer manual printed on lightweight, water-soluble paper, when divers need the complex instructions underwater. Printing that same booklet on lightweight sheets of plastic mounted in a wrist or forearm holder, however, would make the instructions usable where they are needed most.

Robert E. Horn (1999) has defined ID as "the art and science of preparing information so that it can be used by human beings with efficiency and effectiveness" (15). Noted communication scholar Saul Carliner has suggested that this distinction works on three levels:

- *Physical:* ID helps users find and use information by providing a meaningful visual and physical design to communication products, including the design of pages and screens, writing, and production.

- *Cognitive:* ID helps users understand information by structuring and presenting it in ways that help solve problems.

- *Affective:* ID motivates users by gaining their attention and convincing them to act. (45)

What these definitions share is a sense that ID involves more than just creating pretty packaging for information. The scope of ID is more profound: it deals with the relationships among people who create the information, those who use the information, and people's cultures, societies, and environment. Information designers create artifacts that define and build these relationships by helping people solve real problems they face while recognizing the potential affordances or limitations of where they will be accessing information.

In this sense, information design is as old as writing, or even older. The first person who carved an arrow into a tree trunk to mark a path for travelers was an information designer. So is a member of a team today who creates a complex website for customers of an international corporation who need to access purchase orders or shipping records. Both took up the task of making information "clear and accessible to audiences." Both used tools and systems of understanding, like writing and visual design, to convey meaning so it can be used by other people to solve specific problems in specific situations.

Document designers work to understand the problems and situations of information users and information providers, then craft documents that help solve those problems within those situations. They also constantly assess whether the products of their efforts are working successfully and try to improve the quality of their information products.

As you will see as we move through this text, however, though we often prefer easy delineations—a document vs. an interactive interface vs. a three-dimensional space, for example—the "documents" we create aren't just limited to what you might initially think. Part of what makes a good information designer is the ability to creatively assess contexts and make informed decisions about what needs to happen to most effectively communicate the desired information.

THE DOCUMENT IN DOCUMENT DESIGN

So document designers design documents, which apparently are everywhere. But what exactly are documents?

In previous centuries, it might have been easier to answer that question: documents were meaningful marks inscribed, incised, pressed, painted, carved, or scratched on something, whether it was stone, clay, leaves, bark, wood, leather, or paper. These marks might be handwriting, numbers, pictures, or all three, but together marks and surfaces were combined to preserve and convey meaning. Documents of this sort include everything from simple records (receipts, bills of sale, deeds, birth, death, and marriage certificates) to more complex ones (letters, memos, reports, proposals, instructions, documentaries, manifestos, regulations, laws) to great works of art (novels, poems, plays, films, music). In a chaotic and changing world, we still use such tangible documents to fix information so it can be retrieved for later use. Documents like these are the glue that holds together most societies today. They are the objects we use to create more stable relationships between citizens and governments, accusers and defendants, producers and consumers, buyers and sellers, employers and employees, students and teachers.

But the rapid expansion of electronic media has made everybody rethink this earlier definition of *document*. Does the word still mean anything when the digital networks that support something like a website are so changeable? Are websites and other electronic forms of communication documents? Is this text any less a document if you are reading it electronically rather than on paper?

Many have argued that because the web is so dynamic and fluid, it does not convey "documents" so much as rapidly changing, rearranged, and repurposed "content." David M. Levy (2001) has countered this idea by emphasizing the similarities of all documents, paper or electronic: "Attractive and popular though it may be, this understanding of the difference between paper and digital forms is, quite simply, wrong. What it fails to grasp is that paper documents—and indeed all documents—are static and changing, fixed and fluid. It also fails to see the importance of fixity in the digital world. There is a reason why text and graphics editors have a Save button, after all" (36). Levy's comments suggest that although electronic documents might collect millions of bits of information on the fly, these random pieces of information are still assembled for a user to see, in a human-sized scope. At that moment, a document is whatever is available and evident to the user's hands, eyes, and even ears (e.g., many vehicles have backup sonar that beeps more quickly as it approaches an obstacle). Whether the user sees a web page or a printed page at the moment of use, many factors behind the scenes combine to present information in a meaningful way. With electronic documents, those factors include the networks of computers, electrical power, and international standards that make it possible to create on one computer what can be viewed on another. With paper documents, those factors include the networks of paper and ink manufacturing, printing, transportation, commerce, and archiving that must work together to bring a document into our view.

When we look at a document, whether paper or electronic, these myriad factors combine temporarily, settling into something we can use. After we obtain the information we are looking for, we might click to a different web page or turn to a different page in a printed document or go to a completely different document, but the visual field for that moment provides users with a framework for consuming information and an interface for working with that information.

Of course, electronic documents are more fluid than documents inscribed on paper, wood, or stone. We expect some web pages, like those dedicated to presenting news, to change from day to day, or even minute to minute, but we expect other sites to remain fairly constant, and we usually expect paper documents to remain the same, at least until a new version of a paper document is printed. The idea of the document, however, remains useful despite this impermanence. This is one reason digital artifacts follow many print conventions, even though they don't have to (what David Bolter and Richard Grusin call *remediation* in their 2000 book of the same name). We commonly refer to many kinds of electronic communication as documents, including web pages, websites, word processing documents, spreadsheets, slide shows, animations, and blogs, and we treat them as if they have coherence and boundaries, just as paper documents are bounded by their covers and shapes. Even when they break from print conventions, digital artifacts such as websites have quickly settled into a relatively stable set of conventions—for example, site navigation links are almost always at the left or top of each web page.

Arguably, in electronic documents, the permanence of digital information has simply shifted from the interface to the database or network. For example, we expect the appearance of the balance column for our online bank statement to be different each time we look at it, but we certainly hope that the bank will permanently document our payments, deposits, and other transactions. There's actually little money in a bank, in terms of cash—just documentation: numbers in this case. Nobody believes that there are little elves hiding on the bank website, waiting to roll quarters and ready to deposit checks. It's just information, until we print it out for some reason. We also assume that while the numbers will change, the overall appearance will not—the logo and disclaimers will be in the same spot, the places we purchased from will remain in one column, the associated expenditures in another, and so on. So while the critical information within the electronic statement may change, the *concept* of a bank statement—a document—remains.

So for our purposes, documents are not bound by any particular medium. Instead, *documents are best understood as a site where one person can mark information for the use of another.* Document design, then, is the process of creating these sites of interaction in such a way that they respond effectively to the needs of both information producers and information users.

Documents are best understood as a site where one person can mark information for the use of another. Document design, then, is the process of creating these sites of interaction in such a way that they respond effectively to the needs of both information producers and information users.

FROM DOCUMENT TO DESIGN

Because all documents, whether fixed permanently on paper or fixed temporarily on a screen, are presented to the user's field of view, in this book we focus primarily on how document designers make meaning through the design and production of the *visual* features of documents.

Consider an example probably very familiar to you: the academic essay. Your instructors probably had specific requirements for some visual aspects of the text, such as the width of margins. For the most part, you probably complied with those requirements without giving them much further thought (except perhaps about the pickiness of teachers).

In document design terms, however, the academic essays you prepared were designed for a specific user: the instructor. Your instructor's requirements for formatting were intended to make their job—reading and evaluating your essays—easier and more convenient, particularly if your work was to be submitted physically (it still happens!). Such requirements were probably mostly visual, including the following:

- *Double-spaced text*, to give the instructor room for corrections and interlinear comments.

- *Margins of a particular width*, to place a boundary between the text and the rest of the world, buffering the instructor's attention and helping them to focus on the text, as well as providing an area to write longer comments.

- *Paragraph indentations*, to help the instructor recognize when you intended to shift to a new topic.

- *Titles and headings*, usually bigger or bolder than the basal text, to allow the instructor to skim for structure.

- *Page headers or footers with page numbers*, to help the instructor keep the pages in the right order.

- *A staple, paper clip, or binder*, to keep all the pages of your document together.

- *A particular typeface or size of text*, to help the instructor get through a stack of grading without undue eyestrain.

These visual features are very familiar, so we might not pay much attention to them in writing or reading essays. But in other kinds of documents, visual features can become very important indeed. Visual features serve as the user interface of a document, helping users realize a number of important things:

- What kind of document they are reading

- How the document is structured

- Where they are in the document

- What's important in the document

- How to find what they need

Together, these visual aspects of a document have a huge impact on how well the document works.

Let's look at a more complex example: the System Timetable (see figure 1.3), a 132-page document published by Amtrak, the national rail service, opened to the page for the Empire Builder train service from Chicago to Seattle from 2005. We'll start with this older example as a way to consider how we, as designers, might think about designing with highly complex information, and to show how we are constantly rethinking designs as technology and culture evolve.

Both the designers and the users of this document faced difficult challenges. The travelers, who would have been using this document in 2005,

Figure 1.3. Communicating information visually. Note how the designers used visual communication features in this early 2000s document that allow users to navigate the document much as they might navigate the country. In addition, the cover and the Empire Builder page help build a sense of excitement and adventure about rail travel, though one with nationalistic rhetoric we should now question. *Source*: Amtrak.

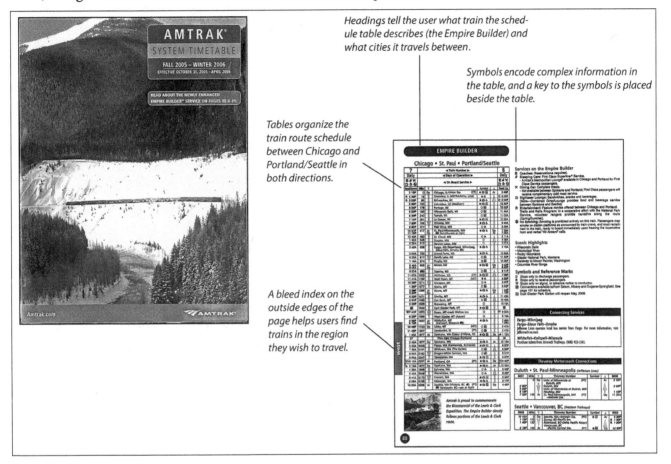

faced the complex task of planning trips by rail. Sure, the internet was alive and well at that point, but complex physical documents were much more prevalent. They needed convenient and economical transportation, but they also needed clear information, including exactly what train they should take, where they could get off, when the train would depart and arrive, and what amenities were available on the train, such as dining and sleeping cars. They wanted to find this information easily, quickly, and accurately. Travel in itself makes people feel stressed and dislocated. If passengers also feel confused or frustrated about the information they receive in this schedule, they might well consider using a different mode of travel on their next trip.

The designer's task was perhaps even more difficult. Train schedules can be pretty complicated, and Amtrak is the biggest passenger rail system in the country. The designer had to take into account all of the many possible users of the document and accommodate the travel plans of many different people. The document designer also had to respond to Amtrak's needs. Amtrak wanted—and still wants—its passengers to ride its trains and pay fares so the company can bring in a profit. Amtrak needs passengers to show up at the right time and place for the right train and to feel good about their trip-planning experience so they will travel with Amtrak again—nothing has changed there, even though many years have gone by. Efficient and cost-effective documents can help accomplish these goals.

A document designer would approach this situation by considering all of these needs (both user and client) and by crafting a document that meets as many of them as possible. The designer must convey accurate information, but, more importantly, they must design that information to meet the specific needs of Amtrak and its passengers as effectively as possible. With these needs in mind, the designer must make decisions not only about the visual design of text, tables, images, and pages but also about the medium and physical size, format, and production of the document. The designer must pay attention to how the document actually works, constantly testing and improving its effectiveness in solving the communication problem. Moreover, the designer must also work within a team of professionals such as advertising designers, writers, and copy editors to produce the document on time and under budget.

In fact, to accommodate the complex needs and agendas of both Amtrak and its passengers, a designer might develop multiple documents in different formats and even different media. Figure 1.4 shows the dynamic, database-driven Amtrak website from 2024, where travelers could input the departure and arrival dates of their trip and develop a personalized schedule. Notice how the design, in moving from paper to screen, changes for modern sensibilities, but the overarching goal is the same—determine from where and to where one wants to go, and when one can get there. There is a subtle lesson here—documents designed for a common, specific purpose tend to adhere to set conventions because people—users—expect to find those particular conventions. This is known as a *genre*: a set of conventions that readily allows users to identify what they are seeing and how to use it (for more information on genre, consider María José Luzón's 2005 article on how we analyze different types of genre, or Carolyn Miller's 1984 argument on the way genres shape social action, among others). In this case, we could call the genre "travel sites" and expect that most, if not all, sites would have a place to input point of departure, destination, date of trip, and so on, and, indeed, looking through travel sites we do in fact see that this is the case.

Figure 1.4. The web version of a print document. On the 2024 Amtrack site, users input critical information (e.g., "From" and "To"). The site then helps them navigate additional considerations, such as price. *Source*: Amtrak, https://www.amtrak.com/.

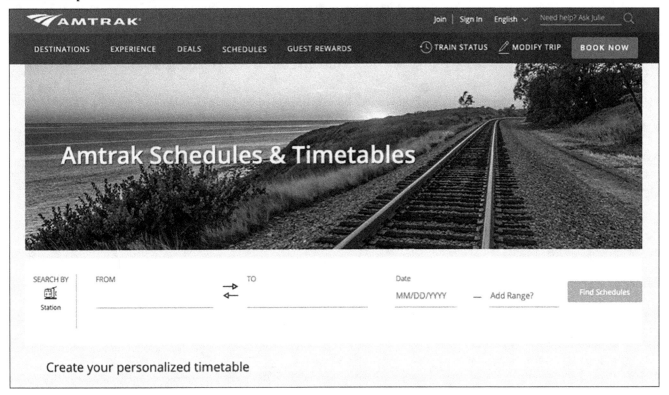

Figure 1.5 shows that same scheduling interface, but now reimagined for a mobile device. Again, think about what the designers have included and what they have left out. As powerful as our mobile devices are—and they are truly exceptionally powerful—their greatest limitation is screen size. Here, everything that we originally needed a massive physical document to display has been further compressed to interactive prompts, thereby letting the technology shoulder a lot of the design work. This approach only works, of course, if the technology works and if the technology has been designed and displayed in such a way as to make the experience of using it pleasurable, or at least functional (an area of research in technical and professional communication commonly referred to as *user experience* [*UX*] *testing*).

Consider, though, that that 2005 *model* of train timetable communication—if not the actual information—is still relevant. Anytime the power goes out, or the internet goes down, someone needs to be able to find out

Figure 1.5. Amtrak's mobile interface for train scheduling. Notice how all of the information previously displayed, from the 2005 print tables to the modern website, has been further compressed. Colored text and a few color bars guide users through decision-making on their handheld device. *Source*: Amtrak.

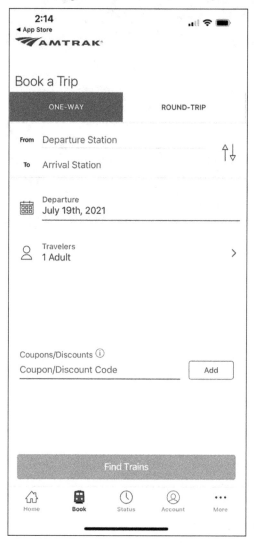

what train goes where, and when. Beyond that, though, physical documents remove the requirement of technological connectivity. Amtrak still prints brochures, and train timetables still exist as physical documents. Users with limited access to technology still deserve to be able to ride trains.

In all of these versions of Amtrak's train scheduling designs, document designers have developed the document's visual interface (whether paper or electronic) and therefore helped shape the user's experience with that interface. This is not to say that all three of these documents work the same way or offer the same capabilities to Amtrak or its passengers, but all three are most definitely documents, crafted to solve particular problems in particular situations.

DOCUMENT DESIGN AS A RELATIONSHIP

As the Amtrak example demonstrates, document design is best understood as a complex relationship between you as a designer and two sets of people: those who ask you (and hopefully pay you) to design documents and those who will use the documents you design. We typically call the former group *clients* and the latter group *users*. The relationship between all three parties is diagrammed in figure 1.6.

CLIENTS

Your client forms one node of this three-part relationship. Sometimes as a designer you might be a permanent employee of the client; other times, you might work as a design consultant working on contract. You might even be both the client *and* the designer, creating something for your own purposes. In any of these situations, it is essential that you understand why, what, and how the client wants to communicate—and to whom.

Clients ask designers to create documents for specific reasons, usually because they want certain things to happen—what we call an *agenda*. Documents cost money, so clients want documents that fulfill their agenda efficiently.

Figure 1.6. Document design is a relationship between the designer, clients, and users. *Source*: Created by the author.

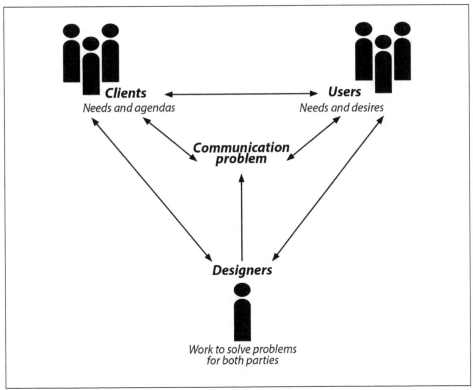

Often, the client's agenda initiates what kind of document will be created. But you will often find that the client's agenda is based on their perception of user needs and desires. For example, the client who suspects that employees won't be enthusiastic about a new training seminar might ask you to design a document that encourages participation and makes it clear to employees why, when, and where to show up.

Client agendas can also be more general and complex. A client might want to reaffirm the user's choice in buying a product and give the user a positive experience so they will recommend the product to others, thus increasing sales. To fulfill this set of agendas, a designer might propose a variety of different documents, from a manual to a quick reference guide to a dynamic customer forum hosted on the client's website.

To gain a clear understanding of your client's agenda, you must ask many questions, especially since clients sometimes aren't entirely clear about their own motivations (see chapter 10 for more on researching client agendas). Asking questions will help you build a picture of what the client wants the document to *do*—an awareness that will make a strong foundation for your design project.

USERS

Finding out about the client may be relatively easy because you can usually identify your clients readily and speak to them frequently. But the second node of the design relationship—users—can be harder to pin down. Nonetheless, every successful document designer invests time to analyze users and their situations, needs, and desires.

Finding out about users is an essential step in creating a successful design. No two users or design situations are the same, so it is important to consider who will use each particular document you create. Perhaps the best way to find out about users is to meet them and talk to them about their interests, biases, and concerns or to watch them and observe the challenges they face. Or you can conduct other kinds of research, such as demographic studies, to find out more about users (see chapter 10 for more about user research).

Users are complex and unique. What works for one user won't always work for another, what works for most won't work for all, and what works one time might not work the next time. To begin understanding designing for users, however, we can start with four general assumptions:

- *Good design helps users solve their problems successfully.* Users are real people with real problems to solve and real situations to face. As a result, users look at the documents we design to fulfill their own goals, not ours.

More about . . . the Term *Users*

Why do we use the term *user* rather than just *reader* or *audience*? Some designers still refer to people who use documents as readers, but that makes it sound as if reading is the only thing people do with a document. Think about all of the other things you do with documents:

- Scroll or flip quickly through a document, checking to see if it is one you want to look at more closely

- Search for particular information you need by looking at the index or table of contents

- Skim paragraphs or sections that look interesting or useful

- Compare one part of the document to another

- Stick a document in your pocket or book bag for easy reference on the road, or bookmark it in your browser for later reference online

People do these and many more things with documents, so they are more accurately classified as users than as simply readers.

- *Good design helps users do what they want to do.* Users do not want to read documents; they want to do things. Users often approach documentation only as a last resort—when they have already failed at all other attempts to complete the tasks they face.

- *Good design helps alleviate that stress by helping users solve their problems quickly and easily.* Users often approach documents already feeling frustrated, worried, or lost. Because documents are often the last resort to solve a pressing problem, users already feel stressed when they pick up a document to look for solutions.

- *Good design helps users gather what they need efficiently and quickly.* When users do read documents, *they rarely read all the way through.* Instead, users tend to be hunter-gatherers, culling visually through a large landscape of data to look for the small pieces of information that will mean something in their current situation.

Taken together, these assumptions should help you design for users as unique, self-motivated people. As a document designer, you will need to spend time *for each document you design*, finding out more about these people and their situations, needs, and desires.

DESIGNERS

Finally, there is *you*: the document designer. It's your job to negotiate the needs of both clients and users. Clients have things they want to happen, problems they hope to solve, people they wish to communicate with through documentation. Users have their own individual problems, needs, and desires, depending on their situations and motivations. As a document designer, you'll need to find ways to make both groups happy with the documents you create. Fortunately, the client's ultimate goal is often to give the user a positive experience, which makes meeting both groups' agendas a little more manageable.

As the mediator between clients and users, you have some significant responsibilities. Although you must strive to please your employers (as all professionals must), you must also stand up for the rights and interests of users. Document designers often struggle through ethical dilemmas in completing design projects. When does spinning the client's product in a positive light become simply misleading users? What information is okay to leave out of a document? How do you decide what things users don't need to know, especially if those things paint the client in a bad light? What responsibilities do you have to make your client look good in situations where users might be getting hurt?

Legal and professional guidelines can provide some help in resolving these dilemmas, but ultimately the designer holds responsibility to represent both clients and users. (See chapter 3 for a broader discussion on ethics in design.)

LEVELS OF DESIGN

Understanding the relationship between clients, users, and you as designer can help you prioritize projects to determine how much effort to put into a design. Without this prioritization, it's difficult to create documents that meet needs efficiently.

In a perfect world, every document would be beautiful. But this is not a perfect world. Design always costs money and always takes time. The less money the client has to spend, the less of the designer's time they can pay for. The tighter the schedule and the smaller the budget, the less elaborate the design can be to meet project deadlines. Despite our best intentions to create beautiful documents, factors like time, cost, and production methods will limit how aesthetically pleasing we can make our designs.

In addition, depending on the client, the users, and the situation, some documents are simply worthy of a higher design investment than other documents. This leads us to the concept of *levels of design*, which we derive from the well-known concept of levels of edit in technical editing practices (see Van Buren & Buehler, 1980). By *a higher level of design*, we mean a document that deserves the investment of more time and more money in its visual design. By *a lower level of design*, we mean a document that does not deserve a significant investment in time or money in its design.

> In a perfect world, every document would be beautiful. But this is not a perfect world.

The level of a design can be described broadly by how elaborate the design is—literally, how much labor is invested in the design. We can define the amount of labor in a document through two sets of indicators:

- *Polished vs. rough production values:* Some production materials are more costly and difficult to work with than others, and some production methods create a more polished document than others. Photocopying a document on standard letter-size paper and binding it with staples has a low production value and requires relatively little labor. Sending a document out to a professional printer for four-color printing and binding has a high production value and thus more labor.

- *Customization vs. consistency:* The more the design takes advantage of off-the-shelf elements, like clip art or image assets from design libraries, or very conventional formats and layouts, the less labor it takes to produce. The more we customize the graphics, colors, or format of a document, the greater the labor required. For example, creating a simple text-based web page requires little customization—the browser does all the work in determining how the text will be displayed. But creating a dynamic, graphically intense website with complex computer scripting to allow user accounts, comments, or user-customized content would take significantly more work.

Most projects will fall somewhere between these extremes, as you can see graphically in figure 1.7.

The difficulty is that we must determine as wisely and strategically as possible what labor investment our particular project deserves in this

Figure 1.7. Levels of design. The greater the level of labor required or justified by a project, the higher the level of design. A particular project could fall anywhere within this field, depending on the situations of the client and users. *Source:* Created by the author.

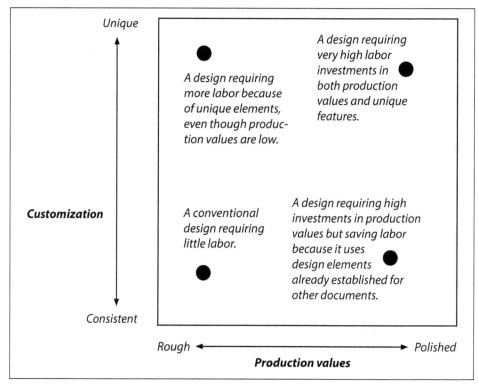

spectrum. Several interlocking factors can help us determine what level of design is justified by a particular project:

- *Ephemerality:* How long is the document supposed to last? If you want it to last a long time, a higher level of design might be justified. If it will go out of date next week, it might not be worth a high design investment. For example, an annual report is usually intended to be distributed throughout a whole year, if not longer, justifying a more elaborate document. A memo or email about the company picnic, however, is likely valid only for a very short time and so justifies a simple document.

- *Relationship to the client's mission:* Is the success of the document critical to the client's mission—their agenda, strategies, or goals? Every document should respond to a real need, but some needs are more important than others. If the document is essential to a client's mission, a higher level of design investment might be justified. If it's less essential, a lower level of design might do.

- *Reach:* How many people will see the document? If the document will reach only a few people, a lower level of design investment might be warranted. If it's going to be seen by many people, it might deserve a higher level of design.

- *User authority:* How much decision-making power will the users of the document possess, and how hard must the document work to persuade them? If the users are important decision-makers, such as purchasers or people who approve or disapprove proposals, the document must persuade them to do something. In those situations, a higher level of design investment might be justified.

Naturally, these factors overlap. A document might have a very small reach—for example, only a few people in a user group—but if those users have a high level of authority and the document must convince them of something that's critical to the client's mission, then a very elaborate design might well be called for.

Accurately determining the level of design is important to the success of any document. If we aim a document at too low a level of design for the situation, we might miss an opportunity to convince an important decision-maker about something essential to the client's agenda, strategies, or goals. But if we exert a lot of effort designing a document that doesn't have much of an impact, we're wasting our client's money.

DESIGN, RHETORIC, AND EMOTION

As this description of the relationships involved in the design process suggests, much of the success of a design depends on how people *feel* about the document. If people feel good about using a document, they're more likely to use it successfully, fulfilling the client's goals—and making the client happy.

As Donald Norman argued in his influential book *Emotional Design* (2004), users do react to documents emotionally, including the documents we design. These reactions can range from delight at seeing a document that answers an urgent question to annoyance at having to look at a document in the first place to being impressed by the professionalism of the organization that published the document. As Norman describes, we react to documents on visceral, behavioral, and reflective levels: When we first encounter a document, we have an immediate, uncontrollable visceral reaction (joy, disgust, boredom, etc.). We then have a behavioral response—the document works (or doesn't) to influence the way we act. Last, we end up considering the document in relation to what the document is, what it should do, and how it contributes to the surrounding culture.

Because people often have such immediate emotional reactions to what they see, design gives us an important opportunity to shape how users will feel about the document. Doing so is an important aspect of usability and the user's experience. Even if a document is accurate and could be used successfully, it might actually fail if users have negative reactions to the document, the authors, or the subject.

Try to think about shaping two kinds of emotional impressions in document design: the user's impressions of the client who is "speaking" through the document and the user's feelings about the document itself and its subject. In rhetoric, the user's impression about the speaker is called *ethos*, and an invocation of the user's feelings is called *pathos*.

The design of a document can very directly imply an ethos for the speaker and pathos for the subject. For example, a company trying to attract the attention of parents of young children might employ a design that makes the company sound fun and interesting, while also being educational. In the website in figure 1.8, LeapFrog used bright colors, images

Figure 1.8. Appealing to an audience of consumers. This web page implies a certain ethos for the company offering the product, learning technologies for children. The designers included children playing with the technology to "speak" for the company, giving it a voice the company hopes users will find appealing. *Source*: LeapFrog, https://www.leapfrog.com.

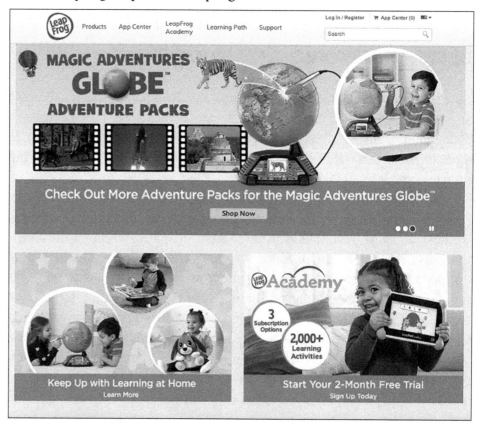

of smiling children interacting with technology, an alphabet background, and more to convey a sense of fun-in-education for their various learning technology toys for young users. The design speaks for the company by implying an ethos that suggests it creates engaging educational products. In turn, the design encourages potential buyers to feel a positive pathos by associating the technology with their desire for their children to have fun—none of the displayed technologies look complicated or frustrating; instead, the emotion conveyed is that children will want to play with these educational toys and technologies, and they will have fun along the way.

However, the same company might want to project an entirely different ethos and pathos for a different audience or purpose, such as with the annual report in figure 1.9. Here, the document puts a different face on

Figure 1.9. Appealing to an audience of investors. This corporate annual report is from the same company that produced the website in figure 1.8. Instead of children, however, the designers chose to include a picture of the chair of the board to "speak" for the company, presenting a different ethos for a different audience: investors. The illustrated office behind the speaker adds a touch of whimsy, but only a touch—this is definitely a business report. *Source*: Vtech.

Vtech, the company that owns LeapFrog—that of the chair of the board—to encourage investors to put their money into a solid, profitable, well-run corporation. This ethos is created visually with more traditional text and page layout, though still modern, as well as with a photograph of the well-dressed chair. The design hopes to encourage users to feel reassured and confident about the subject of the document: the financial probity and success of the company. In an application of pathos, invoking this feeling of confidence responds to the user's desire for financial security.

VISUAL DESIGN

As a document designer, your primary tools for helping clients create a positive experience for users are visual. Document designers create clear visual patterns that help readers see the relationships between different pieces of information. The rest of this book concentrates on this visual aspect of document design. Chapter 2 gets you started by outlining some consistent principles of design.

EXERCISES

Chapter 1 introduced you to the concept of a document; the various roles of a designer in creating documents; the complex set of interactions that occur between clients, users, and designers; and ways we might think about the user's needs and expectations and how to variously address those needs and expectations. These exercises should help you dig a little more deeply into those concepts.

1. Find a document whose client was someone you know or can talk to. For example, you might look for a document produced by your employer or at your school. Interview the client about the document, focusing on these questions:

   o What were your goals in having this document designed? What did you want the document to *do*?

   o Who did you think the audience of this document would be?

   o How was the document intended to fulfill the audience's needs?

   o What parts of the document worked best? What parts didn't work so well?

2. Find someone who might have used the document you examined in exercise 1 and interview them. Consider asking questions like these:

   o What did you find useful about the document?

   o What did you find not so useful about the document?

   o What things, if any, do you think the document designers did not understand about people like you who would be using this document?

3. Find a document you can examine and discuss in class. Form small groups and share your documents with each other. Select one document and choose two members of your group to role-play to the rest of the class—one acting as the document's client, one as the document's user. The client should explain to the user what they wanted the document to do. The user should respond by explaining how the document actually works (or does not work) for users. Afterward, discuss the dialogue as a class:

   o In what areas do the client and user seem to have overlapping agendas or goals?

   o What disagreements do you see between the two roles?

   o How could the document designer have better mediated this relationship?

4. Working with a group, find several versions of the same document from across several years. If possible, find versions of the document in print and online. To help you find documents as they have changed over time, you might want to use the Wayback Machine at archive.org. Consider the following:

   o How has the document evolved over time?

   o What has changed about the document?

   o What has stayed the same?

   o How do those changes (or lack of changes) match up with any cultural changes or events you can discover? For example, has a new application been developed that lets designers work in new ways? Is there a new trend in site design?

REFERENCES AND FURTHER READING

Bolter, J. D., and Grusin, R. 2000. *Remediation: Understanding new media*. MIT Press.

Carliner, S. (2003). Physical, cognitive, and affective: A three-part framework for information design. In M. J. Albers & B. Mazur (Eds.), *Content and complexity: Information design in technical communication* (pp. 39–58). Routledge.

Hart, G. (1996). The five w's: An old tool for the new task of audience analysis. *Technical Communication, 43*(2), 139–145.

Horn, R. E. (1999). What is information design? Information design as an emerging profession. In R. Jacobson (Ed.), *Information design* (pp. 15–33). MIT Press.

Levy, D. M. (2001). *Scrolling forward: Making sense of documents in the digital age*. Arcade.

Luzón, M. J. (2005). Genre analysis in technical communication. *IEEE Transactions on Professional Communication, 48*(3), 285–295.

Miller, C. R. (1984). Genre as social action. *Quarterly Journal of Speech, 70*(2), 151–167.

Norman, D. (2004). *Emotional design*. Basic Books.

Potts, L., & Salvo, M. J. (2017). *Rhetoric and experience architecture*. Parlor Press.

Salvo, M. (2014). What's in a name? Experience architecture rearticulates the humanities. *Communication Design Quarterly, 2*(3), 6–9.

Slack, J. D., Miller, D. J., and Doak, J. The technical communicator as author: Meaning, power, authority. *Journal of Business and Technical Communication, 7*(1), 12–36.

Tham, J. (2021). *Design thinking in technical communication: Solving problems through making and collaboration*. Routledge.

Van Buren, R., & Buehler, M. F. (1980). *The levels of edit*. Jet Propulsion Laboratory.

2

PRINCIPLES OF DESIGN

LEARNING OBJECTIVES

Upon completing this chapter, you will be able to:

- Define design objects and discuss how object-driven designs promote visual thinking

- Apply visual variables (shape, orientation, texture, color, value, size, and position) to bring order to design objects

- Use design principles (similarity, enclosure, alignment, contrast, order, and proximity) to create meaning

~

When we use documents, we often don't consciously pay attention to the actual marks on the page or screen. Instead, we usually rely on our experience and understanding of conventions to understand what a document means. As you read this chapter, for instance, you probably aren't thinking consciously about the individual letters on the page or the relationship between the different parts of the page spread—headers, footers, page numbers, headings, words, sentences, paragraphs, margins, and so on. You're probably concentrating more on what you're reading than on the visual display of the page.

As designers, however, we must develop a heightened awareness of these individual marks, as well as the power of combining them. To help you develop that awareness, look at figure 2.1. Of course, you'll notice right away that the document in figure 2.1 is upside-down. We placed it that way to short-circuit your experience as a user of documents—your practiced, conventional, unconscious ability to decipher meaning from marks on a page. But because you can no longer read the marks on the flyer easily, there's a good possibility that you can see the marks as they are in themselves—shapes and arrangements of shapes on a visual field. Designers often use this technique of looking at a document from a different perspective to help them see the separate elements. Developing a sensitivity to marks as marks will help you become more conscious of how you use marks to create meaning.

In this chapter, we'll discuss two important concepts based on this awareness of the rich and varied characteristics of the marks we can make on a page or screen: *design objects* and *principles of design*.

Figure 2.1. Looking at a document upside-down. Don't try to read this upside-down document. Instead, try just to look at it as a collection of marks arranged into patterns in a visual field. What do you notice when you look at the document upside-down? What objects seem prominent or important? How do the objects relate to each other visually? *Source*: Texas Tech University Health Services, 2006.

DESIGN OBJECTS AND THEIR CHARACTERISTICS

You're probably already familiar with the concept of design objects if you've ever used a computer drawing program, a page layout app, or even the drawing function in a word processing app. In such apps, you can typically access a toolbar of different objects that you can insert or drop into a document. For example, the drawing and insert toolbars from Microsoft Word allows users work with a variety of objects, including different styles of lines, various shapes, and text boxes.

Design objects include any mark or group of marks that can be seen and manipulated on a page. They can be simple marks such as dots, lines, shapes, or background shadings, each of which carries some kind of meaning or function in the design. For example, designers often use a simple printer's bullet (•) to mark the beginning of each entry in a list.

> Design objects include any mark or group of marks that can be seen and manipulated on a page.

Please note that design objects also include the marks we use to make *text*, including letters, numbers, and punctuation. We sometimes think of text as primarily verbal, but in document design, text is just a system of visual marks that stand for linguistic expressions or sounds. Text is just as visual as any other kind of design object.

Design objects can also be *combinations* of individual marks that are treated as one visual unit, as seen in figure 2.2. For example, single letters can combine to create words, words combine to create a line of text, and lines combine to make a paragraph or a column of text. In the same way, a combination of lines, shapes, and letters could be manipulated as a single graphic—a bar graph, for example. Similarly, a combination of composite objects (paragraphs, graphics, headers, footers) can make up a whole page or a screen.

All of these objects mark a *positive space* on the page, but a design object can also be formed by a *negative space*. Positive space includes the marks actually made on a page or screen, such as the text or a photograph. Negative space is the space *between* and *around* positive space. For example, the margins on a page are a negative space. Positive and negative space can be equally important to users for conveying meaning and fulfilling a function in the design. For example, the negative space between a graphic and the text wrapped around it marks a boundary, encouraging users to look at the two objects separately.

Figure 2.2 also clarifies what kinds of meaning design objects can carry. The negative space around the dog images can hold a relatively simple meaning, conveying to users only that the image and its surrounding

Figure 2.2. Rather than thinking of text as simply letters, and images as complete, individual photos, drawings, charts, etc., it might help to rethink of anything you put on the page as made up of smaller and smaller design objects. In this image, for example, the word *text* is text—we can read the *t*, the *e*, and so on, but each letter is really just a picture of a dog's head filling positive space put in the appropriate proximity and alignment contrasting against negative space. *Source*: Created by the author.

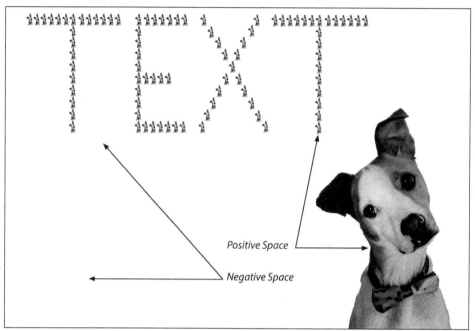

text should be considered as two separate, though related, objects. Other design objects are richer in meaning—the graphic itself, for example, might convey complex data visually. The objects carrying the most complex meanings are likely the marks we use to convey language—letters, words, sentences, paragraphs, and images.

A considerable and growing tradition of object-oriented thinking, derived from cognitive psychology and computer science, supports this way of thinking about design objects. In essence, cognitive psychologists have suggested that we tend to think of our surroundings in terms of objects and their qualities.

Of course, the term *design objects* as we use it here is a metaphor—but a very useful one. Many computer programs can make it look as if you are picking up and moving objects as if they were children's blocks on a table. The common term for this capability is drag-and-drop.

However, when you drag-and-drop an object in a computer app, you're not actually moving anything but your mouse or finger (for touch technologies). Most pages really have only two functional dimensions: width and height (physical pages may play with depth as well, using techniques such as embossing to make elements stand out in relief). We don't "move" the cursor over a button; we just use mouse or finger to trigger a brief animation of a series of pixels. But design objects allow us to *think* of pages as three-dimensional—letting us arrange, combine, or overlap different objects for different effects.

Thinking of marks as design objects also allows us to manipulate their qualities and characteristics with great precision. Semiotic theorist Jacques Bertin (1983) suggested that there are seven *visual variables* we can use to manipulate objects in designs:

- Shape

- Orientation

- Texture

- Color

- Value

- Size

- Position

These variables intersect and combine to give design objects their individuality, recognizability, and meaning.

SHAPE

Shape refers to the two-dimensional area covered or enclosed by an object. As Bertin points out, "the universe of shapes is infinite," even in two dimensions (95). Shapes can be regular or symmetrical, like geometric figures. But they can also be irregular or asymmetrical—letters, for example, are complex, mostly asymmetrical shapes.

Even the tiniest objects have a shape if we look closely enough. Periods, for example, are probably the smallest mark used in everyday typography. But different typefaces have differently shaped periods, as you can see in figure 2.3.

Shapes can carry meaning either by convention or by resembling something. The letters of the English alphabet don't actually look like sounds, but by training in standard conventions, we associate them with sounds. Similarly, some shapes are associated with conventional meanings

Figure 2.3. Differently shaped periods in different typefaces. The possible variations in shape are nearly endless, even with something as prosaic and conventional as a period, shown here at 72 points. *Source*: Created by the author.

Times New Roman   *Verdana*   *Trattatello*   *Farah*   *Goudy Old Style*

(see figure 2.4). Other shapes actually resemble something we might recognize (see figure 2.5), and some shapes might do both, such as the hand shapes in figure 2.6 (though their meaning is culturally specific).

One important shape often used in document design is the *line* (also known in traditional printing as a *rule*). In geometry, lines have only one dimension: length. But in design, lines must have two dimensions to be seen: length and width. By definition, lines are longer than they are wide—but that still leaves a lot of room for variation (see figure 2.7). Lines can be wide or thin, consistent or irregular, curved or angled. Curved and angled lines often imply connection and dynamic relationships. Lines can

Figure 2.4. Objects with conventional meaning. Some objects carry meanings based on convention—that is, society implicitly agrees on what these objects mean by using them repeatedly in similar ways. *Source:* Created by author from https://stock.adobe.com/.

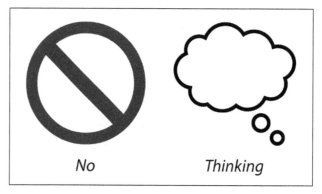

No   *Thinking*

Figure 2.5. Objects with representational meaning. Objects can have representational meanings in that they actually look like what they're meant to indicate. *Source:* Created by author from https://stock.adobe.com/.

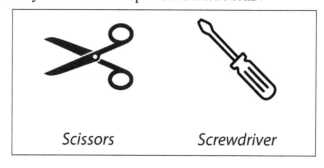

Scissors   *Screwdriver*

Figure 2.6. Objects with both representational and conventional meaning. Sometimes even representational shapes can have conventional meanings. These are representations of hands, but by convention in North America a thumbs-up hand means "good," while a thumbs-down hand means "bad." *Source:* Created by author from https://stock.adobe.com/.

also differ in the shape of their ends. Such lines are often used to imply a relationship or connection between different objects, as on a flowchart. Arrowed lines, for example, imply direction, causality, or chronology.

Lines can also be implied rather than drawn. For example, most of us easily recognize a dotted or dashed line, such as the curved one in figure 2.7. Dotted or dashed lines, however, are composite objects formed by small shapes placed close enough to each other in a row so that they *imply* lines. All of these differences in lines are based on the variable of *shape*. By the same token, any kind of design object—letters, words,

Figure 2.7. Lines with variations in shape. This diagram from a manual on aviation uses many different shapes of lines—straight, curved, dashed, arrowed, and different widths—to show pilots what kinds of markings to expect on an airport runway. Interestingly, an actual runway is also a document with lines that tell pilots how and where to land. *Source*: Federal Aviation Administration, FAR/AIM 2005.

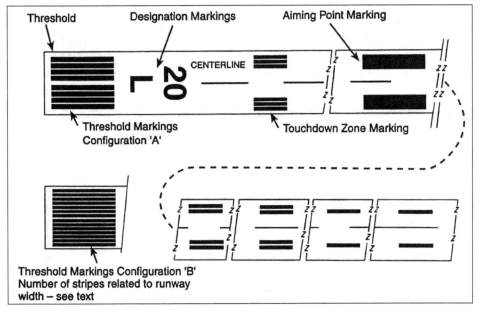

ORIENTATION

If lines are important and useful objects, then the directions in which they point—their *orientation*—must be important too. Each of the lines we have just seen clearly has an orientation, but other shapes can have orientations as well. Text in English is arranged in a horizontal orientation that leads users to read along each line left to right. Text in Japanese, on the other hand, traditionally has a vertical orientation to aid reading top to bottom. And individual shapes, like the hand in figure 2.8, also can have implicit orientations.

Figure 2.8. Implicit orientation. Some shapes imply an orientation inherently. Most people would describe this pointing hand as being oriented or pointed to the right. *Source:* Created by author from https://stock.adobe.com/.

TEXTURE

Texture refers to any pattern applied to an object. Obvious repeating patterns such as cross-hatching are often used for information graphics printed in black and white—for example, to distinguish between bars in a bar chart (figure 2.9).

Figure 2.9. Using texture in a bar chart. This bar chart uses textures to indicate that the different bars stand for different sets of data. *Source:* Created by the author.

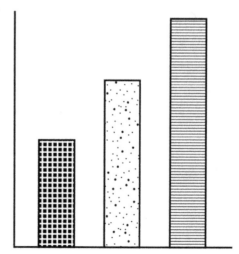

But textures can also include other kinds of patterns, such as applying an image to an object. Examples include using a background image or watermark on a page, a PowerPoint slide, or the desktop interface of your computer (figure 2.10). Such applications of textures can reduce readability, so they are not used as much as previously.

COLOR

Color is an increasingly common variable in design objects because in recent years color reproduction technologies have become cheaper and more convenient. For paper documents, color inkjet and laser printers are readily available. And for electronic documents, nearly all computers and mobile technologies now use color monitors. These factors have increased the likelihood that designers will use color to modify design objects. See chapter 8 for more about using color.

VALUE

Value refers to the *relative* lightness or darkness of a design object compared to its surroundings, including other objects near it and the background. High-value objects are those that contrast most with their surroundings; low-value objects are those that blend in with their surroundings.

On a mostly light background, such as white paper, a dark object typically has a higher value and greater visual density than a lighter object, so the darker object will likely be more prominent. This effect is

Figure 2.10. Using texture in a PowerPoint slide. Default PowerPoint slide templates often have a background texture to give a three-dimensional effect to the design. *Source:* Microsoft PowerPoint 2022.

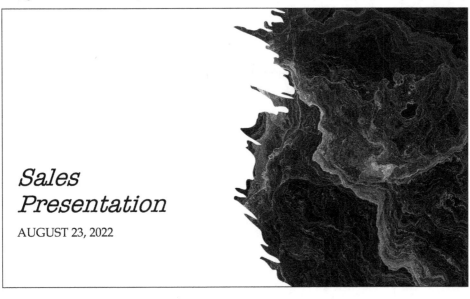

Sales Presentation

AUGUST 23, 2022

reversed on mostly dark backgrounds, such as we often see on websites (see figure 2.11).

Value is often used for emphasis, as high-value objects tend to get more immediate attention than low-value objects.

In printing, drafts of products are often accompanied by their registration marks, or make-ready marks, such as those seen in figure 2.12. These marks help printers ensure that content is printing as intended.

SIZE

However, the impact of value for emphasis also depends on the object's *size*: a very small dark object on a white page will carry less emphasis than a larger dark object on the same page.

Figure 2.11. Contrast and value. Value is determined by the contrast between objects and their backgrounds. The more contrast, the higher the value. *Source*: Created by the author.

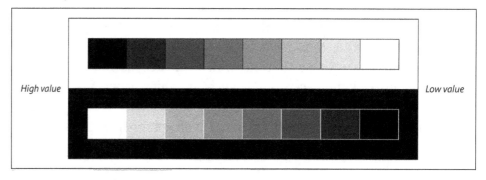

Figure 2.12. Example of marks printers use to line up content for printing and set color and value. *Source*: Created by the author.

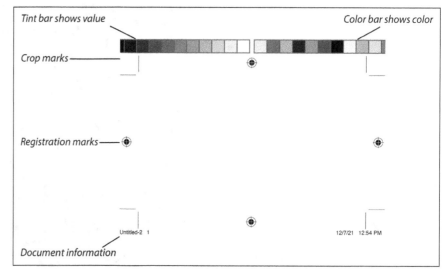

We can control the size of objects with many variables, but two aspects of size are most important: *absolute size* and *relative size.*

Absolute size refers to the dimensions of an object in relation to a standard, such as the centimeter or the inch. Professional print shops still sometimes use these measurement units, as well as less familiar ones such as *picas* and *points.* A pica is $1/6$ of an inch; a point is $1/12$ of a pica or $1/72$ of an inch (1 inch = 6 picas = 72 points). Using accurate absolute measurements is essential in design because it allows us to be precise and consistent in our use of design objects.

To users, however, absolute size doesn't really mean much; we don't often see them pulling out a ruler to measure a design object. But users can visually discriminate between the size of one object and that of another—that is, they can recognize objects as being relatively *bigger than* or *smaller than* with great sensitivity. Designers frequently use this discrimination of relative size to convey two important meanings: that one object is more important than another or that one object contains another. Users often assume that a larger object is more significant than a smaller object—for example, the title of a document is usually larger than its first-level headings, which in turn are larger than the second-level headings. But larger objects can also imply a relationship of ownership over the smaller objects they "contain." A single line in a graph, for example, is considered as "belonging to" the larger object that contains it (the graph itself).

Finally, the size of design objects is also relative to the boundaries of the visual field. These boundaries are easy to recognize: they're the edges of the paper or the screen that define the two-dimensional space in which we work. This might seem pretty obvious, but working within the available space can be remarkably difficult, requiring a careful balancing of different elements, especially in terms of their size. On most pages, some elements are more important than others and so deserve more space. But there's only so much space to go around, and even minor objects—if they're important enough to include—must get their fair share.

POSITION

We can control the position of objects on the plane in two dimensions: width and height. Most layout and drawing programs let us specify the position of an object with great accuracy.

In keeping with the object metaphor, we can also position objects on top of each other—for example, positioning a pull quote on a shaded field to distinguish it from the surrounding text. Of course, the objects aren't *really* on top of each other. The printer or

We can also position objects on top of each other—for example, positioning a pull quote on a shaded field to distinguish it from the surrounding text.

screen actually stops printing gray when the black of the text begins, but we assume that the gray continues under the text, forming a consistent field. (See chapter 3 for more information on this phenomenon.)

SIX PRINCIPLES OF DESIGN

Carefully controlling the qualities of individual design objects is an essential first step in creating good designs. But how do we show relationships *between* design objects? Using a welter of terms and ideas, many designers and scholars have tried to articulate the basic principles by which we create these relationships. As you can imagine, nobody entirely agrees on what the principles of design are; instead, they propose many possible ways of describing basic principles. One fascinating book—William Lidwell, Kristina Holden, and Jill Butler's *Universal Principles of Design*—proposes over a hundred design principles! Despite this ambiguity of terms, a limited set of basic design principles can help you focus your design thinking. Principles of design serve as general rules to help you make good decisions about your designs.

In this book we'll use six basic principles of design that we think govern most visual relationships between design objects:

- Similarity

- Enclosure

- Alignment

- Contrast

- Order

- Proximity

To make this list of design principles easy to remember, just think SEACOP.

These principles are not exclusive. Frequently, objects will be related by multiple principles at the same time, the principles working together visually to create complex meanings. And different designers might use different terms to describe these ideas, depending on their training and background. But employing each of these principles consciously will enable you to create designs that guide and help users through using a document. Moreover, you can use these principles to create a *design system*—a coherent and consistent set of visual elements in a document that help users use the document in ways that fit their needs.

Understanding the principles of design will make you more conscious of the visual relationships between objects when you design. Ideally, the *visual* relationships between design objects should echo and reinforce the *logical* relationships between them. The principles of design work together to build a system of relationships so users can see how design objects work together.

Ideally, the *visual* relationships between design objects should echo and reinforce the *logical* relationships between them. The principles of design work together to build a system of relationships so users can see how all of these design objects work together.

SIMILARITY

Use similarity to show that design objects are alike in kind or in function. Users naturally assume that design objects that *look* similar *are* similar—that they do something alike, that they belong together somehow, or that they are parallel in function or kind. The more consistently a design applies similarity to similar objects, the faster users will be able to discern patterns and use the document efficiently. Similarity is essential for showing relationships of connection, such as coherence and consistency.

We can control similarity very precisely by paying attention to the qualities of design objects: shape, orientation, texture, color, value, size, and position. Objects can be similar in any combination of these variables. For example, print documents often use similarity in their design of page numbers. To work efficiently, page numbers need to be similar in these visual variables:

- *Shape:* all in the same typeface

- *Size:* all in the same size

- *Color:* all in the same hue

- *Value:* all the same relative density (all boldface or none boldface)

- *Position:* always in a consistent, predictable position on the page—for example, the bottom outside corners of each page

Any inconsistency in these visual variables will make it more difficult for users to predict where to find the page numbers—or more difficult for them even to recognize the numbers as page numbers. Users may still be able to use the inconsistent document successfully but less efficiently, and probably less happily, than they might if the page numbers were all similar.

ENCLOSURE

Use enclosure to show separation and to group complex objects. We can use positive or negative elements to enclose objects or groups of objects, setting them apart from their context. Positive elements of enclosure include lines, borders, and shadings that surround an individual or complex object and distinguish it from the surrounding objects on the page or screen. Figure 2.13 includes a number of examples of enclosure.

In this web page design, enclosure is mostly formed positively, with lines and borders. But enclosure can also be formed simply by negative space. People who haven't thought much about design sometimes assume

Figure 2.13. Using enclosure to show separation. This website uses enclosure in several ways to separate areas of interest. *Source*: EPA, https://www.epa.gov.

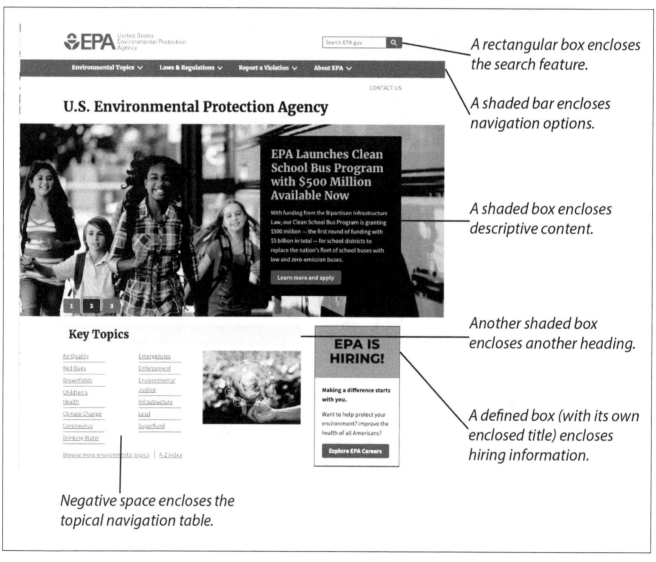

that borders are essential to enclose objects—for example, when designing tables, they might include borders by default. But many well-designed tables use negative space instead of positive borders to enclose each cell, row, and column and set them off from the rest (figure 2.14). Relying on negative space instead of positive borders can make a design look more open and less confusing.

Figure 2.14. Two versions of a table. The first table relies on borders to enclose the cells and form rows and columns; the second relies on negative space (and alignment) to do the same thing. *Source*: Created by the author.

Bacteriological	MCL	Highest Detected Level	Synthetic Organic Chemicals	MCL	Highest Detected
Total Coliform Bacteria	5%	< 5%	2,4,5-TP (Silvex)	50 ppb	ND
Radiological	**MCL**	**Highest Detected Level**	2,4-D	70 ppb	ND
Gross Alpha	15 pCi/L	0.462	Alachlor (Lasso)	2 ppb	ND
Radium-228	5 pCi/L	0.548	Atrazine	3 ppb	ND
Turbidity	**MCL**	**Highest Detected Level**	Benzo(A)Pyrene	200 ppt	ND
Turbidity	TT (NTU)	0.29	Carbofuran	40 ppb	ND
Inorganic Chemicals	**MCL**	**Highest Detected Level**	Chlordane	2 ppb	ND
Antimony	6 ppb	ND	Dalapon	200 ppb	ND
Arsenic	10 ppb	ND	1,2 Dibromo-3-Chloropropane	200 ppt	ND
Barium	2 ppm	0.0279	Di(2-Ethylhexl)Adipate	400 ppb	ND
Beryllium	4 ppb	ND	Di(2-Ethylhexl)Phthalate	6 ppb	0.41
Cadmium	5 ppb	ND	Dinoseb	7 ppb	ND
Chlorine	4 ppm MRDL	1.53****	Diquat	20 ppb	ND
Chromium	100 ppb	0.58†	Endothall	100 ppb	ND
Copper	AL = 1.3 ppm	90th percentile value = 0.1701	Ethylene Dibromide (EDB)	50 ppt	ND
Cyanide	200 ppb	ND	Endrin	2 ppb	ND

Bacteriological	MCL	Highest Detected Level	Synthetic Organic Chemicals	MCL	Highest Detected
Total Coliform Bacteria	5%	< 5%	2,4,5-TP (Silvex)	50 ppb	ND
Radiological	**MCL**	**Highest Detected Level**	2,4-D	70 ppb	ND
Gross Alpha	15 pCi/L	0.462	Alachlor (Lasso)	2 ppb	ND
Radium-228	5 pCi/L	0.548	Atrazine	3 ppb	ND
Turbidity	**MCL**	**Highest Detected Level**	Benzo(A)Pyrene	200 ppt	ND
Turbidity	TT (NTU)	0.29	Carbofuran	40 ppb	ND
Inorganic Chemicals	**MCL**	**Highest Detected Level**	Chlordane	2 ppb	ND
Antimony	6 ppb	ND	Dalapon	200 ppb	ND
Arsenic	10 ppb	ND	1,2 Dibromo-3-Chloropropane	200 ppt	ND
Barium	2 ppm	0.0279	Di(2-Ethylhexl)Adipate	400 ppb	ND
Beryllium	4 ppb	ND	Di(2-Ethylhexl)Phthalate	6 ppb	0.41
Cadmium	5 ppb	ND	Dinoseb	7 ppb	ND
Chlorine	4 ppm MRDL	1.53****	Diquat	20 ppb	ND
Chromium	100 ppb	0.58†	Endothall	100 ppb	ND
Copper	AL = 1.3 ppm	90th percentile value = 0.1701	Ethylene Dibromide (EDB)	50 ppt	ND
Cyanide	200 ppb	ND	Endrin	2 ppb	ND

ALIGNMENT

Use alignment to show connection and coherence. We can show connection by using explicit lines (as we saw in figure 2.7), and this technique works in many situations. But we can also use simple alignment to show relationships implicitly between design objects. For example, consider the two lists of features for a software program in figure 2.15.

Both lists contain the same information, but the list on the left takes more time and effort to read because none of the features align. The bulleted list on the right, however, aligns each feature on the same implied vertical line, making it easier for users to skim down the entries to see if they're interested in buying the product and to compare different features. In fact, the bulleted list uses several different vertical alignments to emphasize its "listness"—the alignment of the heading, the alignment of the bullets, and the alignment of the list text all contribute to the effect. Each level of vertical alignment (or indentation) implies a further step down the hierarchy of the information, from main idea ("product features") to subitems (the features list).

We can see alignment applied more fully in figure 2.16, from usability.gov, a website dedicated to helping designers think more fully about usability and accessibility. The design of this site aligns several design objects both horizontally and vertically to show their relationship as *belonging to* or *parallel with*. Used carefully, alignment can help organize the different parts of a page, making it easier for users to manage the many different design objects that can appear by creating a unified system of page design. Typically, designers use a page layout grid to manage such alignments (see chapter 5 for a discussion of grids and page layout).

Figure 2.15. Alignment and bulleted lists. The alignment emphasizes each element of the list and makes it clear that they are all parallel. *Source*: Created by the author.

Product features: active updating; synchronous data transmission; updated user interface; accessibility kit; backup software; import and export functions

Product features
- Active updating
- Synchronous data transmission
- Updated user interface
- Accessibility kit
- Backup software
- Import and export functions

Figure 2.16. Using alignment to organize information. Note how this website uses alignment to organize different parts of the page. *Source*: GSA, https://www.usability.gov.

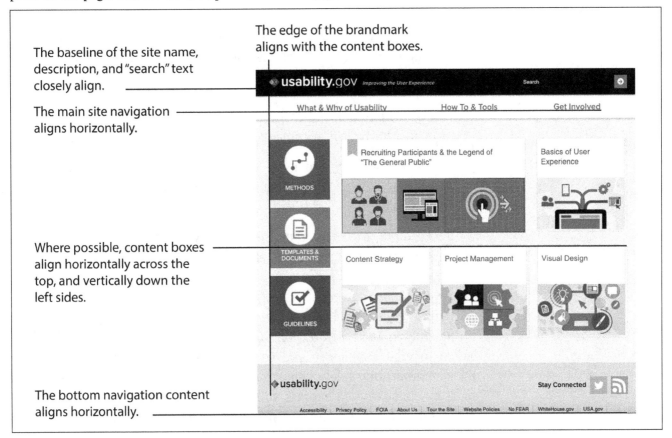

CONTRAST

Use contrast to show difference and create emphasis. Just as objects can be similar in terms of the visual variables, design objects can differ from or contrast with other design objects by having different visual characteristics. Users typically assume that a difference in appearance means a difference in function or meaning.

Similarity and contrast are closely connected: designs often employ both in concert as two sides of the same coin. In fact, it's impossible to show contrast without a comparison to something else; contrast means something only in the context of similarity. For example, a word in bold can stand out only in a field of words in regular text; if all the words are in bold, none stand out individually.

Contrast is a valuable tool for emphasizing important design objects, such as notes, safety warnings, or cautions. Without adequate contrast from the surrounding text, users might not notice such important information.

So designers typically make these objects stand out from the regular or *basal* text of the document in these ways:

- *Shape:* a different typeface; a conventional icon such as a stop sign or warning triangle; a border surrounding the warning

- *Size:* a larger point size than the surrounding text

- *Color:* limited use of color (often red or yellow) in an otherwise black-and-white document

- *Value:* a greater density or higher value than the surrounding objects, created by bolding text or adding a gray field behind the text

- *Position:* separate from the text—for example, a consistent position in the margin of the document

However, to be successful, safety warnings must also be consistently recognizable: all safety warnings should look the same throughout a document. So once again, contrast and similarity work hand in hand. Taken together, similarity and contrast can also be very powerful tools for communicating relationships of *structure* between objects. One of the most obvious ways to express similarity and contrast in this way is with headings, which imply a hierarchical structure for a document and thus echo its logical organization. The key is to make the levels of heading different enough to recognize easily while retaining consistency in multiple iterations of the same level of heading.

For example, the first heading design in figure 2.17, although very consistent, doesn't include adequate contrast between the different levels of heading and basal text, making it hard to make out the logical hierarchy of the document. The second heading design, however, uses size, value, shape, and position to identify different heading levels, while maintaining consistency each time a heading level appears. This heading system visually implies a set of important relationships. Each higher level is different enough (bigger, denser, differently shaped) to convey the relationship of ownership over the subsections it "contains." At the same time, the visual similarity of headings at the same level—for example, the level 2 headings—implies that these headings and the text that follows them are parallel, equal parts of the document.

Finally—and at the most basic level—contrast is essential simply for distinguishing objects as separate from the background. For example, websites with dark or patterned backgrounds sometimes use relatively

Figure 2.17. Contrasting headings. Notice that the contrast and similarity must work as a coherent system. Different levels need enough contrast to seem visually distinct, but not so much that the first-level headings end up too large or the basal text too small. *Source*: Created by the author.

Level 1 Heading	**LEVEL 1 HEADING**
Basal text…	Basal text…
Level 2 Heading	**Level 2 Heading**
Basal text…	Basal text…
Level 2 Heading	**Level 2 Heading**
Basal text…	Basal text…
Level 1 Heading	**LEVEL 1 HEADING**
Basal text…	Basal text…
Level 2 Heading	**Level 2 Heading**
Basal text…	Basal text…
Level 3 Heading	*Level 3 Heading*
Basal text…	Basal text…
Level 2 Heading	**Level 2 Heading**
Basal text…	Basal text…

dark type as well—like the purple commonly used to indicate followed links. In those designs, users have trouble distinguishing between the background and the foreground objects (the text) because they are too close together in value or pattern. This effect is also known as *inadequate figure-ground contrast*.

Don't be afraid to use contrast to make something stand out. Of course, there are limitations—you still want the design to work well as a whole system—but for the most part, more contrast makes for increased usability. To show contrast strongly, consciously apply differences *in multiple variables*—instead of just making something you want to emphasize bigger, make it bigger, denser, a different color, *and* a different shape (figure 2.18).

Figure 2.18. Building contrast. The example on the left employs contrast in just one variable: font size. The example on the right demonstrates that we can build contrast in multiple visual variables simultaneously: not just font size but also color, typeface, and position. *Source*: Created by the author.

ORDER

Use order to show sequence and importance. Order refers to most users' tendency to assume that what they see first on a page is more important than things they see later on. We can use this tendency to provide information in an order that makes sense for the user and for the information a document design presents.

Most users have encountered thousands of documents in their lifetimes, so they're very experienced and practiced in using them. This experience creates certain patterns of use that readers apply to new documents they encounter. As we will discuss in more depth in chapter 5, one of the most powerful patterns in Western Hemisphere readers is that we tend to read languages written from left to right and from top to bottom. When we encounter a page, we tend to look first at the top, then our eyes move in a Z-shaped or an F-shaped pattern across the page. That is, in some cases we read from left to right across a page, then our eyes move down and back to the left, then across to the right again, forming a Z pattern. In other cases, our eyes move most of the way across the first few lines, or skim most of the way across the page closer to the top (eye tracking shows that our eyes stop, or fixate, more often closer to the top of the page) then our eyes return to the left, move down, read or skim part-way across, return to the left, then proceed rapidly to the bottom with increasingly limited across-the-page skimming or reading. This F pattern, described by the Nielsen Norman group in 2006 and reexamined by them in 2017 (and also examined by many other researchers as well) suggests that we tend to focus more immediately, frequently, and intently on the top and the left side of a page than on the bottom or the right side.

Designers take advantage of this tendency in Western reading practices to begin at the upper left and move across and down to order pages so that the most important information is positioned at the top, followed by information of decreasing importance as the reader moves down the page. (Unscrupulous designers sometimes use this technique to emphasize positive information and deemphasize negative information. Sadly, the small print is usually at the bottom of the page rather than at the top.)

This is not to say that readers always create order on the page from top to bottom and from left to right—there are, of course, many other patterns, often affected by the designer's choices of content and placement. Eye-tracking research suggests that users tend to scan the page holistically, hunting for what seems most important to their current task first and then moving on to less important or more detailed items. And more complex designs might encourage users to navigate through a page or screen in a different order than simple top-down, left-right. We can use other principles of design such as alignment and contrast to draw the user's eye to the most important information first and then lead the user through the document in a logical order. In fact, designers can encourage a more productive type of scanning called the "layer cake pattern" by using visually engaging headings and subheadings, which allows users to more mindfully move through content (see work by the Nielsen Norman group, authored by Kara Pernice).

We should also recognize that users create their own order as they encounter documents, depending on their own purposes and situations. If a user is looking for a picture, they might not pay attention to the text, even if it's at the top of the page. Still, ordering design objects carefully can help create meaningful relationships between those objects.

PROXIMITY

Use proximity to show grouping and belonging. Users typically assume that design objects that appear in close proximity belong together. Conversely, design objects that are relatively distant from each other imply that they don't belong together.

We can see this principle at work in figure 2.19. Chances are, you see the group of squares on the left as vertical columns and the group on the right as horizontal rows. Why do you think you see the objects this way? The answer has to do with *proximity*. If the squares are closer together vertically than horizontally, we tend to see groupings that make vertical composite objects—columns. If the squares are closer together horizontally than vertically, we see groupings that make horizontal composite objects—rows. Note also that the squares are grouped not just by *absolute* proximity—the measured distance between them—but also by *relative* proximity. In other words, they build relationships not just by being close but also by being closer in one dimension than in the other.

Figure 2.19. Proximity relationships determine grouping. Because the squares on the left are closer together vertically than horizontally, we see them as four vertical lines, or columns. Conversely, because the squares on the right are closer together horizontally than vertically, we see them as four horizontal lines, or rows. *Source*: Created by the author.

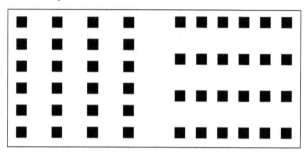

We can use this principle to make larger, composite design objects out of smaller, individual ones, in effect grouping related design objects to show their logical relationship of belonging together. For example, designers often mark paragraphs by including more vertical space between paragraphs than between individual lines of text. As a result, each paragraph appears as a single composite unit. Using the same technique, we typically place a heading closer to the section that follows it—the section it introduces—than to the section that precedes it. That way, relative proximity makes the heading visually connect to the paragraph that follows it (see figure 2.20).

Figure 2.20. Using proximity for headings. Headings should be in closer proximity to the text they introduce than to the text they follow. *Source*: Created by the author.

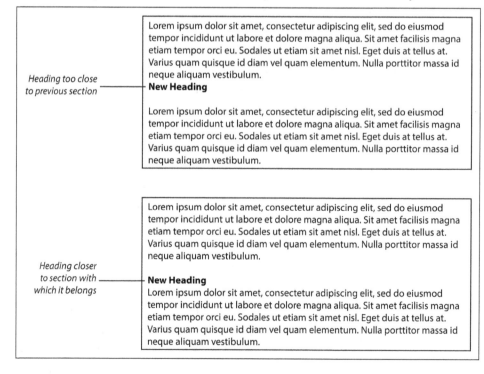

Proximity also makes a natural tool for grouping like concepts together. It's common in information graphics to place related objects near one another to show their relationships, for example, placing labels near the things they explain.

USING DESIGN PRINCIPLES

You can use these principles to combine design objects and create a visual system of meaning in documents. Doing so will require paying close attention to the visual details of what you design. But that attention to detail will give you an intimate awareness of how users make sense of what they see on a page. It will also help you control how your designs meet readers' needs and fulfill your clients' agendas.

EXERCISES

Chapter 2 defined design objects and discussed how object-driven design promotes visual thinking. You read about visual variables (shape, orientation, texture, color, value, size, and position), design principles (similarity, enclosure, alignment, contrast, order, and proximity), and how those principles can be operationalized to create meaning. The following exercises should help you more deeply engage with the concepts you learned in this chapter.

1. Collect a handful of documents so you can explore their designs. You should be able to find documents easily. Try looking at home, at school, at work, or at any public facility, such as a doctor's office, a museum, a library, or the local chamber of commerce. Look not just for advertisements and magazines but also for documents that convey important information to users, such as pamphlets, booklets, manuals, newsletters, and sets of instructions. Discuss how the designs you found use the various variables and principles discussed in this chapter.

2. Interview someone who designs documents as part of their job, or even as a profession. Ask what principles the designer generally uses in creating document designs. Do the principles seem similar to those described here? What terms does the designer use that are different from those here?

3. Take one of the documents you gathered in exercise 1 and look at it more closely.

   a. Choose one page of the document. How many different kinds of objects can you identify? Remember, sometimes objects are composite objects made up of other objects. Describe six of these objects in terms of their visual variables: shape, orientation, texture, color, value, size, and position.

   b. Examine the same page for its use of principles of design: similarity, enclosure, alignment, contrast, order, and proximity. How many uses of the principles can you see? (Not all of the principles may be used or used effectively on the page you chose.) What principles would you apply to improve the page's design?

4. Design your own business card. Be sure to employ as many of the design principles as possible in creating your design. Include at least the following information on your card:

   o Your name

   o Your phone number

   o Your email address

   o Your mailing address

   o Your slogan (make one up!)

   You can use a word processing program, a drawing program, or just a sheet of paper and a pencil.

REFERENCES AND FURTHER READING

Berger, A. A. (1998). *Seeing is believing: An introduction to visual communication* (2nd edition). Mayfield.

Bertin, J. (1983). *The semiology of graphics*. University of Wisconsin Press.

Johnson, M. (1987). *The body in the mind: The bodily basis of meaning, imagination, and reason*. University of Chicago Press.

Kostelnick, C., & Roberts, D. D. (1997). *Designing visual language: Strategies for professional communicators*. Longman.

Lidwell, W., Holden, K., & Butler, J. (2003). *Universal principles of design: 100 ways to enhance usability, influence perception, increase appeal, make better design decisions, and teach through design*. Rockport.

Nielsen, J. *How to conduct a heuristic evaluation*. https://www.nngroup.com/articles/how-to-conduct-a-heuristic-evaluation/

Pernice, K. (2017). *F-shaped pattern of reading on the web: Misunderstood, but still relevant (even on mobile)*. Nielsen Norman Group. https://www.nngroup.com/articles/f-shaped-pattern-reading-web-content/

Pernice, K. (2019). *The layer-cake pattern of scanning content on the web*. Nielson Norman Group. https://www.nngroup.com/articles/layer-cake-pattern-scanning/

Schriver, K. A. (1997). *Dynamics in document design: Creating texts for readers*. John Wiley and Sons.

Swann, A., & Dabner, D. (2003). *How to understand and use design and layout* (2nd edition). HOW Design Books.

White, A. W. (2002). *The elements of graphic design*. Allworth.

Williams, R. (2015). *The non-designer's design book* (4th edition). Peachpit.

3

THEORIES OF DESIGN

LEARNING OBJECTIVES

Upon completing this chapter, you will be able to:

- Describe how we receive and perceive visual information to make meaning

- Place various theories of design within a framework engaging with visual perception, visual culture, and visual rhetoric

- Produce documents within that theoretical framework to motivate action and understanding

Sometimes visual design seems simple: We look at a document and feel immediately whether it looks good or has something wrong with it. With practice, some designers develop what is called a "good eye" for design: an intrinsic, habitual, instinctive feeling for what will probably work, based mostly on their own experience.

But, as designers, our feelings about what looks "good" may not match the users' feelings. Rather than relying on feelings, a master designer should be able to *explain* what works and why. Theory is an attempt to provide those explanations—ultimately, to discover and convey fundamental principles of the human experience that can be applied to new design situations.

Finding these explanations can be a difficult but illuminating challenge, primarily because there are so many ways to explain how people interact with documents. Consider the example in figure 3.1.

We can look at this document through a myriad of angles, but three general approaches are most common:

- Visual perception

- Visual culture

- Visual rhetoric

Theories of visual perception concentrate on using psychology and biology to explain how all human beings experience the world through their sense of vision. Understanding how users perceive documents helps us create easy-to-use documents that attract users' attention. So, for example, the leaflet in figure 3.1 is printed in bright colors designed to attract a child's attention, and it asks children to interact with the document and their surroundings by matching what they see on the card to what they see in the museum. Theories of visual culture tend to explain visual design by recognizing how societies and groups establish what visual perceptions *mean*. In the case of the museum leaflet, the images employed represent the European cultures that created the art the

Figure 3.1. Explaining how a design works. This leaflet introduces young museum visitors to art by asking them to be "art detectives." We can explore how this document works through visual perception, visual culture, and visual rhetoric. For example, one panel, perforated into four clue cards, asks children to find the detail pictured on the front by following the clues on the back. Doing so requires perceptual skills such as pattern recognition. The repeated use of magnifying glasses takes advantage of common cultural awareness about detective fiction, going all the way back to Sherlock Holmes. *Source*: The J. Paul Getty Museum, "Be an Art Detective."

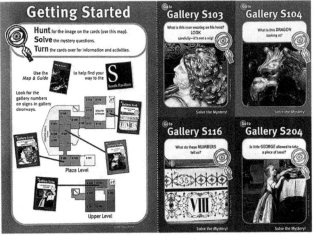

museum preserves. The concept of being a detective is also reinforced by the visual mark of the magnifying glass repeated throughout the leaflet; people know the link between detectives and magnifying glasses from their cultural experiences. Finally, theories of visual rhetoric borrow from both perception and culture to create visual designs that meet specific needs for specific people. This leaflet encourages children to be interested in and excited about the details of art by guiding an interactive activity with the art collection.

Among these three general approaches, scholars and designers have proposed many individual theories. Naturally, these theories do not all agree with each other, and they each have their strengths and weaknesses. Researchers and practitioners from many fields—including physiology, psychology, rhetoric, communication, art, and design—are still trying to figure out exactly how visual experience works.

So rather than trying to reconcile all of these different viewpoints, we will present a framework for understanding them, examining how they work together and where they disagree. Besides, recognizing the breadth of competing theories about visuality will make you a more thoughtful designer than simply choosing one theory and applying it formulaically.

VISUAL PERCEPTION

Humans are highly visual—in the way we experience the world, describe what we experience, and think about our experiences. In fact, our *visual perception* is so central to our personal experience that we often don't think consciously about what we see.

Imagine yourself binge-watching a new show on your tablet. You're watching characters you've come to know and love—or hate. You laugh. You cry. You get angry. At some level you know that you are watching pixels flicker on and off on the screen, but you try to forget that you are watching actors on a soundstage, and that you're watching a video recording created and displayed with technology. But are you consciously aware of all that? And most of us *do* laugh, or cry, or get angry even though we are only seeing patterns of lights on a screen. As a society, we wouldn't be so fascinated with movies and shows if they didn't engage us, didn't move us in some way, didn't entertain us. Somehow our visual perceptions allow us to take these shifting lights and turn them into a compelling visual experience.

This ironic gap between what we *sense* as biological entities and what we *experience* as seeing humans is a central question in understanding visual perception. Gestalt psychologist Kurt Koffka summarized his and others' attempts to understand this irony with the question "Why do things look as they do?" This question may seem simple, but it has proven remarkably difficult to answer, leading to widely different theories of how we perceive visually.

> "Why do things look as they do?"
> — Kurt Koffka

In this section, we'll discuss some of the most influential theories of visual perception. Understanding these different approaches to Koffka's question will help you better understand how users create meaning out of the documents we design.

NEUROPHYSIOLOGY

Neurophysiology explains visual perception by exploring how organisms, and particularly their nervous systems, respond to visual stimuli. In making this exploration, neurophysiologists have discovered features of the human vision system that have a direct effect on how we see the world.

One example is the neurophysiological explanation of how we see colors. For centuries, scientists and philosophers such as Sir Isaac Newton and Wolfgang Goethe theorized about ways to systematize our perception of colors, producing remarkably persuasive accounts of how colors work together. It was only in the twentieth century, however, that scientists could finally explain how the biology of our nervous system creates the sensations we perceive as colors. They did so primarily by subjecting the nervous system to experiments, such as shining a pure light into a subject's eyes and measuring the chemical or bioelectrical response. Collectively, experimental techniques like this are called *psychophysics*.

Neurophysiologists discovered that humans have specialized nerve cells on their retinas known as *rods* and *cones*. These cells have different but complementary responses to light. Rod cells start a nerve response related to the amount of light cast on the retina through the lens of the eye. As a result, they are very sensitive to the edges and shapes of objects. Cone cells, however, invoke a nerve response related to the wavelength of light received. Together, the responses of rods and cones create the physical sensations we think of as color. From the responses of our cones, we perceive different wavelengths of light as different *hues* of color, and from the responses of our rods we perceive different intensities of light as different *saturations* or *brightnesses* of color. (For more information on color perception, see chapter 8.)

Neurophysiology offers powerful explanations of many aspects of visual perception because it concentrates on human beings as biological creatures who have developed through evolutionary changes. For example, neurophysiologists discovered that rods are more numerous and sensitive than cones. This discovery explains why as light recedes at twilight, the color seems gradually to fade away from the objects we view. Under these conditions, our rods are still working at detecting brightness, edges, and shapes, but the smaller number of cones have too little visual stimulus to invoke the hue response. This mechanism also makes sense in evolutionary terms. Early humans who could make out the shadowy forms of predators in dim shadows or the dark of night survived more often than those who had less sensitive perceptions of edges and shapes. The survivors passed on this skill; as a result, evolutionary forces encouraged the development of a multitude of rod cells for sensitive light response. Color, however, became more useful in daylight as a means for early humans to discriminate between what was edible and what was not. Good color perceivers were more likely to distinguish an edible mushroom from a poisonous one or ripe fruit from unripe and thus to survive and pass on this ability to their offspring.

This kind of research into the basic building blocks of sensation is neurophysiology's specialty. Neurophysiologists call these building blocks *primitives*, including basic sensations such as brightness, hue, edges, and patterns. Primitives combine into more complete sensations called *percepts*—for example, the recognition of shapes.

These neurophysiological aspects of experience are important to document design because they help us understand how users respond biologically to what they see. Neurophysiology helps us understand how users detect the shapes they see on pages, and so it is the basis of many technologies of document production, including color monitors and color printing. For example, the realization that the neurophysiology of color perceptions is based on experiences of red, green, and blue led to the development of color monitors, which use red, green, and blue light to create a wide range of colors.

Despite its usefulness and explanatory power, neurophysiology still has its limitations. Neurophysiology tends to focus on biological *mechanisms* of perception rather than on the *psychology* of perception. It explains how visual stimuli affect our nervous system, but it doesn't explain very well how we make meaning out of these stimuli. It makes perception seem a passive biological response, rather than something people do actively. Neurophysiology also implies that our experience of the world is mostly indirect—almost as if our consciousness is hidden deep inside our neural system, weighing and assembling what we sense but never coming into direct contact with anything.

GESTALT

One approach that counters neurophysiology's focus on isolated neurological responses is *Gestalt theory*. While acknowledging what neurophysiology tells us about human perception, Gestalt attempts to create a more comprehensive version of human perception, contending that experience is more than the sum of its parts. Gestalt suggests that we don't just see edges, colors, brightnesses, and shapes—we see fruit, or a tree, or a dog. In other words, we experience what we see as a *whole*—which is also a meaning of the German word *Gestalt*.

The primary figures in Gestalt psychology's heyday (between about 1890 and 1940) were the Frankfurt University psychologist Max Wertheimer (1880–1943) and his students Wolfgang Köhler (1887–1967) and Kurt Koffka (1886–1941). These scientists argued that we should try to answer Koffka's question "Why do things look as they do?" by concentrating on the phenomena of perception as experienced by a naive, uncritical observer. They identified several phenomena that are stable across different people and cultures; these phenomena have become known as the *Gestalt laws of perception*, the most significant of which are these:

- Figure-ground discrimination

- Laws of grouping

- Good figure

These laws have been particularly influential on document design. In fact, you'll probably notice several connections between the following sections and the principles of design in chapter 2.

FIGURE-GROUND DISCRIMINATION

One of the most basic laws Gestalt psychologists identified was that we tend to organize the visual world into two categories: figure and ground. *Figure* is what we perceive as an object, and *ground* is what we perceive as the object's context. Figure-ground discrimination describes how we tell the difference between the two. This Gestalt law has a direct application to document design: it helps us predict how users will recognize design objects (figures) on a page (the ground).

We use a variety of cues to determine what is figure and what is ground. For example, we typically recognize the *smaller* continuous parts of a visual field as figures and the *larger* parts as the ground. Consider the first graphic in figure 3.2, a white square and a black circle. Simply by recognizing these marks as "objects," you have made some judgments about figure and ground. If we take the square as the visual field, then most of us would identify the black circle as a figure on that ground because it is smaller than the surrounding white space. But if we take the entire page as the visual field, then we recognize both the square and the circle as figures on the ground of the page. We also tend to see the black circle as being "on" the paper, when it could just as easily be a hole through the paper (see the discussion of continuation when we discuss grouping later in this chapter for more about why we make these sorts of judgments).

One application of this principle to document design is the concept of *figure-ground contrast*. If the figure and the ground look different, users have an easier time distinguishing one from the other than if the figure and the ground look similar. For example, in figure 3.2, it's easier to make out the black circle on the white square than it is to make out the light gray circle on the dark gray square. As these graphics suggest, figure-ground contrast can be formed with brightness (different values), hue (different colors), or saturation (different purities of color). In addition to the contrast of the figure to the ground, good figure-ground contrast relies on the consistency and distinctness of the figure's edges. Figures with fuzzy edges are harder to recognize because they have poor figure-ground contrast. The last graphic in figure 3.2 shows the same size circle and square but with a pattern applied to each that makes it hard to see where the figure ends and the ground begins.

One of the more general applications of figure-ground contrast is in camouflage, which works by blurring the distinction between figure and ground. But camouflage is rarely a good idea in document design,

Figure 3.2. Figure-ground contrast. Black and white are highly contrastive, making it easy to see the black circle against the white square. The gray square and circle contrast significantly less. The patterned gray circle is difficult to make out against the patterned gray square. *Source*: Created by the author.

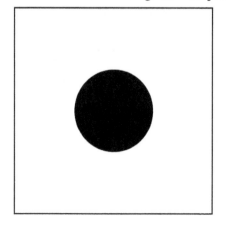  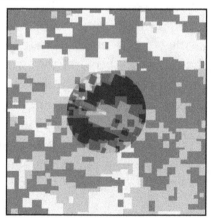

where *inadequate* figure-ground contrast is a common problem, making documents harder to read. Documents that use background images, patterns, or watermarks often provide too little contrast between the ground and the design objects placed on that ground. Text, because it's made up of relatively small and intricate shapes, can easily get lost on a busy background.

GROUPING

Gestalt theory postulates several ways we group figures once we have recognized them through figure-ground discrimination. The most important grouping indicators are proximity, similarity, continuation, and common region.

Proximity and similarity. The natural tendency of users to group figures can be very useful in document design. Designers rely on proximity and similarity as two of the six principles of design you read about in chapter 2 (see figure 2.19). To summarize, the law of *proximity* suggests that we tend to group figures that are closer together, and the law of *similarity* suggests that we perceive similar figures as belonging together.

Continuation. The law of *continuation* suggests that we will assume a connection between figures that are lined up, even if we don't have any direct evidence that they are connected. Consider the graphics in figure 3.3. Most of us would recognize the first graphic as a circle obscuring the top edge of a rectangle, assuming that the rectangle's top edge continues behind the circle. But there is no real evidence that this perception is accurate. In fact, there could be a semicircular gap in the rectangle that the circle merely covers up. The law of continuation describes our tendency to see that upper line of the rectangle as continuing "behind" the circle, despite the lack of visual evidence that this is so. Continuation also explains our tendency to see shapes where they do not exist. For example, the "Pac-Man" shapes in figure 3.4 might seem unitary and

Figure 3.3. The law of continuation. Our sense of continuation "behind" an object is an important tool for recognizing objects, but it can also be misleading. *Source*: Created by the author.

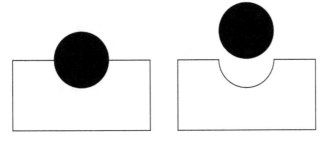

Figure 3.4. Creating shape through continuation. Continuation can create shape perceptions and groupings. Here, the three shapes can be arranged to imply a triangle through continuation, even though there is no triangle explicitly marked. This continuation also makes the three shapes look connected to each other. *Source*: Created by the author.

separate when arranged randomly. But when arranged differently, the same shapes make us see something that's not explicitly there: a triangle. The triangular arrangement, now one of the most common illustrations of continuation and the illusion of brightness enhancement, was first formally described in 1955 by psychologist Gaetano Kanisza and is often referred to as the "Kanisza triangle." Continuation also makes us see things as groups, even when there isn't any positive connection between them. We see the three black objects that form the implied triangle as "belonging together" because the acute angles they contain seem to form lines that continue across the gap.

Continuation is the foundation of alignment, another principle of design discussed in chapter 2. When we align design objects on a page, we suggest to users that those objects are connected by an implied continued line. For example, we almost always align individual letters of text on a consistent, implied line (the baseline). Readers naturally group the letters along this implied line into lines of type, words, sentences, and paragraphs.

Figure 3.5. Variations in text size affect readability. *Source*: Created by the author.

Doing otherwise makes reading harder.

Common region. The law of common region is the foundation for the design principle of enclosure, as discussed in chapter 2. *Common region* claims that people see objects sharing a common space as belonging together (Palmer, 1992). For example, consider the three sets of objects in figure 3.6. The first set of objects seems to be grouped several ways:

Figure 3.6. Forming a common region. A common region can be formed explicitly by drawing boundaries completely around grouped objects or less explicitly by using dividing lines or even negative space. *Source*: Created by the author.

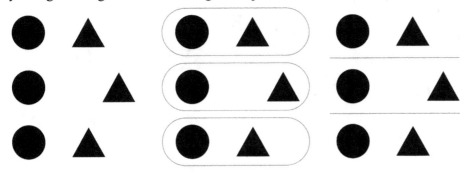

by continuation (rows of shapes), by similarity (circles and triangles), and by proximity (the triangles are closer to each other than to any circle, and vice versa). All of these groupings suggest two distinct groups: the line of circles on the left and the line of triangles on the right.

Yet if we draw boxes around some of the shapes, we form a common region that now pairs each circle with a triangle. This enclosure overpowers the principles of continuation, similarity, and proximity in grouping the objects. Even simple lines drawn between each pair of shapes will invoke a common region because they imply a horizontal connection between the object pairs. Negative space can imply common region, too, if it is not overpowered by other principles.

Document designers use common region all the time—for example, to set off design objects from the text, such as graphics, notes, caution and warning statements, headers, and footers. In electronic documents, common region leads designers to create buttons for important links or commands, such as those in a website navigation bar.

GOOD FIGURE

Gestalt suggests that figures that use proximity, similarity, continuation, and common region consistently and unambiguously are stronger, more stable, and more recognizable than those that don't. For example, examine the two sets of objects in figure 3.7. Both use the laws of proximity and continuation to imply a complex object (lines) built out of simpler objects (diamonds). But because the line on the bottom is straighter (continuation), more regular (similarity), and more consistent (proximity), it makes a better figure than the line on the top.

Users typically recognize good figures in document designs more readily than weaker figures. Logically, if we want users to recognize

Figure 3.7. Creating a good figure. We perceive good figure particularly when Gestalt laws are followed completely. Here, for example, the aligned row of diamonds makes a better figure than the haphazard row. *Source*: Created by the author.

something important, we should make it a strong figure. This application leads many document designers to create simple, rather minimalistic documents that focus on visual clarity, good alignment, and obvious groupings of related information. With the graphical possibilities of software, it's easy to create fancier, more visually complex designs that don't form good figures. These designs might get more initial attention, but they might also be less easy or efficient to use.

GESTALT AND DOCUMENT DESIGN

As you might have guessed by now, Gestalt had an enormous influence on document design, if only because its principles are so easy to apply to documents. In fact, many Gestalt experiments involved asking subjects to look at shapes printed on cards—very similar to what we do when we look at a document on paper or on a screen. Although this parallel has been a boon for document designers, some critics claim that Gestalt works better in the laboratory than in the real world.

In addition, despite how convincing and useful Gestalt principles may be, Gestalt psychologists have never been able to explain successfully *why* Gestalt principles work. After the first half of the twentieth century (when Wertheimer, Koffka, and Köhler were active), perception studies proposed alternative ways of understanding perception that more successfully explained why perception works as it does, but these newer theories don't necessarily work well with Gestalt. For example, Gestalt does not mesh well with neurophysiology, which focuses on finding the explanations for perception that Gestalt fails to supply. Finally, although the Gestalt principles provide good indications of how a *general, universal human being* might respond to a document, they don't always predict how individual users or particular groups of users might respond. Users come from many different cultures and backgrounds and with many different abilities; individual users or particular groups of users often respond differently to the same documents.

CONSTRUCTIVISM

Another explanation of perception is constructivism. Imagine you are backing your car out of a garage (see figure 3.8). Before you start to move, you glance quickly in several different directions: in the side mirrors to your left and right, and in the central rearview mirror. You even crane your neck around left and right to check for anything that might be behind the car. Once you're satisfied that the coast is clear, you back out, still constantly adjusting the course of the car with additional glances forward over the fenders, as well as backward in the mirrors and over your shoulder. If you're an experienced driver, you perform all of these actions quickly, efficiently, and almost unconsciously.

So what have you done here perceptually? You haven't looked at *every* aspect of the world around you—even the mirrors and your backward glances can't reach everywhere behind the car. Instead, you have used your fragmentary perceptions to *construct* a model of the visual world that fills the gaps in your perception. You also constantly test that model against additional observations, building a more and more accurate idea of your surroundings.

Figure 3.8. Constructing a visual reality. We actively construct a visual reality by mentally combining many quick views of our surroundings. *Source*: Created by the author.

Constructivist theories explain this kind of scenario particularly well. Like Gestalt, constructivist theories find neurophysiology unsatisfying because it implies that people are mostly passive receivers of stimuli. But unlike Gestalt, which posits that we understand the whole visual scene at a glance, constructivist theories propose that human beings perceive the world in fragments from which they actively construct a visual reality. Constructivist theories focus on explaining how we use the limited building blocks of sensation and perception to create bigger concepts of the visual world we live in.

Most constructivist research focuses on how we build a visual world by *sampling* the visual ecology by taking in perceptions a bit at a time—what John M. Findlay and Iain D. Gilchrist (2003) have called *active vision*. When we look at an object, constructivism suggests, we don't see it all at once; instead, we take multiple visual snapshots, looking at different aspects of the object. These snapshots are called *fixations*, and the fast eye movements between fixations are called *saccadic movements*. According to constructivism, we use repeated fixations on different parts of the visual field to build up a picture of what we perceive.

As a result, we never actually see the entire visual field precisely or completely—rather, we put together our perceptions dynamically, filling in the gaps with memory and experience. Even in the perception of a single visual field, we must use short-term memory to remember what we saw in the previous fixation and add it to what we see in the next. Long-term memory and general experience affect our perceptions as well. One experiment, for example, found that when children were asked to estimate the physical size of coins shown to them, they overestimated the size of the coins of higher value: the higher the value of the coin, the greater the overestimation of size. The children's knowledge of the relative monetary values of the coins, in other words, influenced their ability to estimate size visually (Gibson, 1979).

Constructivist theories suggest that humans start with hypotheses about how the world works. For example, the children in the coin experiment started with the general idea that bigger things are often more valuable than smaller ones. But we constantly seek out and absorb additional information about the world, and we just as constantly readjust our hypotheses. In fact, constructivism suggests that we have to stay on this constant search for visual information because our visual apparatus doesn't really provide enough data for our minds to create a sense of the world. Instead, our minds supply most of our mental model of our surroundings, using incomplete and fragmentary sensations to build more and more accurate mental models. This explains how we can make our way through a familiar room blindfolded, but in an unfamiliar room, we must watch carefully where we're going, taking in lots of visual snapshots to build a mental model of our surroundings. In this regard, while acknowledging the importance of neurophysiology, constructivism moves the locus of perception even further into cognition and psychology, but away from biology.

Constructivism has been influential in document design, particularly as researchers have used eye-tracking equipment to record how users look at a document. As we mentioned in chapter 1, eye-tracking research has found that for most printed documents, readers of English and other left-to-right languages fixate first at the upper left-hand corner of a document. Then, in a series of saccadic movements and fixations, users move their eyes in an F or Z pattern down the rest of the page, scanning for the information they need. The more consistently the text on the page fits these patterns, the more likely that users can make clear connections between their fixations as they build a picture of the visual field and its contents. (See chapter 5 for further discussion of this observation.)

Constructivist theories of perception reveal the importance of giving users clear visual cues about the structure and content of documents. For example, obvious headings, labels, and titles for text and graphics give users' eyes somewhere to rest momentarily as they build a mental model of the page, spread, or screen. Without these cues, users must slog their way through the entire document, rather than using their natural fixations and saccadic movements to build up a concept of the document's structure and contents efficiently and dynamically.

ECOLOGICAL PERCEPTION

Like neurophysiology, however, constructivism implies a separation between people and the world that some scientists and philosophers find troubling: it suggests a fragmentary and incomplete world we can never fully comprehend. Ecological perception counters constructivism by suggesting that we perceive a full and rich world of sensory data directly and immediately, without the intervention of discrete neurological responses or prior hypotheses—a phenomenon called *direct perception.*

Psychologist James J. Gibson, the major proponent of ecological perception and direct perception, argued that rather than perceiving the world as tiny, discrete sensations or percepts that we build into bigger concepts, we experience the surrounding world or ecology simultaneously, in a rush of available data that we interact with dynamically and unconsciously (1979). But unlike Gestalt, which focuses on a mostly stationary viewer, ecological perception notes that we do not stand still while we look at our ecology. Without much thought, we cock our heads to the side to get a better angle of view, we move closer or farther away from an object, and we walk around and look at the back side of whatever we're examining.

According to ecological perception, we also consider the relationship of objects to our own bodies and recognize how we might use objects. For example, we might recognize an object as something we can grasp, twist, pull, push, climb on, walk across, pick up, or throw. The qualities of objects that allow us to use them in some way are called *affordances* (figure 3.9). The object's relationship to our bodies determines what extent of graspability, sitability, or walkability it affords.

In document design terms, we must provide affordances in documents so users understand what they can use to work with the document. In designing the physical document, we can provide tabbed indexes, convenient bindings, and other physical features that make a document easy to use. In terms of page design, we can provide clear and definite navigational cues, such as links, buttons, and form fields that users can easily recognize as design objects they can *do* something with. This recognition of the importance of affordances is one reason why the cursor in some web browsers turns into a small, pointing hand when we place the mouse over a link. The hand implies that the link is something that can be *touched*, if only remotely through a mouse-click.

This focus on affordances in the world that surrounds us fits in particularly well with discussions of *usability* in document design. Usability studies can take into account all aspects of the user experience—not just the surface of the page, but the back of the page, its size and shape, even its materials—to find how users interact with the objects in their ecology (for more information on usability, see "Visual Rhetoric as User-Centered Design" later in this chapter). The idea of affordances has also been very influential on *interface design* for software and websites, and in physical

Figure 3.9. Examples of affordances. We recognize affordances as tools for interacting with things around us here by pulling a chain, turning a doorknob, or flipping a switch. Documents can include similar affordances, helping users manipulate and work with information. *Source*: Created by the author.

design as well, as described by Donald Norman in his seminal work *The Design of Everyday Things* (2013). Designers create visual-spatial cues to suggest to users what affordances the interface offers—that is, what can be clicked, dragged, pushed (as in buttons), or otherwise manipulated. So promising is this connection that designers have begun to think of design much more holistically as part of the user's ecology, labeling their efforts *interaction design (IXD)*, or even more broadly, *user experience (UX) design.*

However, if constructivism implies that too many things (i.e., values and preconceptions) come between people and the world, ecological perception implies that there is nothing between people and the world. Ecological perception's emphasis on direct perception without the intervention of cognition moves us farther away from the mind and back into the senses. For this reason, although many have found the concept of affordances useful, Gibson's concept of direct perception has found fewer supporters.

THEORIES OF VISUAL PERCEPTION INFLUENCE DOCUMENT DESIGN

Because these theories concentrate on explaining human beings' basic interactions with the world, they have had a great influence on document design. Documents, after all, are part of the world we live in and interact with. If our designs do not take the basic dynamics of users' visual experience into account, they may be unsuccessful or even inaccessible to users.

All of the theories of visual perception we have discussed attempt to explain how we recognize objects. Object recognition is simultaneously all of these things:

- A fundamental part of our neural apparatus for seeing (neurophysiology)

- A technique we use to mark the boundary between one object and its surroundings (Gestalt figure-ground discrimination and good figure)

- A way to recognize how we might interact with our environment (ecological perception, constructivism)

All of this thought about object recognition helps us design usable documents. For example, consider figure 3.10, an excerpt from Hugh Cleveland

Figure 3.10. A document lacking visual clarity. This text from a scanned PDF document (and a close-up) shows how theories of perception can help explain problems of reading. *Source:* Created by author from Ross, H. C. (1952). *Substitute for fear.* House-Warven, p. 54.

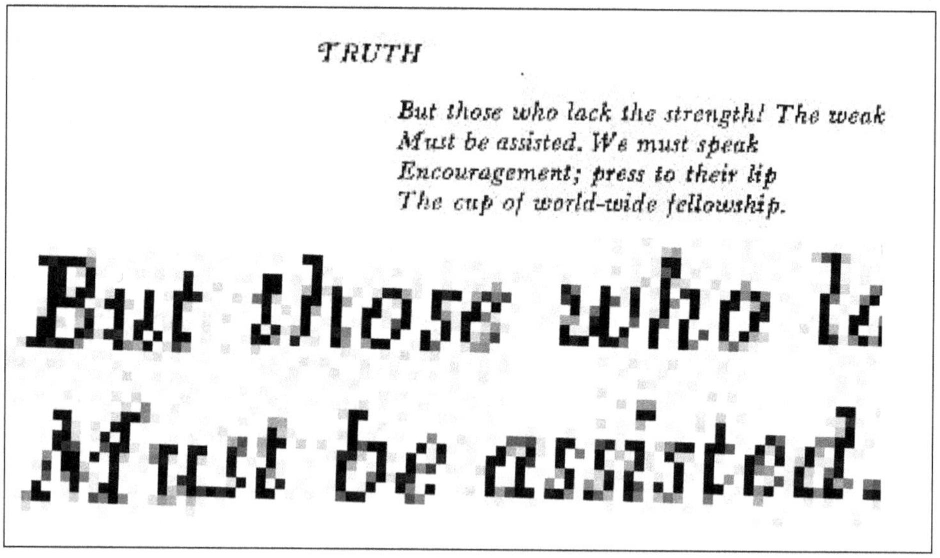

Ross's *Substitute for Fear* (1952), which has been scanned and saved as a PDF (Portable Document Format). We might not read text like this very often anymore, but in your role as a document designer, you absolutely must be able to consider what will happen when you move objects back and forth between the physical and digital worlds. In this case, when the book was first designed as a paper document for a print-based publication, this text was easy to read. Since someone needed a digital copy, and the book had been released as print-only, the pages were scanned to PDF. Because of the limitations of the scanner, or the software, or the settings when the document was scanned, the document lost a lot of clarity for on-screen viewing, and it becomes even fuzzier when it is printed out again, as it is here.

It's easy to see what's happened here. In the translation of the printed text to PDF for distribution online, the page image had to be downsampled to a lower resolution, both to prepare it for viewing on a screen (which has coarser resolution than a printed page) and to make it small enough as a file to transmit quickly. Unfortunately, these constraints lead to a document that users will find hard to read, whether they attempt to do so on the screen or print out the document. The letters have blurry edges, and they are not particularly dark, so they don't have good contrast against the white paper; they also have halos of extraneous gray smudges around them. The technology you now have access to might keep this from happening—or it might not. Changes between digital and physical states might have almost no impact, or they might result in dramatically altered documents.

Theories of visual perception provide us with good explanations why the characteristics that appear in this case make a document hard to read. Neurophysiology tells us that our rod cells are sensitive to boundaries between one pattern and the next. If the boundary isn't clear, we'll have to look at it more closely to see it. Gestalt suggests that we can distinguish figure from ground more easily if the figure is "good"; the letters in figure 3.9 are slightly hazy and not clearly aligned, making it hard to tell figure from ground. Both constructivism and ecology suggest that we gather information from the environment to interact with it; if the information is vague or hard to discern, how can we do so effectively?

However, as we've noted particularly for Gestalt, most theories of visual perception focus on human beings in a general, universal sense. This focus is a strength because it gives us basic principles, like the Gestalt laws, that should apply well to most communication situations. But it's also a weakness because it doesn't allow well for the effect of our *nurture* as well as our *nature*. In other words, theories of visual perception don't always give adequate weight to the influence of our cultural and social experience on what we see.

VISUAL CULTURE

If you've ever made a reference to a movie, song, or book that was met with complete lack of understanding and blank stares, you've run into a cultural barrier to understanding. We are all products of our culture—we were raised in a particular way, by unique people, around other unique people. We watched certain shows growing up that came to mean a lot to us, or read certain books, or listened to certain music. There's a group of people in the world who get all of the inside jokes and references to *Spongebob Squarepants*, but those same people may not know *Ren & Stimpy* or *The Animaniacs*. Some folks can tell you the registry number of the USS *Enterprise* (NCC-1701), or that *Serenity* is the ship captained by Malcolm Reynolds in *Firefly*. Some of us can immediately picture Gilligan from *Gilligan's Island*, others know all about Meredith Grey in *Grey's Anatomy*. Some of us see a 112-ounce can of pudding and immediately associate it with Carl in *The Walking Dead*. We can do the same thing with music, or art, or sports teams, or even politicians and political scenarios. The point is that we form cultural associations with particular elements that have meaning to some people but not to others.

When we use those associations in the form of visuals in our designs (a face, a team jersey, a cartoon, sometimes even just a particular color combination) we are drawing on *visual culture* to help form a narrative.

The images from shows, movies, music videos, sports reels, and more draw on images created in studios and in the world and shared via electronic signals and flashing lights, so they start with our visual perceptions, but they focus more fully on the connections we make between those perceptions and our culture or the culture of those who created the program. Visual culture influences how we ascribe *meaning* to what we see. When we watch something as visually complex as a sitcom, we don't rely just on our visual perceptions—we use the web of associations, connotations, and connections that our visual perceptions evoke.

The same thing applies to the documents we use, which are complex cultural objects as well as physical and perceptual ones. Visual culture is particularly important to document design because documents are not just part of our natural visual environment. Instead, someone who comes

Figure 3.11. Carl Grimes, from *The Walking Dead*, eating a 112-ounce can of pudding on a roof. If you know the story, you associate a particular set of emotions with this image—joy, empowerment, an odd sense of childlike peace, a kid getting to be a kid in a terribly difficult world, perhaps. If you don't, it's because you weren't part of this particular aspect of popular culture. As a designer, it's your job to be aware of the types of cultural associations your audience will likely draw (or not) from the visuals you use. *Source*: AMC, *The Walking Dead*.

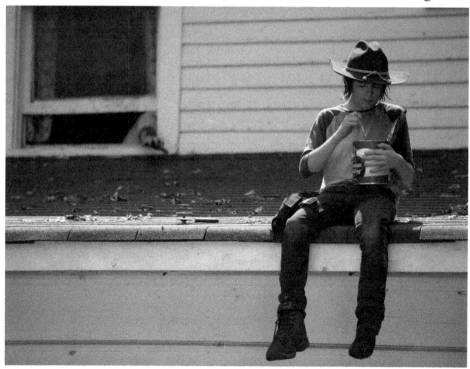

from a particular cultural background designs the documents. And the documents are used by people with their own unique and varied cultural backgrounds. Both the designer and the user rely on a complex field of conventions to convey meaning. For example, the size (or *font*) of the type used for different design objects in this book conveys a particular meaning. We use the largest fonts for chapter titles and headings; we use a smaller font for basal text; and we use an even smaller font for captions to the figures. The contrast between these sizes helps convey the book's structure. Simply put, in hierarchical terms the bigger parts (chapters, sections) are marked with bigger type, while smaller parts are marked with smaller type. This use of fonts to show structure isn't an accident—it's a result of strong visual conventions with which designers consciously guide users.

VISUAL LANGUAGE

In some ways, visual communication can convey as much as written communication, forming in effect a *visual language*. For an example of this, let us consider driving a car.

When we drive, we rely not only on our own perceptions to guide the vehicle but also on the "rules of the road." We must learn these rules before we are permitted to drive, and we are penalized when we break those rules. Driving laws govern not only our own actions but also those of other motorists to keep them from running into us. After we have been driving awhile, we internalize these rules and do not have to think much about them. This internalization reduces our cognitive load, allowing us simply to follow the conventions without giving them much thought.

But if we visit a country where the rules are different, our cognitive load increases dramatically. Then we must constantly remind ourselves of the simplest things, such as what side of the road to drive on (see figure 3.12).

This example parallels in some ways our use of language. When we are young children, we think about our language a lot, constantly acquiring new words and rules for combining them into expressions. After a while, though, we stop thinking about the words themselves and simply speak. It's only when we try to learn a new language that we must think about language rules again. In other words, languages work by conventions, such as the shared agreement on what words mean (semantics), how they are spoken (phonetics), and how they can be combined (syntax).

Similarly, visual language supposes that people within a culture have well-developed general agreements or *conventions* about what images can mean, how they can be presented, and how they interact. For example,

Figure 3.12. Considering cultural conventions. All conventions have a cultural element—some drive on the left side of the road, some on the right. The conventions of visual design are no exception. *Source*: Traffic jam Bangkok, photo by Sky 269 (originally color), Creative Commons Attribution-Share Alike 4.0 International license.

Charles Kostelnick and Michael Hassett (2003) have discussed at length the visual convention of the arrow and similar pointers (see figure 3.13). Arrows are exclusively *visual* marks—we can't read an arrow out loud the way we can read a written word. Yet users from many cultures recognize arrows as meaning something—usually "look here," "go this way,"

Figure 3.13. Considering visual conventions. Pointer marks can come in many shapes and sizes, but visual conventions help us understand that they mean similar things. *Source*: Created by author from https://stock.adobe.com.

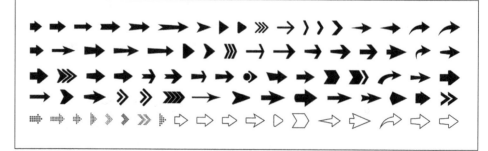

or "do something with this." Some scholars have labeled our ability to "read" visual images as *visual literacy*. According to Paul Messaris and Sandra Moriarty, visual literacy is "the viewer's awareness of the conventions through which the meanings of visual images are created and understood" (2005, p. 481).

INTERCULTURAL COMMUNICATION

Visual conventions aren't determined just by popular culture but by societal culture—our intercultural expectations. One often-used example is the meaning of *red*. In most Western cultures, red signifies danger—that's why our stop signs and stop lights are red. But in Japan, China, and other Asian countries, red often means happiness, joy, and celebration.

So in document design terms, a document designed for one culture's conventions of visual language might not work for another culture. With this realization comes a big problem, since we live in a global community with international businesses and organizations. How can a global business create a document that will work successfully in the visual languages of all of the cultures with which it must communicate?

Two general approaches hope to solve this problem of intercultural communication: *globalization* and *localization*. Designers using a global approach try to create documents (usually in simplified English, the most common language on the planet) that they hope most people can understand. By avoiding colloquialisms, metaphors, local expressions, and local visual conventions, globalization attempts to create a document that will succeed for most people in most cultures. Globalized documents, however, can never hope to speak successfully to people of all cultures, and this approach often produces oversimplified, unsuccessful documents.

Localization takes the opposite approach: rather than creating one very basic document, it creates different versions of a document for each set of users in different cultures. Localization often involves translating documents into a different language, and it can also involve creating new designs for documents using local visual conventions. However, good localized designs often go beyond simple translation. They *articulate* the parts of a document to meet the needs of an intended audience (see, e.g., Slack et al., 1993). Articulation involves revising so that images, expressions, and linguistic constructions are relevant to the target audience in a way that translation-only localization might not be.

When localizing a document, designers typically try to take into account all they know about the particular culture that will use it. Ideally, localization helps designers create culturally sensitive documents. But localization inevitably costs more than globalization, making some organizations reluctant to invest in this practice except for their most mission-critical communications.

When designing for audiences outside of your own culture, it's important that you do your research. Designers often fall into the trap of basing designs and language on cultural stereotypes—you probably know many, and, regrettably, there is no shortage of them online and in the news. When localizing a document you may make mistakes, but you can decrease the likelihood of mistakes by talking with folks from the target audience, looking at designs that have worked in the past, and generally becoming as familiar as possible with the culture for which you are designing.

SEMIOTICS

Given the complexity of visual communication (to say nothing of intercultural visual communication), theorists have tried to find more accurate ways to describe how people ascribe meaning to visual language. One common approach comes from *semiotics*, the study of signs and sign-systems. Its most influential early proponent, the anthropologist Ferdinand de Saussure (1857–1913), suggested that a *sign* is actually a complex formed by the relationship between two components: the *signifier* (the mark we see) and the *signified* (the concept the mark points to). The signifier can be any visual mark, including images and letters. However, this relationship is ambiguous and unsettled; it depends on people's willingness to accept it to make it real. In other words, we rely on convention to link a signifier (i.e., the written word *bird*) and its signified (the concept of bird) and make meaning.

Charles Saunders Peirce (1839–1914) proposed a way to understand signs with even greater discrimination. Although Peirce (pronounced *purse*) used different terms than Saussure, he suggested that there are three kinds of signs, depending on the tightness of the signifier's connection to the signified:

- Icons are signs that look like the thing they signify. Most representational images—photographs, for example—are icons.

- Indexes are signs that have a clear connection to whatever they signify (i.e., they indicate something). Peirce's example was a signpost, which indicates the road to the destination written on the sign.

- Symbols are signs that have an arbitrary relationship to what- ever they signify; Peirce described them as signs that "have

Figure 3.14. Examples of icon, index, and symbol. C. S. Peirce's concepts of icon, index, and symbol can help us understand how people find and create meaning from what they see. *Source:* Created by author from author photo and https://stock.adobe.com/.

become associated with their meanings by usage." Examples include written letters, words, and texts, as well as signifiers that don't have a direct relationship to whatever concept they signify. For example, signifiers such as #, *, and $ have only conventional relationships to the concepts "number," "note," and "money."

So in figure 3.14, the photograph of a house is an icon because it *looks like* or *represents* a house. The skull and crossbones carry meaning as an index because it *indicates* that anything associated with the image is dangerous. And an arbitrary mark like the international "no" or "prohibition" sign is a symbol because it has no direct connection to anything except the connection we've collectively decided upon.

Of course, the category into which a sign falls depends entirely on how we use the sign. A picture of a house can be an icon when it's used to mean *this particular house*. But it could also be used more generally to signify the general concept of home, which is why we still see it used in the toolbars of most web browsers to mark a link to the user's home page (figure 3.15). Used this way, this mark would be an index because it points to a *metaphorical* home on the web: the base page the user wants to see upon first opening the browser. However, this mark doesn't

Figure 3.15. Indexes vs. icons. In this context, a web browser toolbar, the image of the house acts as an index rather than an icon because it points a user to their home page, not an actual house. *Source:* Created by the author.

represent any particular house (as an icon would), and it isn't entirely arbitrary (as a symbol would be).

Moreover, semiotics reminds us that conventional signs are not stable or unitary. Rarely can we assume that a mark means only one thing; more often, it can mean several or many possible things. This parallels spoken language, where a word might have one *denotative* meaning (like a dictionary definition) but many *connotative*, or implied, meanings—a concept known as *polysemy*. Denotatively, the word *father* means an organism that offers genetic material to create offspring; connotatively, *father* can have many implied meanings, from positive (warmth, nurture) to negative (control, discipline). Roland Barthes (1915–1980), in particular, brought to our attention the relationship between signs and connotations. The constellation of possible meanings that surround a sign relies almost entirely on culture, society, and usage.

An awareness of the complex, imprecise, and sometimes arbitrary nature of signs helps document designers because it makes us sensitive to the many ways users might interpret the marks on a page. Often, users latch on to a denotation or connotation we didn't intend. As document designers, we must be on the watch for marks that users might interpret differently than we intended. We also need to *ask* users what they think something means by testing the usability of our designs.

VISUAL CULTURE AND POWER

As we've seen, signs can act as a visual language; we even use signs, in the form of letters, to signify our spoken language. And the meaning of those signs is mostly conventional and culturally derived. But this observation begs a question: *Who* decides what signs mean?

In some cases, signs come to mean what they do organically, from the habit of usage by many people across the years. Alternately, the meanings of signs are established by an authority, such as a government or an international standards body. For example, the design of stop signs in the United States is determined by the US Department of Transportation.

But in many cases, power plays a role in how meaning gets attached to signs. People who have the money and freedom to speak louder and broadcast their ideas through the media are more likely to establish the meaning of a sign than those who are poor, isolated, or silenced. One example of this dynamic is the use of corporate logos, such as the Nike symbol, the Coca-Cola logo, or McDonald's golden arches. Each of these corporations has spent billions of dollars on advertisements that constantly stress positive associations between these symbols and the company and its product. They've been so successful in doing so that it's hard to find a place in the world where someone *won't* recognize these symbols as meaning what the corporations wish them to mean. If these corporations didn't have the money to advertise their logos on such a large scale, we would not recognize them so easily. The Nike symbol would be a meaningless squiggle, we might think Coca-Cola was a brand of hot cocoa rather than soda, and we might take the golden arches to just look like a stylized letter *M* (which, of course, it is—polysemy again!). In the meantime, these corporations have so successfully wielded their brand power that they have probably pushed out of existence many smaller businesses that might make a better pair of shoes, a tastier drink, or a more palatable hamburger.

Scholars interested in cultural studies focus on these issues of power as they relate to visual communication. The power of some people or organizations to set the meaning of signs in such a society can amount to what Antonio Gramsci (1891–1937) called *hegemony*, the tendency of particular ideas, usually associated with powerful people, institutions, or even general cultures, to limit and even control the ways we interact and communicate—in a sense, to write and enforce the rules of discourse. This power has positive benefits. For example, we wouldn't want people to make up their own rules for voting in public elections. But it also makes individuals conform in their meaning-making. For example, many US citizens find the term *socialism* unappealing, even though some US allies, such as Britain, Sweden, and Italy, have more or less socialist governments. In its most basic form, socialism refers to the collective ownership of something. Even in the United States, we use socialism for some aspects of our system, such as publicly owned electric power plants, air traffic control systems, and national parks. But the common wisdom that socialism is dangerous, emphasized repeatedly by US politicians and powerful social groups, limits our ability to use the term without raising people's fears through the specter of Stalinist communism, fascism, and war.

Of course, corporations and governments can never entirely control language or visual communication. Even powerful organizations must try to speak to users in a language they can understand and respond to. As a result, people also have a certain amount of power to determine the terms in which they're spoken to.

VISUAL CULTURE AND ETHICS

From this recognition of the relationship between visual culture and power, document designers have grown much more aware of their ethical role in working with both clients and users. As we discussed in chapter 1, most document designers serve as the representatives of clients who can pay them for their work (we hope). But when you rely on the client for your income, you may encounter pressure to act in ways that are good for your client but not so good for the people who use the documents you design.

As a document designer, you will often encounter situations in which you must decide how to balance the agendas of your clients with the needs of users. You must also find appropriate compromises between your own ethical values and the client's objectives. For example, imagine you are hired to work on the website design for a firearms manufacturer, like the one in figure 3.16. Depending on your ethical perceptions about guns, you might have a variety of ethical qualms about working on this project. If you don't agree with the prominence of guns in US culture, you clearly must find a way to negotiate between your ethical values and the client's need to sell the product. You may ultimately find that you cannot work on the project, even if it means losing money or even your job. On the other hand, if you do not object to guns, you might still have ethical problems with the image of the waving US flag on the website. Does this design cheapen the meaning of the flag by using patriotism to sell

Figure 3.16. Raising ethical questions. Design projects always involve ethical questions, regardless of the designer's personal views. Depending on your social values, professional values, and moral sense, you might consider the red, white, and blue color scheme and a waving American flag either acceptable or unethical. You might also question the imagery and color palette of a gun designed exclusively to attract women, or even the implication that pink is inherently feminine. *Source*: Accu-Tek Firearms, http://www.accu-tekfirearms.com/.

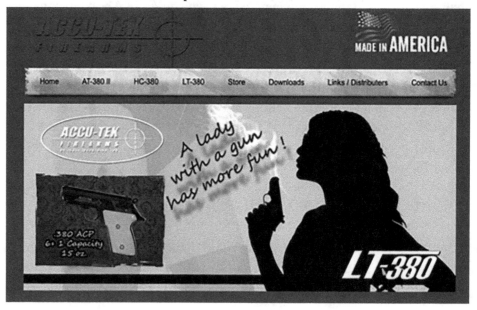

a product? Does it conflate patriotism with the right to bear arms? You might also question the color palette (a pink background) and imagery (the gun has a pink grip) associate with a gun-marketing campaign aimed exclusively at women: Does "a lady with a gun" really have more fun?

Consider a different scenario: You're asked to produce a brochure for a pharmaceutical company. The client wants a cover photo of happy, healthy people to emphasize the company's commitment to community service. But you discover that the company has been indicted in lawsuits for rushing unsafe drugs into the market. Is the planned design an unethical manipulation of users' feelings? If you continue to work on the project, are you involving yourself in that manipulation? These questions are thorny and difficult to resolve. What is most important is to make a conscious, informed decision about your actions in document design, striving to create coherence between the ethical standards you hold and the work you do.

Design Tip:
Making Ethical Decisions in Document Design

Thinking of visual design in cultural terms requires that we also think about the ethics of our designs. Document designers must make decisions about issues such as these:

- How do we design documents that meet the needs of users when those needs might be at odds with those of our clients?

- In emphasizing the positive features of the client's agenda in our designs, are we obscuring the more negative aspects?

- Do our designs represent the world (whether information about a service, program, product, or policy) accurately, without distortion? For example, in what ways might our charts and graphs distort the data they're meant to convey—either by presenting data inaccurately or by omitting data that contradicts the client's aims?

- Does our work forward ideologies or ends with which we can't agree? How do we manage to make a living while standing up for our own ethical principles?

None of these questions is easy to answer. And the answers depend on many competing sets of moral values, including your society's values, your profession's values, and your own values. The answers you come up with will also depend very much on the particular situation you find yourself in, making it difficult to make black-and-white pronouncements about what's right or wrong.

VISUAL RHETORIC

Thus far in this chapter, we've discussed theories of visual perception and theories of visual culture. But you may have noticed that both sets of theories tend to focus on *describing* how people respond to the visual world they live in—either the natural ecology or the artificial one created by culture. These descriptions are valuable, but how do we use them when we design a document?

One way to address this problem is to focus on a third set of theories, which we'll categorize under *visual rhetoric*. You may have heard the word *rhetoric* used before to describe empty or cynical communications—for example, one politician saying that another's words are "just rhetoric," rather than honest facts. But rhetoric has a much broader scope: Paolo Valesio has defined rhetoric as "all of language, in its realization as discourse" (1980, p. 7). Generally, most theories of rhetoric use the term to discuss how people try to influence each other through human communication. Kenneth Burke, for example, called rhetoric "the use of language as [a] symbolic means of inducing cooperation" (1950, p. 43). More specifically, Wayne Booth argued that rhetoric is most concerned with a "rhetorical stance" negotiated among three essential elements: "The available arguments about the subject itself, the interests and peculiarities of the audience, and the voice, the implied character, of the speaker" (1963, p. 141). Rhetoric, in other words, focuses on the relationships among a speaker or *rhetor*, the rhetor's goals, and the people to whom they are speaking.

Rhetoric has long been used to analyze these relationships in written communication, especially the relationships between writers, their goals, and the audiences to whom they write. This analytical process also helps rhetors to plan and create communications that will fulfill rhetorical goals with particular audiences.

Visual rhetoric extends rhetorical theory by taking into account the visual design of a document, as well as its words. In this regard, it synthesizes and applies what we have discovered about visual perception and visual culture. Visual rhetoric is a way of understanding the relationship among clients, users, and designers. It takes advantage of what theories of visual perception tell us about the nature of human viewing in general and what theories of visual culture tell us about how people from different cultures view created objects. Visual rhetoric applies our understanding of both perception and culture to speak convincingly and effectively to users.

In the sections that follow, we'll discuss two aspects of visual rhetoric. First, we'll discuss it as a collection of strategies for persuading users. This aspect of visual rhetoric focuses on the client's goals and the designer's attempts to represent those goals visually to an audience. It also includes the question of visual ethics. Then, we'll discuss visual rhetoric in terms of *usability*, a term that brings the focus of the rhetorical stance to the user and his or her needs.

VISUAL RHETORIC AS PERSUASION

Visual rhetoric applies to the design of all documents, from websites, like we saw with Accu-Tek Firearms; to manuals, reports, and proposals; to advertisements, posters, and flyers; and to anything else designed to motivate action or engage users. Of these, proposals (in all their various forms) seem to rely the most on persuasion. When we write a proposal, we are by definition putting forward an idea to someone who can make a decision to accept or reject it. That proposal might be to convince someone to allow us to do a variety of things:

- To perform a service, such as offering to design a website (a services proposal)

- To provide a product at a particular cost and under certain conditions (a sales proposal)

- To do a project that someone finds worthy of financial support (a grant proposal)

In all of these situations, we must convince the decision-maker to allow us to do the work or provide the goods. We must also convince the decision-maker that our proposal is better than that of any competitor. What we say in our proposal is obviously important, but the visual design of our proposal can play a role in convincing the decision-maker that our proposal is the best. A proposal with a clear, usable, smart-looking design is often more convincing than one that looks confusing, awkward to use, or clumsy. In other words, a well-designed proposal reflects on us, making us look like we know what we're doing.

But do we need to convince a user of anything when we design a report or an instruction manual? Often, we do. Corporate annual reports, for example, often have a persuasive mission as well as a reporting mission. They not only report on the financial performance of the company, but they also make stockholders feel confident about the corporation so its stock price goes up. And corporations also hope their annual reports will encourage potential stockholders to buy stock in the corporation.

Even something as seemingly value-neutral as a set of product instructions must also convince users of a variety of things. For example, the instructions must convince users that the instructions are worth reading. The instructions must convince users that the instructions are authoritative and trustworthy. And they must convince users that the product the instructions explain was worth the cost of purchase. Finally, a well-designed set of instructions can persuade users that they should buy more products from the company.

It's particularly easy to see visual rhetoric at work with websites designed to sell products. On any number of sites you'll see designs selected to invite users in. Typefaces are selected to make pages look open and inviting, color directs users to important information, and images direct us to think or feel a particular way about products. We can easily analyze sites by looking at the rhetorical strategies at work through the three aspects of rhetoric first identified by Aristotle:

- *Ethos* refers to the sense the user gathers of the speaker's character (or in our terms, the client's character).

- *Pathos* refers to aspects of the document intended to evoke an emotional response in the user.

- *Logos* refers to the logical or factual information conveyed by the document.

This rhetorical framework makes it easy to analyze how a document's visual rhetoric works and how well it works.

Of course, the Aristotelian approach is only one model. There are many other alternatives, some more suited to specific document designs and situations. Kenneth Burke's rhetorical theory, for example, works particularly well in helping us understand how images can influence users. Burke analyzed rhetoric in terms of what he called the pentad of act, scene, agent, agency, and purpose.

For example, figure 3.17 shows the privacy policy pamphlet from the banker Wells Fargo, with a highly cropped photograph on the cover. From the photograph—and from knowing about Wells Fargo's origins as a shipping agent in the nineteenth-century American West—we understand that the *scene* is next to a wagon (which also serves as the company's logo). The *agents* are the two people handling the strongbox. The *act* is that of the agent on the ground handing the strongbox up carefully to the agent on the wagon. The *agency*—the means by which the act is completed—is the strongbox itself, with its sturdy construction, prominent lock, and equally prominent label. The purpose for the agents' care in handling the box is clearly to safeguard whatever the box contains—keeping it locked up and handing it directly to someone trustworthy.

This application of Burke's rhetorical theory helps us understand what the designers were thinking when they chose this image for this

Figure 3.17. Influencing users through a photograph. This leaflet uses visual rhetoric to convince customers that the bank will keep information as secure as it once kept money and shipments in the old West. *Source*: Wells Fargo and Company, "Keeping Your Information Safe and Secure."

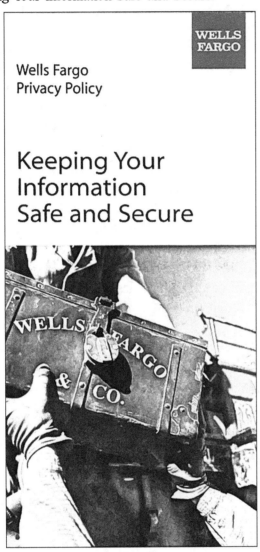

design. Together, the Burkean elements in this dramatic scene imply that Wells Fargo employees today will guard our privacy from theft as carefully as their predecessors handled letters, papers, and currency shipped through the company across the dangerous Western frontier. In fact, this rhetorical analysis reveals that the designers probably *created* the image specifically for this purpose. The photograph is not a nineteenth-century original but a photograph staged for this document (or similar documents that hinge on the idea of safekeeping). The analysis allows us to think about how we can create or use images in our own designs to reinforce the rhetorical purpose of the documents we create.

VISUAL RHETORIC AS USER-CENTERED DESIGN

User-centered design recognizes that successful visual rhetoric requires not just a convincing visual presentation, but a design that helps users fulfill their own needs and agendas rather than just those of the client. This focus on the user is inherently rhetorical. As Robert R. Johnson (1998) argued, rhetoric is closely tied to user-centered design because it foregrounds the need for communicators to understand their audience.

Document designers have increasingly come to realize their responsibility to represent the interests of users in the visual rhetoric they create. But user-centered design is also smart from the client's perspective: It's easier for a client to convince a user of something if designers create a design that meets the user's needs and expectations. A successful user-centered design is said to have good usability.

To understand user-centered design more fully, we can turn our attention to one of its most prominent proponents: Jakob Nielsen. Although Nielsen's work has focused mostly on website design, it applies equally well for other kinds of document design. Nielsen (2000) suggests that usability can be defined through five variables:

- Learnability: The quality of the document that allows users to figure out how to use it. Users are likely to find more success with documents that are easy to learn.

- Efficiency: The ease and speed with which users can use the document. A document users find efficient to use is one they'll probably use more effectively.

- Memorability: How well the document helps users become habituated to its means of use. A memorable document is more likely to be used and used effectively.

- Error avoidance: How well the document helps users avoid errors—particularly navigational errors—and recover gracefully when they happen.

- Subjective satisfaction: How well the document convinces users of its utility. If the document fulfills a need, users will trust it and use it repeatedly.

These variables are only one way of looking at usability and user-centered design, but they foreground the importance of designing documents that meet users' needs and expectations.

Finding out what those needs and expectations are, however, can be a challenge. The most important impact of the user-centered focus in document design has been a growing attempt to understand users better. Designers do so by using two primary approaches: user research and usability testing (for more on these topics, see chapter 10). They conduct user research to find out more about the users, employing a variety of techniques. For example, user researchers (or document designers wearing their "user-research hats") might conduct surveys to find out about the users' background, education, and other demographic information. Or they might conduct focus groups by gathering a group of potential users and asking their opinions about different subjects related to the document or about a draft of the document itself.

Rather than just asking users what they *think*, designers conduct *usability testing*, asking users to perform a series of tasks with a document—for example, finding the customer service phone number on a website or following one procedure in an instruction manual. To provide more data for analysis, usability researchers often ask users to narrate their thoughts as they perform the tasks. The researchers record the user's attempts and study the recordings to find out where the document went wrong—that is, where the user had trouble completing the task. The goal is not to test the user but to test the document against the user's real-world responses.

The goal of user research and usability testing is to explore how real users might use and feel about the documents we design. Employed together, these methods help us gauge how to respond to users' situations, in terms of both their visual abilities and their attempts to use a document within the context of all the other documents they have used—in other words, in terms of both visual perception and visual culture. For visual perception, user research in particular can help us understand whether potential users of a document have a perceptual limitation. If you're designing a document for a group of senior citizens, for example, you might want to find out how well most senior citizens can see. Or if you're designing a document primarily for men, you might want to know what effect color blindness might have on their perceptions (about 8 percent of men have some impairment of color perception, as opposed to 0.5 percent of women). User research can also help us understand the cultural attitudes and conventions user groups bring to their interactions with documents. Usability tests can reveal whether the designs we create work, visually, as we intend them to. They can help us recognize where our designs fit in with users' cultural or conventional expectations and how we need to redesign documents to better fit their expectations.

FROM THEORY TO PRACTICE

In this chapter, we discussed a variety of ideas about how users perceive their visual environment, both through their biological and psychological equipment and through their cultural and social expectations. Together, these two aspects of understanding how users see the world help us make good design decisions that fit both our clients' objectives and our users' needs.

Of course, we have scarcely touched upon many important theories. No one chapter—in fact, no one book—could provide a full account of all the theories of visual perception, visual culture, and visual rhetoric. But we encourage you to read further about these and other theories that might give your work as a document designer a stronger foundation.

EXERCISES

This chapter has covered a lot of information! You've learned a bit about how we receive and perceive visual information, and you've been introduced to theories of visual perception, culture, and rhetoric. Considering perception, you've learned about neurophysiology, Gestalt psychology, constructivism, and ecological perception. In terms of culture, you've considered our visual language(s), intercultural communication, semiotics, power, and ethics. And, considering rhetoric, you've thought about persuasion and user-centered design. Now it's time to practice. These exercises should help you better understand what you've just read.

1. Find a document that you think is visually interesting for some reason. Then ask two people from different cultures to look at the document and tell you their responses. What do they like? Dislike? What elements do they find interesting or boring? Listen carefully to what they say and how they express it. Then write a report in which you describe the background of your two "viewers," outline what they noticed, and write a reflective comment in which you think about how their comments fit into (or don't fit into) visual perception, visual culture, and visual rhetoric.

2. Choose a document or a website and discuss it from two of the perspectives we examined in this chapter. For example, you might analyze a web page from the perspective of visual perception and visual culture, or from the perspective of

visual rhetoric and visual culture. What aspects of the design would the perspectives you have chosen explain differently? What aspects of the design would the perspectives you have chosen explain similarly? What insights did you gain from looking at this artifact from different perspectives?

3. This exercise will help you start thinking visually. Using only one page and no more than three abstract shapes, create a sign that expresses an idea. Try *not* to use any letters or words. Here are some ideas, but feel free to think of your own: descend, flip, slow, accelerate, celebrate, move, dance, attend, avoid, enter, no, yes, pick up, drop, throw, push, pull. After you've created your sign, write a cover memo for your classmates or your instructor in which you explain how your design will influence users, basing your argument on the concepts about perception presented in this chapter. How would people of different backgrounds or cultures respond to your design? In what situations would your design work best?

REFERENCES AND FURTHER READING

Barthes, R. (1977). *Elements of semiology*. Hill and Wang.

Booth, W. C. (1963). The rhetorical stance. *College Composition and Communication, 14*(3), 139–145.

Boothe, R. G. (2003). *Perception of the visual environment*. Springer.

Brumberger, E., & Northcut, K. (2016). *Designing texts: Teaching visual communication*. Routledge.

Bruner, J. S., & Goodman, C. C. (1947). Value and need as organizing factors in perception. *Journal of Abnormal and Social Psychology, 42*(1), 33–44.

Burke, K. (1950). *A rhetoric of motives*. Prentice-Hall.

De Saussure, F. (2006). *Writings in general linguistics*. Oxford University Press.

Debord, G. (1995). *The society of the spectacle* (D. Nicholson-Smith, Trans.). Zone Books.

Findlay, J. M., & Gilchrist, I. D. (2003). *Active vision: The psychology of looking and seeing*. Oxford University Press.

Forgacs, D., & Hobsbawm, E. J. (Eds.). (2000). *The Antonio Gramsci reader: Selected writings, 1916–1935*. New York University Press.

Gibson, J. J. (1979). *The ecological approach to visual perception*. Houghton Mifflin.

Gordon, I. E. (1989). *Theories of visual perception*. John Wiley and Sons.

Hall, E. T. (1971). *Beyond culture*. Anchor/Doubleday.

Handa, C. (2004). *Visual rhetoric in a digital world: A critical sourcebook*. Bedford/ St. Martin's.

Hill, C. A., & Helmers, M. (Eds.) (2004). *Defining visual rhetorics.* Lawrence Erlbaum.

Hoffman, D. D. (1998). *Visual intelligence: How we create what we see.* Norton.

Jenks, C. (Ed.). (1995). *Visual culture.* Routledge.

Johnson, R. (1998). *User-centered technology: A rhetorical theory for computers and other mundane artifacts.* State University of New York Press.

Kostelnick, C., & Hassett, M. (2003). *Shaping information: The rhetoric of visual conventions.* Southern Illinois University Press.

Meggs, P. B. (1992). *A history of graphic design* (2nd edition). Van Nostrand Reinhold.

Messaris, P., & Moriarty, S. (2005). Visual literacy theory. In K. Smith, S. Moriarty, G. Barbatsis, & K. Kenney (Eds.), *Handbook of visual communication: Theory, methods, and media* (pp. 481–502). Lawrence Erlbaum.

Mirzoeff, N. (Ed.). (1998). *The visual culture reader.* Routledge.

Mitchell, W. J. (2002). Showing seeing: A critique of visual culture. *Journal of Visual Culture, 1*(2), 165–181.

Nielsen, J. (2000). *Designing web usability: The practice of simplicity.* New Riders Publishing.

Norman, D. (2013). *The design of everyday things.* Basic Books.

Palmer, S. E. (1992). Common region: A new principle of perceptual grouping. *Cognitive Psychology, 24*(3), 436–447.

Peirce, C. S. (2006). What is a sign? In N. Houser et al. (Eds.), *The essential Peirce: Selected philosophical writings* (Vol. 2, 1893–1913, pp. 4–10). Indiana University Press.

Ross, H. C. (1952). *Substitute for fear.* House-Warven.

Slack, J. D., Miller, D. J., & Doak, J. (1993). The technical communicator as author: Meaning, power, authority. *Journal of Business and Technical Communication, 7*(1), 12–36.

Valesio, P. (1980). *Novantiqua: Rhetorics as a contemporary theory.* Indiana University Press.

Zanker, J. (2010). *Sensation, perception and action: An evolutionary perspective.* Bloomsbury Publishing.

UNIT 2: PROCESSES

4

THE WHOLE DOCUMENT

LEARNING OBJECTIVES

Upon completing this chapter, you will be able to:

- Discuss how a document's ecology relates to user perceptions

- Articulate how perception, culture, and rhetoric shape a document's use and impact

- Make decisions about a document's design based on conventions, human factors, transformation, and cost

- Discuss multiple format types and articulate their decision-making process based on desired audience, purpose, and contexts

~

Imagine you've just bought a car. You probably have questions about how to work some of the features, like the stereo, lights, and so on. To find answers, you'll probably refer to the owner's manual at least a few times as you get to know your car.

But think for a moment about where that owner's manual must live and work. Normally, it's kept in a pretty dark and dangerous place (for documents, at least): the glove box, where it has to contend with not only gloves but a flashlight; a wad of napkins, straws, and ketchup packets from the last drive-thru you visited; a handful of assorted pens and pencils; and perhaps even a screwdriver and a pair of pliers—all sorts of objects tailor-made to poke, bend, rip, crease, and otherwise mess up a typical document. Then there are the conditions under which it's used in the car. It has to stand up to the abuse of being repeatedly pulled out and tossed into the glove box. Plus, it has to help you find information quickly, when you need it. It can't be too big to use while you're sitting in the driver's seat. But it has to be big enough to contain a lot of important information. And ideally, because we sometimes drive at night, it should be something you can make out easily by a dome or map light. On top of all that, it has to serve a rhetorical purpose: to make us feel good about having bought the car so we might do so again someday or recommend the brand to our friends. Altogether, that owner's manual must do a tough job in a difficult environment, and it must do it all as a physical document (in many cases, at least), because there's no telling when you may need it in a place where you have no connectivity.

Figure 4.1 shows how one car manufacturer addressed this difficult job. This set of manuals for a Mazda is a handy size, easy to flip through, and convenient to use in the car, and includes a tabbed, quick tips guide (at the bottom left of figure 4.1). Rather than the quick tips guide being bound conventionally on the left, it's bound at the top edge to encourage easy flipping. To facilitate that pattern of use, it features a special trim on the bottom (outside) edge of each page to create a tabbed index. Owners of the Mazda MX-5 can use the tabs to flip exactly to the point they wish to read. To manage that tough environment of the glove box, the set of documents then fits into a protective case (in itself a document, as it lets us know the brand and logo). The quick tips guide is also printed on heavily coated and varnished paper—it's likely to see the most use and also has to stand up to potentially being left out of the case in the glove box.

Figure 4.1. A set of documents designed to work in a difficult environment. The designers of this had to think carefully about the conditions in which it would be stored and used. *Source*: Mazda.

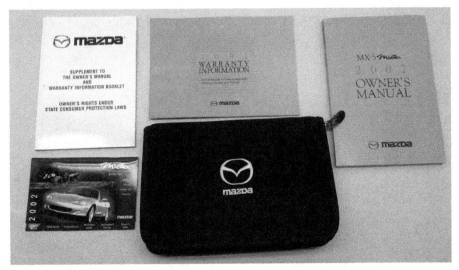

This example suggests that designing a document requires thinking not only about two-dimensional layout, colors, typefaces, and so on, but also about how the document will perform as a physical, three-dimensional (3D) object in a particular environment. As Charles Kostelnick (1996) has pointed out, aspects of the *whole document* (which he calls *supra-textual elements*) have a significant effect on the way users interact with documents. So we need to think about the whole document and its place in users' lives.

What do we mean by the *whole* document? In chapter 2, when we discussed design objects as a way to think about the marks we put on pages or screens, we noted that thinking of these marks as "objects" is simply a useful metaphor. But the *whole* document really is a physical object. Thinking about the whole document adds a third dimension to design: the document as a tangible, physical object with a front, back, depth, and weight, as well as two-dimensional width and height. In this chapter we consider the whole document as the physical aspects of any document design: the size of the document, its binding, the technologies used to produce it, the paper on which it's printed, or the screen on which it's presented.

You might think it would be better when producing a document to begin with something smaller than the whole document, such as typography or page design. And sometimes designers *do* start with one feature—a color scheme, a typeface, a logo, an image, or even elements that show brand identity with the client's other products or publications. But the decisions you make about the whole document provide the context for all the decisions about the smaller parts. Addressing the big decisions first can help make those smaller decisions more manageable and efficient.

> The decisions you make about the whole document provide the context for all the decisions about the smaller parts.

We'll begin this chapter by discussing the whole document through the perspectives of perception, culture, and rhetoric. Then we'll discuss two important whole-document design decisions: *medium* and *format*. Throughout, we'll provide plenty of examples to help you make these decisions effectively in your own document design projects.

THREE PERSPECTIVES ON THE WHOLE DOCUMENT

PERCEPTION

People experience documents not just as two-dimensional visual fields, such as pages or page spreads, but also as 3D objects they can touch. Printed documents have a weight and a relationship to the hand that can make them feel either natural or awkward to use. Even electronic documents take on the physical features of the hardware that embodies them, from the glowing pixels of the monitor to the software interface to the keyboard, mouse, or trackball we use to navigate the document.

Considering the whole document requires that we pay attention to senses other than just our vision: touch, hearing, and even smell. (Taste doesn't often come into play once users get old enough to read documents rather than to chew on them!) The sensation of touch can be affected by the quality and kind of paper and binding. The sense of hearing can help users navigate flipping through a printed document, and auditory clues (like clicks and beeps) can give users feedback in electronic documents. Even the smell of a printed document can have an effect on users, as some find that certain inks and papers have a strong odor.

Creative designers often find ways to combine sensual elements: you may choose your room deodorizer based on not only the packaging's description of the odor, but the colors that suggest that odor, and often even a scratch-and-sniff component so you can experience the odor firsthand prior to purchase. The band Cake, for example, once released a limited-edition version of their *B-Sides and Rarities* album that featured five different scented packaging versions: a red version that smelled like roses, brown to smell like leather, green for fresh-cut grass, purple for grape, and yellow with banana scent. Some greeting cards include small sound players that activate when the card is opened to play a tune or even a recording of the sender's voice. These examples, and more, help us start to think about how engaging with a document is, or can be, a full-sensory experience.

The physical features of a whole document are integral aspects of our experiences with documents. To use what some scholars call *visual-spatial thinking*, users need these—at least some of these—physical cues to understand the structure and content of a document (see Johnson-Sheehan & Baehr, 2001). You've probably used this kind of thinking when looking for something you had read earlier in a document. You might not remember the exact page number, but your spatial senses help you remember that it's near the end of the document or on the top of a right-hand page. Or you might even have changed your document physically for easy reference: adding a sticky note, dog-earing a page, or highlighting a passage in a document or bookmarking a page on a computer screen.

CULTURE

As 3D, physical objects, documents have had a significant effect on culture. Entire academic fields, such as history of the book, bibliography, and print history, focus on exploring the history of printing, its key technologies, its products, and their effect on culture. Oxford bibliographer D. F. McKenzie (2004) referred to these studies as the "sociology of the book"—an examination of the web of people surrounding the creation and consumption of books, including authors, designers, printers, booksellers, and readers.

But a book is only one kind of document, and perhaps not the most culturally influential one today. For most people, the number of books they read pales against the number of other kinds of documents they encounter every day—what they receive in the mail (either snail mail or email); what they use to understand how to do their work; what they read as they browse the internet. In many ways, Western culture has transitioned from the age of the book to the age of the document, a broader category that includes the many other kinds of documents we live with. These documents have a significant and growing effect on what we do and on how we see ourselves, both individually and as members of a culture.

As a result of this ubiquity, even people who seldom think consciously about documents can have a very well-developed sense of document conventions. Although they might be hard-pressed to name what characteristics make a document a technical manual, users typically know a technical manual when they see one. Conventions about the whole document can be very strong, which works both to our advantage and to our disadvantage. If we create a technical manual that firmly meets users' conventional expectations for that genre of writing (see chapter 1 for more on genre), they'll immediately recognize the manual for what it is, but it might not stand out very much. An unconventional manual, one that doesn't necessarily meet genre expectations, might attract users' attention more quickly, but they might consider it annoying or off-putting because it flouts established conventions, and they might even resent having to learn how to use something new.

RHETORIC

The rhetoric of the whole document arises from its place in the user's life and experiences. In this sense, what matters isn't just the document itself but the ecology in which it's used—the physical environment of the user. Two ecological factors are especially important:

- *The occasion of use:* When will users use the document? Will they refer to it repeatedly, read it through only once, or use it some other way? What circumstances will motivate them to use the document?

- *The conditions of use:* Will the document be used in an office, a car, an airplane cockpit, a factory floor? Will the conditions

be dry, wet, windy? Can users lay the document on a flat surface, or must they hold it in their hands? Will the document
be fixed to a surface, displayed on a stationary screen, or
appear on a portable screen such as a handheld mobile device?

Understanding these factors will help us create a document that is both
convincing and usable in the user's specific environment.

More broadly, the physical design of the whole document can have
a significant effect on how users respond both to the document itself
and to the client who commissioned its creation. A document bound
expensively with valuable materials gains an ethos of luxury, exclusivity,
and professionalism. One bound cheaply—perhaps only stapled—may
carry a rhetoric of disposability or informality. Sometimes disposability
is appropriate, depending on the level of design the document merits. Or
a designer might want to sacrifice production costs and the hidden costs
of corporatized labor practices to make a physical, rhetorical argument
with the design of their product. Author and designer Mike Monteiro
(2021), for example, offers a zine format (see figure 4.2) version of his

Figure 4.2. Zines, usually small circulation, self-published, or independently published magazines, have been a staple of fringe, or outsider, communities since
their rise to popularity alongside the punk movement of the 1970s. Many play
with size, format, and layout. *Source*: Photo by the author, personal collection
of D. G. Ross.

book *Ruined by Design*, as he argues, "Zines [small circulation, usually self-published magazines] were, and amazingly continue to be, markers of fringe communities. They're the social networks those communities use to communicate with each other," and he wants his book to have that sort of *ethos*. But on other occasions, a fancier whole-document design is crucial to convincing the user that it's valuable and important.

MAKING DECISIONS ABOUT MEDIA

In simple terms, *media* (singular: *medium*) refers to the means by which a document is conveyed to readers. However, media are not neutral conduits; the medium of a document places significant constraints on its design, as each medium has different capabilities and limitations.

In document design, the two most significant media are *print* and *screen*. Print incorporates any document marked in some kind of ink on a physical page, usually made of paper, cardboard, or plastic. Screen incorporates any document conveyed to readers through an electronic viewing interface, such as a computer monitor or digital projector.

Making good choices about what medium to use for a given document design project can be a complex task. In this section, we'll discuss the following factors that affect decisions about medium:

- Conventions

- Human factors

- Transformation

- Cost

We'll also discuss single-sourcing, a technique for using more than one medium.

CONVENTIONS

When choosing the medium in which to produce a document, one factor to consider is the experiences users have had in receiving similar documents. These experiences can build very strong conventional expectations. If we break those conventions thoughtlessly, we risk creating a document that will make users uncomfortable, or even a document that users won't recognize.

Consider the humble memorandum, the long-standing embodiment of interoffice communication. For a century, memos were printed on paper and routed through an organization's mail system. To expedite writing memos, companies sometimes used a memo form, with blanks for the sender (From) and the receiver (To), as well as the date and subject. This convention of using the paper medium for memos is still common today, even though most companies use email instead. However, email encountered resistance when it was first introduced, primarily because it presented interoffice communication in an unfamiliar, unconventional medium—the then-high-tech green or amber text on a black computer screen. Email system designers overcame this problem by creating email program interfaces that follow the conventions of traditional paper memo forms. Look at your own email application and you'll likely see the traditional fields for routing information—To; Cc, meaning "carbon copy," a throwback to writing or typing documents over a sheet of carbon paper placed over an additional blank page so that the typed or written matter would be directly duplicated; and Bcc, meaning "blind" carbon copy, an exact replica sent but hiding the emails of recipients—and a subject line, followed by a space in which to write the memo (figure 4.3). Only the

Figure 4.3. Email message form. An email message form contains many conventions left over from paper memo forms, such as blanks for the receiver and the subject. The date and the sender are filled in automatically, making these blanks unnecessary. *Source*: Created by the author.

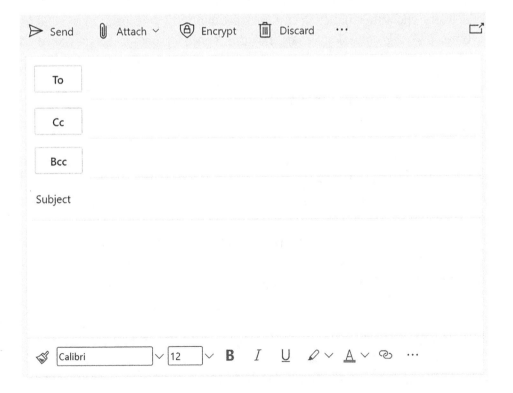

> Design documents to meet your users' expectations.

Date and From lines are omitted because the computer supplies this information automatically.

If you do choose to use a medium different from what users expect, be prepared to explain the advantages of the new medium so users will be willing to give it a try. Users are often uncomfortable with documents that have transferred from print to screen (or screen to print), particularly if they have become accustomed to the original medium. For example, users may complain about documents on paper if they have become used to the additional functionalities of screen documents, such as search capability. In most cases, design documents to meet your users' expectations.

HUMAN FACTORS

User conventions are just one aspect of making a decision about medium. Sometimes the physical qualities of a medium make a big difference in users' abilities to use the medium effectively in their ecology.

Users experience a document's medium physically, as part of their tangible lives. We often think of media as being somewhat ephemeral, transitory, hard to grasp in a physical sense; pages decay and screens change. At the very least, we might think of them as entirely two-dimensional spaces with only width and height. But print and screen media both involve users in multiple dimensions, incorporating depth (or the illusion of depth) and the passage of time. Paper has a thickness and a heft that users respond to, and users encounter paper documents sequentially, one page or spread at a time. Users' terminology reaffirms this sense of physical experience in time. In paper documents, we speak of a section of a document as being "before" or "after" another section; in websites, we use "back" and "forward" buttons to move through the sequences of pages we have viewed.

The same physical multidimensionality applies to screen documents because users encounter these documents on interface devices such as computers, which have depth and weight (as anyone who has juggled a notebook computer through an airport security check can attest). Good designers pay attention to the *human factors* of use regardless of what medium they choose.

Many human factors come into play in document design, but three particularly important ones can significantly affect our choice of medium: the medium's interaction with light, its resolution, and its manner of transmission.

INTERACTION WITH LIGHT: REFLECTIVE OR TRANSMISSIVE

All vision involves light, but how and where light is generated affects the different ways we interact with print and on-screen documents. In print documents, light comes from some other source (the sun, a desk lamp, fluorescent overhead lights) and *reflects* off the page to our eyes. In screen documents, the screen itself *transmits* light to our eyes by turning on a pattern of glowing pixels.

This distinction is important because we must take into account the situations in which users will use a document. In low-light conditions—in a darkened room, for example—a document that transmits its own light might make more sense than one requiring a separate light source. In brightly lit conditions the opposite might make more sense: a paper document is easy to read in bright light, while too much light can make seeing a screen more difficult. In addition, not all users have the same visual capacity; some can see better than others. Depending on their vision, users might prefer paper documents to screen documents. Or if their technological setup permits digital magnification, the opposite might be true. We must accommodate users by providing the document in a medium they can see easily.

RESOLUTION: HIGH TO LOW

To make good decisions about media, we also must consider the different resolutions of print and screen. *Resolution* is the fineness of detail that the medium can provide to viewers. Screens usually have a much lower resolution than print. Most screens are typically limited in their resolution by the size of the glowing, square pixels that make up the screen image. Computer monitors typically have a pixel size of less than 100 pixels per linear inch (ppi), making each pixel $^1/_{100}$ of an inch easily visible to most human eyes. (Look closely at a computer monitor when it's displaying an image; you should see a grid of tiny squares, each of which is an individual pixel.) As a result of this relatively low resolution and the square grid of pixels, screen images are often jagged around the edges, particularly on curves or angled lines. Even the technological approaches to solving this problem have their own usability issues. Some computer systems, for example, smooth (or *dither*) the edges of letters to provide a gradient between the letter and the surrounding space, but in the process they can make text look indistinct or even fuzzy. This fuzziness makes text less legible because it provides inadequate figure-ground contrast and poor edge discrimination.

Printed documents, even those produced on a garden-variety laser or inkjet printer, typically have a resolution of 300 dots (of ink) per inch (dpi) or better. So a printed document has a resolution almost ten times higher than the typical screen document: 300 dpi equals 90,000 dots in a square inch (300 wide × 300 tall), whereas a screen resolution of 96 ppi equals only 9,216 pixels in a square inch (96 wide × 96 tall). The resulting dots in a printed document are usually difficult to see with the naked eye. Because the dots are made from microscopic blobs of ink, they also aren't square, allowing them to blend together more smoothly than do pixels on a screen.

The higher resolution of print is one of the biggest reasons users print screen documents before reading them, even if the documents aren't intended to be printed. A common user complaint with electronic documents is eyestrain, particularly for documents intended to be read at length. Designers also often choose print over screen when conveying information that requires fine detail, such as high-quality photographic images. But when it comes to documents that will be used only for brief periods, such as reference materials, users will gladly give up higher resolution in favor of convenient electronic searching or hyperlinked content. (For more information on image resolution, see chapter 6.)

TRANSMISSION AND ACCESS

Deciding whether to produce a document in print or on-screen involves accepting widely different means for getting the document to users. For the most part, print documents must be physically carried from one place to another, usually through shipping, mail, or direct distribution. Screen documents, however, can be broken down into electronic bits and bytes and transmitted through the internet.

Each medium's method of transmission has advantages and disadvantages. But the biggest distinction is that documents produced in print are usually more *portable* than those produced for the screen, whereas those produced for screens are more easily *networked*.

> Documents produced in print are usually more *portable* than those produced for the screen, whereas those produced for screens are more easily *networked*.

By *portability*, we mean the quality of a document that allows users to take it with them, to consult at any time they want. Books are a technology thousands of years old, but one of the reasons they remain popular is their portability. Users often appreciate a document they can stick in a pocket or shove in a briefcase, especially documents they might use as a reference. This is why we still call some documents *handbooks* or *manuals* (literally, "by hand").

This quality of portability means that print documents must be designed to fit comfortably in people's hands—not only the user's but also those of others who must handle the document, including sellers and shippers. For example, many print documents must fit into a package for shipping, and many leaflets and brochures are designed to carry postage and address information so they can be delivered through the mail. This dynamic gives many print documents a natural affordance to the human body: they're designed to fit *us*. Fortunately, print documents are also very durable, and they can still be read easily even after all the wear and tear they suffer because of their portability.

However, not all information fits us easily. What about a document that's too long to carry around physically or to find information in easily? That's where screen documents come in. Screen documents specialize not so much in portability as in *networkability*—the quality that allows users to access them over long distances through a network. This quality also allows screen documents to be linked to one another. A paper document might refer to another paper document, but the user must follow the reference manually—by physically getting the other document, either personally or by proxy (asking someone else to send it). But because most screen documents today are digital, they can be associated with each other easily, usually through a system such as the web that allows searching, automatic indexing, and hyperlinks.

This networkability comes at the expense of portability and therefore of access. And although it might sound obvious, obtaining networked documents requires access to a network with its entire infrastructure, including sources of electrical power and wired or wireless data transmission. (How many times have you heard "Sorry! The network is down"?)

Even though, according to Pew surveys, only 7 percent of Americans don't use the internet (Perrin & Atske, 2021), and internationally "only" around 10 to 13 percent of the world's population do not have access to reliable electricity (International Energy Agency, 2023; World Bank, 2019), if you are designing electronically (and ethically), you still need to consider issues like reliable and convenient internet access. Even in the most heavily networked areas of the world, an electronic fire response document, for example, that needs to be used in a power outage might be useless, or a storm evacuation plan that is inaccessible during a hurricane might not be worth the weight of the waterlogged computer it's stored on. In contrast, print documents are always on, always ready to be used.

TRANSFORMATION

These physical factors (light, resolution, and transmission) are important considerations when determining what medium is best for a given document. But screen and print documents also have an essential difference in their ability to transform.

Documents in print are relatively *static*; once they're printed, they stay that way. For some kinds of documents, that static nature fits the purpose of the document—for example, a company's sexual harassment policy document should probably stay the same for many months or years. The static nature of paper documents can also convey a sense of solidity to the document, building an ethos of stability for the client. Although annual reports are now often distributed as both print and electronic documents, the glossy, printed annual report will be around for a long time yet.

Documents produced for the screen, however, are quite *dynamic*, with the potential of changing thousands of times per second. This dynamic nature gives screens the illusion of movement—for example, by showing a few dozen frames per second of a video clip, an animation, or a progress indicator. It also allows electronic documents to change quickly according to user input. So a website, or your web browser, might remember your information from the last time you filled out a form and already have many fields completed before you fill it out again.

One advantage of this dynamic nature is that electronic documents can be updated quickly and inexpensively because an electronic document is typically saved in one location (a network or internet server) and simply accessed by many users through their computers. Updating a printed manual requires printing the manual over again, if however many copies are required, updating an online help system requires making the change only once and resaving the file that users will access through the network.

However, this distinction between static and dynamic isn't firm. Even some print documents are dynamic in that they are designed to be changed or modified after production. Some printed signs are designed so letters, words, or entire sections can be updated or replaced, and printed forms are designed so users can add their own information. Others ask users to somehow manipulate the document in order to gain additional information. A card about natural gas safety (figure 4.4), for example, includes a place for readers to scratch so they can experience what a natural gas leak smells like, or a magazine insert may include perfume samples that are "unlocked" by unfolding part of the document. We also know that users often change print documents for their own convenience by folding, tearing, or marking them. And nearly all print documents created in business, government, or education today at least *start* as electronic documents—that is, as electronic files that are printed onto paper as a last step in the production process.

Conversely, users can and often do print documents designed for the screen, even if the designers never intended the documents to be printed. Users can also save networked electronic documents locally on their own hard drive, creating a static version of a dynamic page by saving it as a PDF or an HTML file.

When deciding in which medium to produce a document, consider how the document should or could change once it's in the user's hands,

Figure 4.4. A dynamic paper document. Paper documents can be dynamic: this natural gas utility card includes a scratch-and-sniff panel so users can learn what a natural gas leak smells like. *Source*: City of Zachary, LA.

as well as how frequently it may have to be updated. Certainly, if the document must be updated regularly—such as a product catalog—consider the screen medium, where documents can be updated simply by making a change and saving it. If solidity or consistency is more important, however, choose print.

COST

In themselves, print documents usually cost more to produce than screen documents. If you want to provide a print document to a thousand users, you must pay for a thousand copies. But if you want to provide an electronic document to a thousand users, you must pay for only one copy, which most users can access on their own computer screens.

But print documents often cost less per copy the more you print. The expense usually comes with setting up printing in the first place; after you've printed a hundred copies, printing a thousand more isn't typically that much more expensive. (Photocopying and laser printing are exceptions: the last copy costs just as much as the first one.) We must also take into account the whole cost of electronic documents—including the expertise required to create them, the constant maintenance they demand, and the infrastructure of power and networks required to get them to users.

Producing a document in color is also much more expensive in print than on-screen. Most computer monitors today can show thousands of colors as easily and cheaply as they can show black and white. But printing multiple colors actually requires printing a document in multiple inks, which increases the cost considerably. (See chapters 8 and 11 for more information on color printing technologies.)

Regardless of the medium, documents cost money. In choosing a medium for your own designs, pay attention to what makes economic sense for the situation.

SINGLE-SOURCING: HAVING IT BOTH (OR MANY) WAYS

Up to this point, we've distinguished media as an exclusive choice designers must make between print and screen. But our technologies allow designers and their clients to create documents that can take on print *and* on-screen forms, depending on the situation and the needs of users. This technique is often called *single-sourcing*.

The basic idea behind single-sourcing is to create content once and then reuse it in multiple ways. For example, you might write the text and prepare the illustrations for a report separately and then use single-sourcing technologies to output that content to a website, a PDF, or a printed document, depending on user needs and demands. The train schedules in chapter 1, whether in print or on-screen, were created from a basic set of consistent information: the actual schedule of train arrivals, departures, and routes.

In the most advanced forms of single-sourcing, the content is stored in a database or text file and assembled on the fly in whatever output medium the designers have planned for. To do so, the content is compiled in an *Extensible Markup Language (XML)* file with *metatags* that describe each piece of content. Metatags are elements of *metadata*—literally, data about data. These behind-the-scenes attributes let designers assign attributes and mark key information for use by various technologies. For example, an XML file might include metatags to mark the title, author, date, headings, illustrations, captions, and paragraphs of a document. Metatags aren't visible to the user; they simply give computers a consistent way of identifying different elements in the document. Finally, a series of digital *style sheets* directs computers how to display each different kind of metatag when the document is ported to page or screen. For example, the style sheets might specify that when the title of a document is displayed on-screen, it will appear as 24-point Arial Bold, but that when the title is displayed in print, it will appear as 24-point

Gill Sans. The advantage of this approach is that when you change the basic information in the XML file, those changes automatically cascade to all of the different kinds of output you have planned for.

As you might imagine, single-sourcing, creating systems of documents relying on metadata, can be a complex task, and a full discussion of single-sourcing technologies is beyond the scope of this book. In fact, single-sourcing is really just one aspect of what we now think of as *content management*, but one very common technique for single-sourcing is to use simple Acrobat PDF files. As we mentioned earlier, PDFs already straddle the boundary between print and screen. They are an electronic file format that users can view on-screen through many different technologies. But they can also be printed, coming out almost exactly like what was represented on-screen. So PDFs make for a nice approach to single-sourcing if your client isn't up to the technological investment of more complex single-sourcing techniques.

MAKING DECISIONS ABOUT FORMAT

Once you have decided what medium is best for your document design project, you must also determine what *format* will work best. By format, we mean all of the possible physical arrangements of documents within the medium, including size, shape, height, width, depth, texture, arrangement of pages and leaves, and so on. A document designer's primary goal in choosing or creating a format is to make the document's format fit its intended use and intended users.

Each kind of format—and each genre of document you create—has its own conventions and history, its own quirks and possibilities, its own manner of use that it suggests to users by its physical design. So in this section, we'll discuss how to make choices about what format might be best for your project.

CONVENTIONS IN FORMAT

Users often have well-developed conventional expectations about format, just as they do for medium. If you provide documents in an unfamiliar format, readers may feel uncomfortable using the document and either avoid using it or use it in a way you didn't anticipate.

Find out what format users expect to see for documents similar to the one you are designing. Then think carefully before choosing a format. Consider, for example, the experience

> Find out what format users expect to see for documents similar to the one you are designing. Then think carefully before choosing a format.

of the mail-order plant nursery White Flower Farm, whose catalog had been produced for more than fifty years in an unusual format: a $10^{1}/_{2}''$ × $8^{3}/_{8}''$ horizontally oriented booklet. The company decided to try a less expensive format, using a shape and size that has become common for many magazines: $7^{3}/_{4}''$ × $10^{5}/_{8}''$ in a vertical orientation. But the earlier, horizontal format had become conventional for White Flower Farm's customers, who were frustrated by the new format.

Here's what the company had to say after it returned the catalog to its original shape and size:

Nostra Culpa

Last year at this time, we mailed our 53rd spring catalogue and the first ever in a vertical format. We made the choice for economic reasons, and, like many purely economic decisions, it turned out to be a mistake. We didn't like the way it looked or felt, and it wouldn't fit on the shelf with its predecessors. Worse yet, it became impossible to find in the deluge of printed materials that swamps our desk. Judging by our correspondence, some of which was what the State Department calls "open and frank," you felt the same. So the natural order of things has been restored. (White Flower Farm, 2004, p. 2)

The title of this apology, which means "our fault" in Latin, suggests what a big mistake the company felt it had made in breaking the conventions it had established with users of its catalog. Though eventually the company had to bow to economic reality and produce its catalog more economically, the customer response to the initial change shows how design of the whole document can become embedded into our expectations. Users had come to expect something out of the company's information products, including a catalog distinguished from the many others in the market by its size and shape. The earlier format was also unusual enough to make it easy to find; the new format looked like all the other catalogs people constantly receive. Moreover, the new format didn't fit into the users' ecologies—the spaces (shelves, desks) where people kept the catalogs in their offices or homes.

Of course, you may determine that switching to an unconventional medium has so many advantages that defying convention makes sense. Earth First!, for example, has changed the format of its primary print publication several times, moving from a simple set of stapled $8^{1}/_{2}''$ × 11" pages to a newspaper layout, then to a magazine format, then to a smaller zine, then back to a larger zine (figure 4.5).

Figure 4.5. Changing formats. A few examples of changing publication formats from Earth First! between 1980 and 2023. Each design is a move to accommodate changing expectations and needs, from typewritten and photocopied sheets, newspaper-like formats, and variations on glossy magazines to more zine-like formats. *Source*: Auburn University Libraries Special Collections and Derek G. Ross's personal collection.

An innovative format might meet users' needs so much better that it's worth the added time and expense to create something cool and new. A newer format might also be able to accommodate changing expectations and needs regarding images, advertisements, and more. There's also something to be said about the power of novelty to attract a user's attention. But be sure to *ask* users through usability testing what they think about an unusual format—preferably during the prototyping stage of the design project so you can make changes *before* producing the document in its final form. (See chapter 10 for more advice on usability testing.) Designs must fit into users' expectations, to be sure, but must also fit into those users' document *ecologies* (you might consider Clay Spinuzzi and Mark Zachry's 2000 work on *genre ecologies* as one way to start an investigation into this line of reasoning). We store our documents in particular ways, often along with other documents, so changing something as apparently simple as a publication size (as described in the White Flower Farm example) ultimately affects the ways in which we are able to access our collections of physical documents.

FORMATS IN PRINT AND SCREEN

With print, the number of available physical formats is vast. For example, in the print medium, we can create documents in formats like these:

- Brochures and pamphlets

- Leaflets

- Books

- Booklets

- Quick-reference cards

- Flyers

- Newsletters

- Signs

Why so many different formats? Because the essence of the print medium is *paper*, a material we can fold, cut, staple, stitch, punch, glue, and bind into a myriad of unique formats within the general categories above. Although most paper starts out rectangular, its flexibility—with its opportunities for folding, punching, perforating, trimming, and binding—makes the format possibilities almost endless.

It is also possible to create screens in many physical formats—that is, in different shapes, sizes, and arrangements. Early television screens were round. But in addition to what we likely view as the most common rectangular screens of our mobile devices and televisions, we now also have screens that can be folded or rolled; immersive rooms with screens on the floor, the ceiling, and all four walls; vapor and fog screens for multidimensional projection and digital manipulation; massive, building-sized projection screens; and more.

But for the vast majority of users today, all screens are in pretty much the same physical format: flat, rectangular, and wider than they are tall. This format places significant limitations on the possibilities for screen designs, constraining designers to what they can show in such a rectangular space. Of course, those limitations are balanced by the many flexibilities offered by a screen's dynamism as a medium, as we discussed earlier. Screens can display websites, digital kiosks, and touch-screen interfaces. They can be made in many sizes, from cell phone screens to jumbotron sports scoreboards.

But because of the limitations of screen formats, in the following sections we'll focus on discussing a few of the many formats of print documents.

VARIABLES IN PRINT FORMATS

Choosing the right format involves weighing a number of factors, some of which are similar to those we considered in making a decision about what medium to use. The more complex your design, the higher the cost and the more time it will take to complete. So in the rest of this section, we'll discuss a number of common formats, moving from simple, fast, and cheap formats to more complex, time-consuming, and expensive ones.

Several variables come into play when designing a format for a document, but the most important are paper, folds, page shape and orientation, trims, and bindings.

More about . . . Physical and Electronic File Formats

Although most screens have only one *physical format*, they can display many *electronic file formats*. An electronic file format is a computer standard for creating and reading files of some particular type, such as a word processing document, a page design file, or an image file. We can usually tell the electronic file format of a computer file by its *extension*—that is, the three or four letters after the period at the end of the file name (e.g., .docx). However, these electronic file formats differ from physical formats in that they're essentially invisible to users and meaningful only to computers.

However, document designers use electronic files to *create* most documents today, whether they're to be printed or displayed on-screen. So to create effective documents through computer technologies, you need to explore the possibilities and limitations of the file formats you'll be working with.

For print documents, common electronic file formats include InDesign files (INDD), QuarkXPress files (QXD), and Microsoft Word files (DOCX). These file types are usually either printed directly or translated to other file types for electronic delivery.

Other file formats are designed to stay electronic—most often, to be delivered through a web browser. The most obvious are web pages, which are formatted as HTML files. Some of these files are written as static documents; others start with a computer script in a file format such as Active Server Pages (ASP) or PHP Hypertext Processor (PHP), which take information from a database and assemble an HTML output on the fly. Electronic file formats also include those that require a browser plug-in or helper application, such as Flash and Shockwave (SWF) animations (though these have largely been discontinued and are no longer supported online except by opensource emulators) and Adobe Acrobat Portable Document Format (PDF).

PDF is also the most print-like of all electronic file formats, since a PDF file holds the electronic description of a printed page that can be sent directly to a computer printer. In fact, many print shops now ask that print jobs be delivered to them as a PDF, since that file format can include all information about typefaces, images, and layout in one file or a series of connected files.

Design Tip: Choosing a Print Format

Consider the following issues when choosing a format for your document.

CLIENT ISSUES

- **Rhetorical purpose.** What agendas does your client want to fulfill with the document?

- **Level of design and cost.** What level of design does your situation call for? Is this a mission-critical document that would benefit from expensive special features like unusual formats, trims, bleeds, or perforations (high-level design)? Or is it something more humble and basic, where extra expense would simply be wasted, calling for a low-level design?

- **Production capabilities.** What production capabilities do you have? Will you be working with just a desktop computer and a laser printer, a copy center, or a full-service print shop? The more complete the production capabilities, the greater flexibility you will have in deciding on a format.

- **Distribution.** Will the document be packaged in a particular-sized product box? Will it be mailed? If so, sizing it in a common postal size will keep costs down.

USER ISSUES

- **Conditions of use.** How, where, and under what physical conditions will users use the document? How will they handle the document? Will they want to flip through it, lay it flat, or hang it on a wall? Will they use it in a wet or windy environment? What else will users be doing while they look at the document?

- **Interaction.** Will users want to write or mark on the document? Will the document incorporate a worksheet? Do you want them to fill out a form or survey and return it to you?

- **Conventions.** In what format will users expect the document to be presented? Expectations about formats can be very powerful; if you present information in an unexpected format, users might not recognize the document as something they need.

- **Storage.** Will users keep the document in a file drawer? Post it on a wall? Throw it in a toolbox?

- **Longevity.** Will users refer to the document frequently or read it only once and discard it?

PAPER

Paper can have a huge impact on document format. New designers often start with a sense that all paper is 8½" × 11" (letter size), but this couldn't be farther from the truth. Although it's more economical to stick with commonly available sizes, documents can be produced in nearly any dimensions a job requires, from business cards to billboards. The only limits are the original sheet size from the manufacturer and your print shop's cutting and printing capacity.

> Keep in mind what paper you intend to use for different purposes from the very beginning of the design process.

We can also take advantage of many different kinds of "paper" for special applications such as cover stock and packaging materials—everything from standard photocopy paper to high-tech plasticized sheets to cardboards.

Keep in mind what paper you intend to use for different purposes from the very beginning of the design process. (For more information on specifying paper for your designs, see chapter 11.)

FOLDS

The size of a document ultimately depends both on the size of the original paper sheet and on how the paper sheet is folded. Sheets can be folded either with *simple folds* (one or more parallel folds) or with *complex folds* (one or more folds across a previous simple fold) (figure 4.6).

Figure 4.6. Simple and complex folds. *Source*: Created by the author.

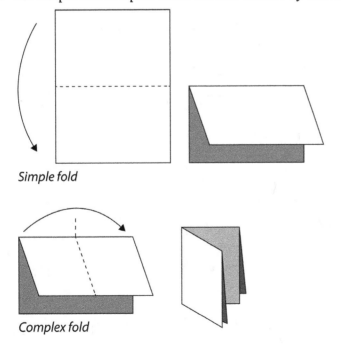

Simple fold

Complex fold

The finished, folded size of documents is often a consideration in design since documents have to fit particular conditions of use and storage. Many documents are sized to fit in standard leaflet racks, for example, or to fit in product packaging. In general, the more folds incorporated into the format, the greater its flexibility as a space for accommodating different kinds of information. But with more folds comes greater complexity, particularly in terms of layout and production.

Folds break up the surface of the sheet into smaller pages or panels, which can be used individually or in combination to fit different kinds of content (figure 4.7). Generally, *page* refers to one side of one *leaf* of a bound format. In a book, for example, each leaf would contain two pages: the front or *recto*, which is visible on the right when the book is held open, and the back or *verso*, which is visible on the left. *Panels* are the areas formed by folds on the same side of a sheet. A single sheet folded in thirds would have three panels on the inside and three on the outside. All of the pages or panels visible when you open a document out flat are collectively called a *spread*.

Pages, panels, and spreads serve as natural areas in which to divide and organize content. Content can be designed to stay within individual panels or to extend over two or more panels of a spread, lending great flexibility to the design.

PAGE OR PANEL SHAPE AND ORIENTATION

Pages and panels are generally (although not always) rectangular. If the dimension of the document running parallel to the text is narrower than the dimension running perpendicular to the text, the document is said to

Figure 4.7. Pages, leaves, panels, and spreads. Each leaf has two pages, recto and verso. *Source*: Created by the author.

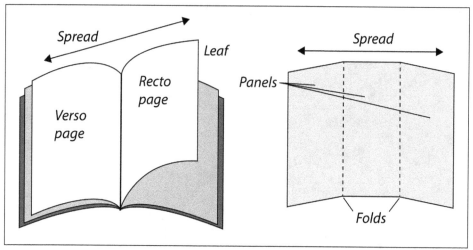

have a *vertical* or *portrait* orientation. If it is wider than tall in relation to the text, we say it has a *horizontal* or *landscape* orientation. (See figure 4.5 for examples of these orientations.)

The relationship between the height and width of a page, panel, sheet, or screen is also referred to as its *aspect ratio*, represented in terms of the numerical relationship between width and height, in that order. For example, a standard 8½" × 11" piece of paper in *landscape* orientation (11" wide × 8½" high) has an aspect ratio of 1.2941, or 11:8.5. Expressed in a way we might more commonly see, we might consider that aspect ratio as a fraction ($^{12941}/_{10000}$), then simplify it. The greatest common divisor here is 2941, so the aspect ratio can also be expressed as 4.4:3.4, but we don't generally express aspect ratios with decimals because we want to as accurately as possible convey the relationship between width and height (the reason we don't see 11:8.5 either), so multiplying by 10, we get 44:34, or simplifying, 22:17. That complicated path led us to the same result as simplifying 11:8.5 (again, working to remove the decimals, in this case multiplying by 2 to arrive at 22:17). Some common aspect ratios you might see are 4:3 or 16:9. A 4:3 aspect ratio, for example, might give us documents at 4" × 3", 8" × 6", 12" × 9", and so on.

It's particularly important to pay attention to page or panel orientation because of the constraints it places on text. Readability decreases if lines of text are either too long or too short. If pages or panels are oriented vertically, they might constrain text enough to decrease readability because the lines are too short. If the pages or panels are oriented horizontally, they might encourage us to use too long a line of text. In that case, you might have to divide the page or panel into columns of text. Large formats, such as tabloids, also need this treatment because even though they might be oriented vertically, they're still wide enough to make line length a potential problem. (See the discussion of grid systems in chapter 5 for more information about laying out text on pages.)

TRIMS

Paper can also be trimmed to a special size before printing, or it can be trimmed after printing to its final size and shape. Most straight trims are made with a *guillotine* or *shear*—electrical or hydraulic machines that can cut cleanly through a thick stack of paper in one stroke.

Patterned trims are called *die-cuts* because they're created with a specially shaped cutter known as a *die* that presses through the sheets. The tabs on the Mazda quick tips guide in figure 4.1 were die-cut so users can thumb to the section they want to consult.

Paper can also be trimmed after folding and binding (see the following section, "Bindings") to create many sheets from one larger sheet. Most books are created this way. Between four and thirty-two pages are printed on a large sheet. The sheet is then folded, bound, and trimmed on the top and front edges of the book to release the individual pages. This technique makes laying the pages out on the sheet rather complex, but fortunately, a print shop will take care of this aspect of your work for you (see chapter 11 for more about this).

BINDINGS

Binding allows us to use multiple sheets in the same document. Binding methods include stitching (stapling), sewing, gluing, and mechanical binding (see chapter 11). Through binding, you can add a sturdy cover, if necessary, to protect the document from handling and extend its service life. Covers are also a great way to inform users about the document's contents and advertise its importance.

COMBINING THE VARIABLES

The variables of print format discussed in the previous sections interact to create the finished document. The document format—its general size and orientation—starts with sheet size but is then modified by how that sheet is folded and trimmed. This multiplies the options for document format tremendously. In the rest of this chapter, we'll discuss some of the more common print formats to give you a sense of some of these options.

To help keep track of these many variables, please refer to figure 4.8, which lays out some common formats in a range of complexities. The formats are grouped into two types: one-sheet formats and multiple-sheet formats.

Of course, this is a simplification of the most common formats—there are many others available, and a number of variations you should have in your toolkit of techniques. So after examining each of the formats in turn, we'll discuss a few special options that can add functionality or interest to document formats.

ONE-SHEET FORMATS

One-sheet formats are simple to design and easy to produce, so there are many common formats in this category. Because of their simplicity and economy, documents in these formats are useful for functions from basic office communications, such as letters and memos, to basic public communications, such as flyers, information sheets, product instruction sheets, and folded leaflets. Because they use only one sheet or leaf of paper, these formats are known as *leaflets*.

One-sheet formats are often produced on laser printers or photo-copiers, so they often start from the standard letter-sized paper ($8^{1}/_{2}$" × 11") or legal paper ($8^{1}/_{2}$" × 14") that fits into these machines. But that doesn't mean that we have to use these sheet sizes with one-sheet for-mats. A full-service print shop can easily make unconventionally sized one-sheet designs. Even most standard office copiers and laser printers can accommodate European sheet sizes such as A4, which is slightly taller and narrower than the US-standard $8^{1}/_{2}$" × 11" letter-sized paper (see chapter 11 for more information on paper sizes).

ONE SHEET, NO FOLDS

Documents formatted as one sheet with no folds are designed to fit either just one side or both sides of the sheet. They might be folded for distribution (e.g., business letters are often folded in thirds to fit into business envelopes), but these folds are not typically part of the visual layout of the document.

No-fold one-sheet formats often work best where the entire sheet will be easily visible, such as flyers, posters, and information sheets intended to be posted on walls, bulletin boards, or other flat surfaces. These docu-ments often serve a marketing purpose as well as an information-delivery purpose, so they can often involve a high level of design to attract the attention of passers-by—as you can see from NASA's poster introducing viewers to Katherine Johnson's important work on early space flights (figure 4.9).

Because they might change frequently, no-fold one-sheet documents can also incorporate a relatively low level of design. And because doc-uments with a low level of design can be produced very quickly, you'll often see them used for temporary signs or notices.

No-fold one-sheet formats are also used for standard business corre-spondence such as memos and letters because they're easy to distribute and require little design effort. These everyday documents also involve a low level of design because they are meant for a limited audience and deliver ephemeral (although sometimes important) information. But some business documents are worth investing in a higher level of design; for example, it's not uncommon for a client to request a two-color or process-color stationery to impress readers.

Figure 4.8. Simple to complex formats. *Source:* Created by the author.

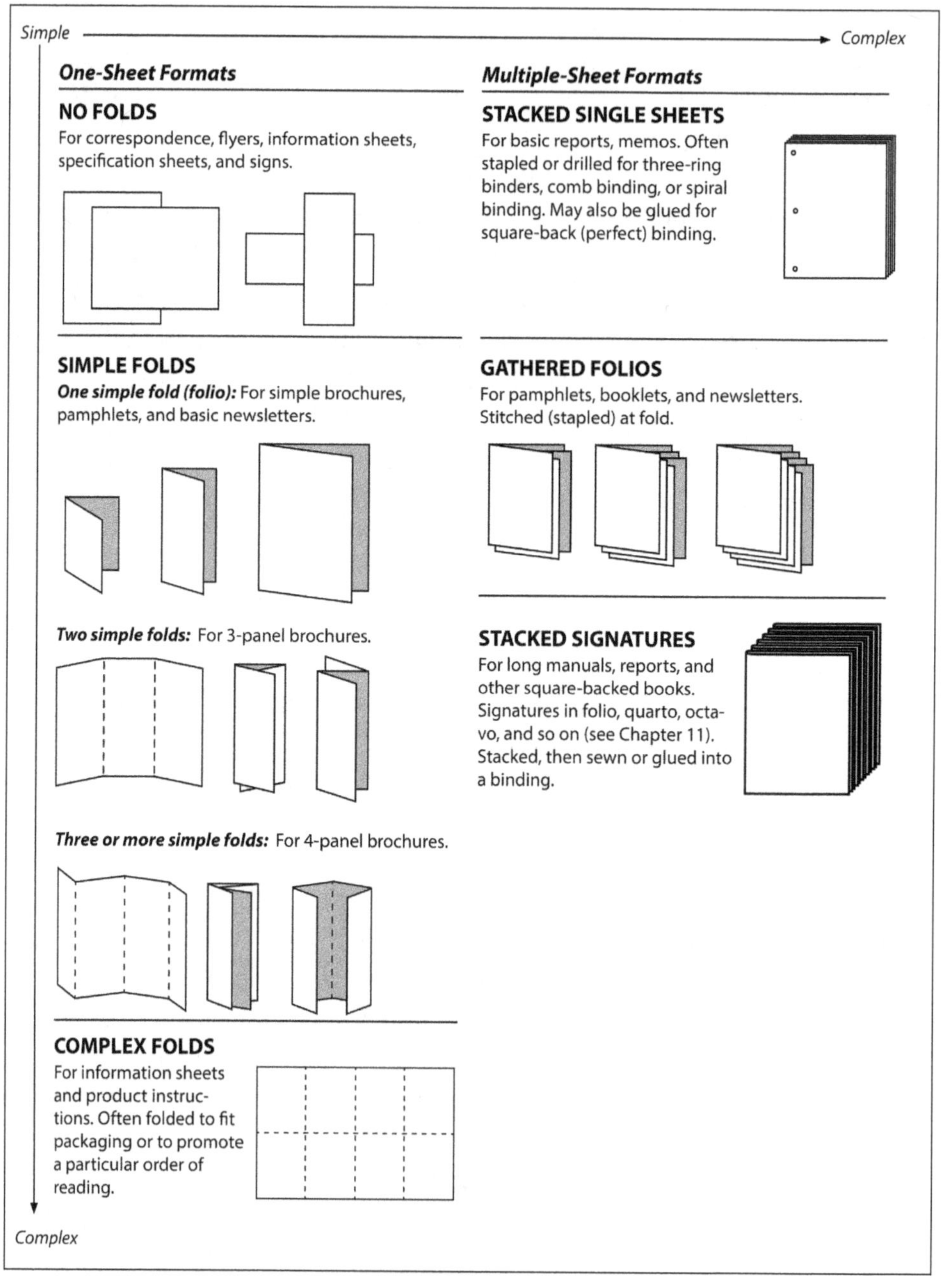

Simple ——————————————————————————————→ *Complex*

One-Sheet Formats

NO FOLDS

For correspondence, flyers, information sheets, specification sheets, and signs.

SIMPLE FOLDS

One simple fold (folio): For simple brochures, pamphlets, and basic newsletters.

Two simple folds: For 3-panel brochures.

Three or more simple folds: For 4-panel brochures.

COMPLEX FOLDS

For information sheets and product instructions. Often folded to fit packaging or to promote a particular order of reading.

Multiple-Sheet Formats

STACKED SINGLE SHEETS

For basic reports, memos. Often stapled or drilled for three-ring binders, comb binding, or spiral binding. May also be glued for square-back (perfect) binding.

GATHERED FOLIOS

For pamphlets, booklets, and newsletters. Stitched (stapled) at fold.

STACKED SIGNATURES

For long manuals, reports, and other square-backed books. Signatures in folio, quarto, octavo, and so on (see Chapter 11). Stacked, then sewn or glued into a binding.

Complex

Just because this format is used for business letters, however, doesn't mean that it's restricted to standard letter-sized paper. Larger sheets can be used to attract attention across greater distances or in specific local situations. The image in figure 4.9 is designed at a larger size to go on classroom walls and draw the viewer's attention, for example, and the image in figure 4.10 markets new mowers to shoppers at Home Depot.

Figure 4.9. One sheet, no folds. This poster introducing viewers to Katherine Johnson's important work on early space flights uses a high level of design to attract users' attention. *Source*: NASA.

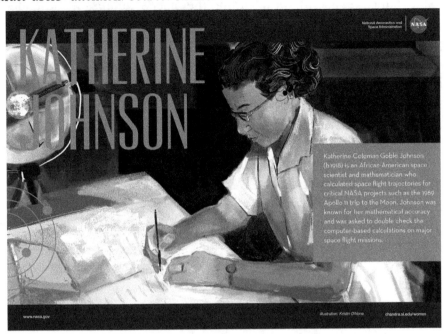

Figure 4.10. Large single sheets can capture attention. This poster at Home Depot is designed to fit into slots under merchandise to show potential buyers important product features. *Source*: Home Depot.

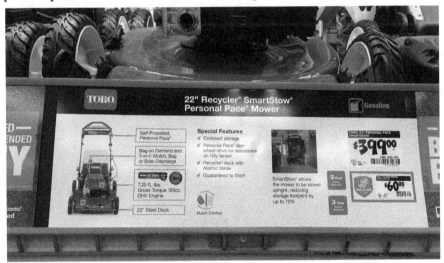

ONE SHEET, ONE SIMPLE FOLD

One-sheet formats with one simple fold are called *folio leaflets*. Folios (from the Latin for "leaf") expand the design and layout possibilities for a single-sheet document, giving us four panels or pages to work with—two on the outside and two on the inside. The front outside or cover panel is typically what the user sees first, so it's a good place for something important—a title or an attention-getting graphic. The back outside panel is useful for printing the user's address (if the document is to be mailed) or some background information about the client organization.

Remember, however, that a folio format requires some thought about the *imposition* of the panels—the way they're laid out on the sheet for printing. Because the sheet will be folded, the outside will hold panels 1 and 4, and the inside will hold panels 2 and 3 (see figure 4.11).

The size of the original sheet also makes a difference to the format, as does the orientation of the fold (see figure 4.12). Folios can also be large or small—anything from a credit card–sized folio that fits in a wallet to a newspaper tabloid size.

Folios can also be relatively tall and narrow or relatively short and broad, like the Dia de los Muertos leaflet in figure 4.13, which is based on a full 8¹/₂" × 11" sheet folded across its length, resulting in a 5¹/₂" × 8¹/₂" finished format.

Figure 4.11. Folio impositions. *Source*: Created by the author.

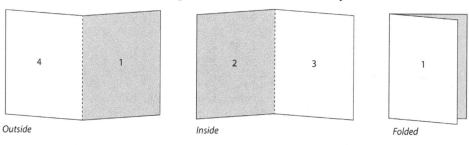

Figure 4.12. Folios in several sheet sizes and orientations. *Source*: Created by the author.

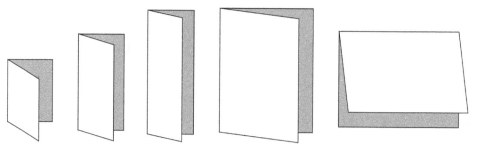

Figure 4.13. A broad folio. Notice how the designer extended content (the "In the Dead of the Night" event information and the picture of clouds) across both of the inside panels, tying them together into a visually unified spread. This document measures $5^1/_2$" × $8^1/_2$" folded, on an $8^1/_2$" × 11" sheet. *Source*: Texas Tech University, "Dia de los Muertos."

Cover

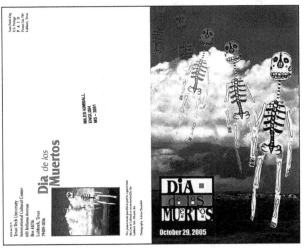

Outside spread

Inside spread

ONE SHEET, TWO OR MORE SIMPLE FOLDS

One-sheet formats can also incorporate two or more simple folds. Because multiple simple folds break up the page into multiple panels, these formats can increase the possibilities for design. For example, these formats allow for some interesting dynamics in how the panels become visible to users as they unfold the document.

Three-panel leaflets. One of the most common such formats is the three-panel leaflet (see figure 4.14)—sometimes called a "trifold" leaflet, even though it actually uses two folds to make three panels on the outside and three on the inside.

Three-panel leaflets can be folded either in a *C-fold* or in a *zig-zag fold* (also known as an *accordion fold*). In a C-fold, the third leaf (panels 4/5) nests into the crease created by the first two leaves (panels 1/2, panels 3/6), so it must be sized a tiny bit narrower than the other two for the leaflet to fold flat. In this format, panels 1 and 6 almost always stand alone as the front and back of the folded leaflet. But the other panels can be combined into larger units: panels 2, 3, and 4, the inside of the leaflet, can form either one large unit or two units (such as one spreading across panels 2 and 3 and a second taking up just panel 4). Zig-zag folded leaflets allow us to design spreads made from panels 2 and 3 and panels 5 and 6 (figure 4.15).

Though a relatively simple and common format, three-panel leaflets can lead to both challenges and opportunities, primarily because we can

Figure 4.14. Imposition of a three-panel C-fold leaflet. Two folds form three panels on each side. Leaf 4/5 must be slightly narrower than the other panels so it can tuck into the fold. *Source*: Created by the author.

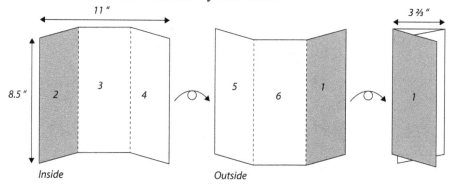

Figure 4.15. Zig-zag fold. A zig-zag fold offers different potential views to readers than a C-fold leaflet. *Source*: Created by the author.

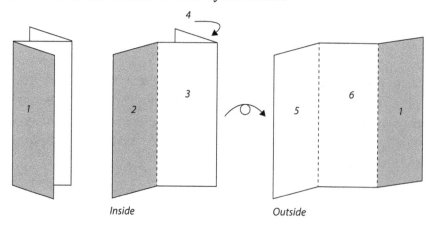

never be sure in what order users will see the panels. Consider a typical C-fold leaflet (figure 4.16). Users will usually look at panel 1 first—the "cover" of the leaflet (first view). But as users turn to look inside the leaflet, before completely unfolding it, they may pause to look at panels 2 and 5 (second view). We can design the leaflet so that 2 and 5 are seen as a unit or as separate content areas. If we design it as a unit, however, it still must make sense when users finally do open the leaflet all the way (third view). This flexibility of reading order provides a lot of opportunities for creative and interesting design, but it requires careful planning.

Figure 4.17 shows how a designer can unify two or more panels in a three-panel leaflet. In figure 4.17, a colored field extends across the top of inside panels 3 and 4, and this design feature continues on the outside with panel 5. All three panels (3, 4, and 5) are thus made to look as if they are conveying information on a similar level—the details of three areas of content: "At Your Home or Office," "On the Go," and "At Your Service."

Four-panel leaflets. Of course, there's no rule restricting us to two simple folds in a sheet; designers often specify three or more folds. One common format is a four-panel leaflet, which starts from an 8½" × 14" legal sheet with three folds, making four panels on each side (figure 4.18).

The advantage of this format when made from a legal sheet is that we get two more panels (one more leaf, front and back) than in a three-panel leaflet, yet the finished, folded size of the document is just a bit smaller than a three-panel leaflet built from a letter-sized sheet (14" ÷ 4 panels = 3½"; 11" ÷ 3 panels = 3⅔"). So both formats can fit in the same standard leaflet rack. Legal sheets can also fit in most office printers and photocopiers, keeping production costs low.

Four-panel leaflets can also be folded several different ways, which gives us a lot of flexibility of design (figure 4.19). For example, the most common fold, *double parallel folds*, can use two different panels for the

Figure 4.16. Revealing a three-panel leaflet. The way we fold a three-panel leaflet affects what readers see as they unfold the panels. *Source*: Created by the author.

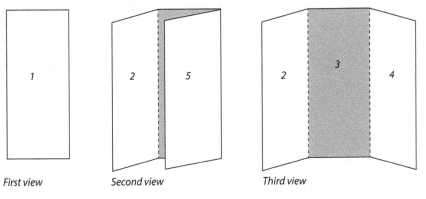

First view   Second view   Third view

Figure 4.17. Design features extending across panels. This United States Postal Service C-fold leaflet ($3^3/4" \times 8^1/2"$) uses a colored field on the inside spread to join panels 3 and 4, extending this field onto the outside on panel 5. The back (panel 6) is left for information about the publication and version information. *Source*: United States Postal Service Publication 611.

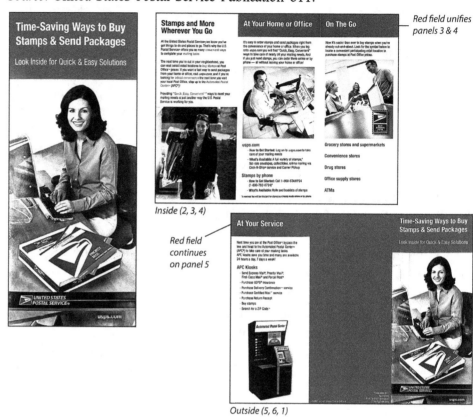

Figure 4.18. A four-panel leaflet. Folds for a four-panel leaflet that will still fit in a three-panel leaflet rack. *Source*: Created by the author.

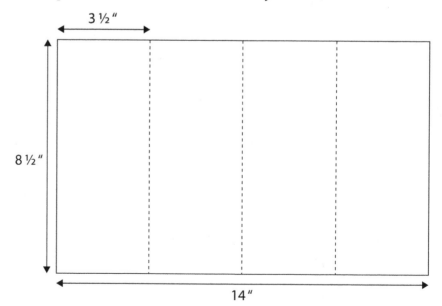

front cover. The first is similar to what we might see in a three-panel leaflet, with the cover set as the first panel on the outside (figure 4.20). This format gives users two options to see the interior of the leaflet: either turn just the first leaf (the cover) or turn two leaves (the cover leaf and the next leaf beneath it). Either way, the user must be presented with a logical display of information. If users turn only one leaf in the example in figure 4.19, they will see two unrelated panels that nonetheless look visually unified ("Payments by Mail" and "Consumer Update Now Available in Spanish"). If they turn both leaves, users will see a coherent spread of two connected panels explaining "10 Easy Ways to Save Your Energy Dollars This Winter." Alternately, double parallel folds allow one of the middle panels to serve as the "cover" panel, as you can see in figure 4.21. In this type of leaflet, users would open the cover panel to see the second view, two related panels providing information about the MallRide program. Users would then open these two panels to see the route map on three panels of the full inside spread. The final, back panel is reserved for contact information.

Of course, with these more complex formats, users are more likely to encounter a document in different ways than we expect. Arrange the panels in the format so that no matter how the user opens the leaflet, it will make sense.

ONE SHEET, COMPLEX FOLDS

Finally, we can use one sheet to make documents with complex folds (folds across a simple fold). You might be most familiar with complex folds from highway maps, but complex folds are also used to format documents such as large instruction sheets that fold to fit product packaging, informational and marketing materials, and more. Besides using space efficiently, documents with complex folds end up with a natural

Figure 4.19. Folding patterns for four-panel leaflets. *Source*: Created by the author.

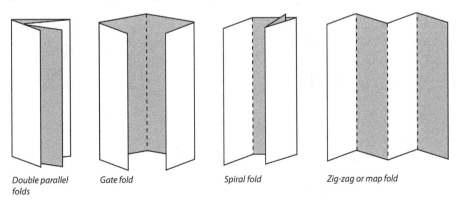

Double parallel folds   *Gate fold*   *Spiral fold*   *Zig-zag or map fold*

Figure 4.20. Leaflet in double parallel folds with cover on first outside panel. This pamphlet is 3³/₈" × 6¹/₂", folded from a 13" × 6¹/₂" sheet. *Source*: Atmos Energy Corporation, "Consumer Update."

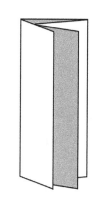

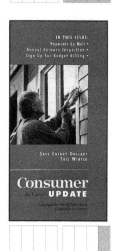

Cover

User opens one leaf to see two unrelated but visually unified leaves.

User opens two leaves to see two logically related leaves.

Figure 4.21. Leaflet with double parallel folds. This format changes the viewing order of the panels, and is 3¹/₂" × 8¹/₂" folded from a 14" × 8¹/₂" sheet. *Source*: Denver Regional Transportation District, "Free Mall Rides" (2608-04.01).

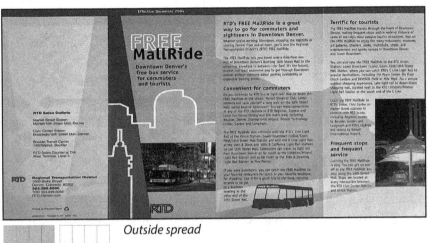

Outside spread

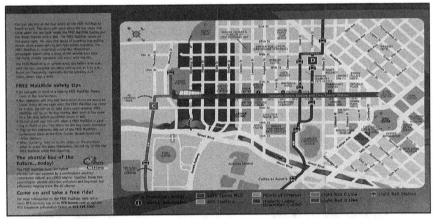

Inside spread

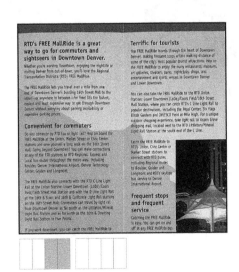

Cover and first view   *Second view*

grid that can be used to organize the content and affect what readers see as they unfold the sheet. (For more on grids in page design, see chapter 5.) For example, the (fictional) guide to the New Oban Express, produced by Six to Start to accompany its Train to Oban virtual race (which is set in a zombie-infested, postapocalyptic environment) in figure 4.22 uses complex folds to guide the user through the document.

The New Oban Express guide starts from a $16^{1}/_{2}$" × $11^{3}/_{4}$" sheet, folded first with three vertical folds, then with one horizontal fold, resulting in eight panels on each side of the sheet. Folded, the document measures $4^{1}/_{8}$" × $5^{7}/_{8}$". The designers draw attention to the cover with a picture of a bullet train speeding across a futuristic environment and the title "New Oban Express." The back of the document invites users to "Escape the London Hordes for Good," along with marketing text. As the user opens the document, the increasingly large design spaces invite different uses. At its 4" × $^{1}/_{8}$" × $11^{3}/_{4}$" size, one side of the document shows the "London to New Oban" route, including a map of England and descriptive text. At $8^{1}/_{4}$" × $11^{3}/_{4}$", we have the "New Oban Express Safety Instructions," along with icons with both conventional and representational meaning

Figure 4.22. A complex fold document. This document measures $4^{1}/_{8}$" × $5^{7}/_{8}$" folded, from a $16^{1}/_{2}$" × $11^{3}/_{4}$" sheet. *Source*: Six to Start, New Oban Express.

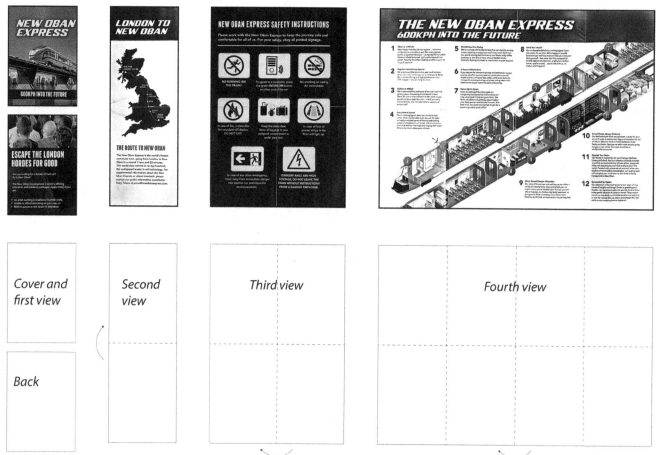

(see chapter 2)—an image of a silhouetted person with a line across indicating "No Running on the Train!," for example. At its full 16½" × 11¾", the document offers a full-page numbered cross section of the train. The logical order of opening the document provides increasingly complex information, and fictional though the New Oban Express might be, it's easy to see how designers can use this complexity to invite users deeper and deeper into a particular topic.

It's equally easy, however, to lead users astray by not paying attention to how they might interact physically with complex fold formats. Consider the example of a university bus route map in figure 4.23. The route

Figure 4.23. A confusing complex fold document. This document measures 3¹¹/₁₆" × 8½" folded, from an 11" × 17" sheet. In the first use view, the user sees the cover of the Citibus route map pamphlet. In the second view, the user opens one flap of the pamphlet to see a bus schedule. In the third view, the user opens the second flap to find that the combined route map is upside down. *Source*: Citibus, "Texas Tech University Citibus Campus Routes."

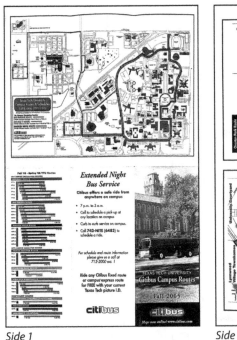

Side 1

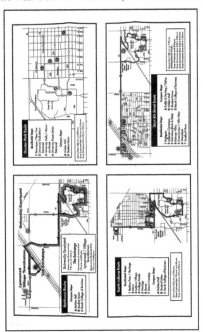

Side 2

Cover

First view: User sees the cover of the Citibus route map pamphlet.

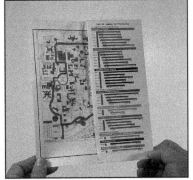

Second view: User opens one flap of the pamphlet to see a bus schedule.

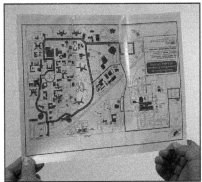

Third view: User opens the second flap to find that the combined route map is upside-down.

map starts with a standard 11" × 17" sheet, folded first in half to 11" × 8¹/₂" and then in thirds across the first fold to a finished size of 3¹¹/₁₆" × 8¹/₂"—small enough to fit in standard leaflet racks. In most aspects, this format works very well. It provides clear information about the available bus routes, incorporating one combined route map and four individual route maps. But as the user unfolds the leaflet to read its contents (figure 4.23, third view), one of the content areas—the combined route map, spreading across three panels—turns out upside down from the user's perspective. This problem might not be fatal, but it makes the user flip the document upside down to read the map correctly.

MULTIPLE-SHEET FORMATS

Despite the wealth of options and advantages of single-sheet formats, some documents are simply too long to be delivered that way in print. There are also physical limits to the convenience of sheet size: we wouldn't want to design a single sheet that when unfolded was bigger than the user's desk or arm spread unless we had a very compelling reasons to do so (e.g., when designing a large sign to be seen from a distance).

That leaves us with the dilemma of how to put together many sheets into one document. Early document designers managed this problem by stitching together sheets end-to-end, storing them either as a scroll or as a zig-zag folded stack. But these arrangements had definite usability disadvantages—particularly if the documents were ever dropped and unrolled!

So the scroll was superseded centuries ago by the *codex* format (plural: *codices*). A codex is simply a document formatted on multiple sheets, all bound together at a common spine at the left, right, top, or bottom of the stack of sheets. The spine serves as a hinge allowing users to flip the individual leaves of paper back and forth. Naturally, over the centuries a lot of effort has gone into the design of this hinge, also known as a *binding*. Together, the codex and its binding form an efficient, compact, and user-friendly package that is still the format of choice for lengthy documents, such as textbooks, long reports, or manuals.

Codices are more commonly termed *books, booklets, pamphlets,* or *brochures* (from the French verb *brocher,* "to stitch," since codex bindings are often sewn or stitched together). These terms are fluid, but generally booklets and pamphlets have fewer pages than books, and brochures have even fewer pages than booklets or pamphlets. Any codex with hard covers (also known as *boards*) is generally considered a book, regardless of length.

In this section, we'll discuss several options for formatting documents as a codex, focusing on the different styles of binding available today.

Some of these bindings are simple and cheap, others elaborate and expensive, but all are focused on meeting particular physical needs of users as they grapple with a long document. Different bindings have different usability characteristics, particularly in terms of durability and the ability to lie flat when open so users don't have to hold the document to use it. Opening flat is particularly important for instructional documents like manuals because the user's hands are usually occupied while following the instructions in the manual.

As we discuss multiple-sheet formats, keep in mind that we're talking about the format as the user handles it, not about how multiple-sheet formatted documents are actually created. In fact, many multiple-sheet formats are actually produced by printing from four to sixty-four pages on very large sheets, which are then folded, bound, and trimmed to the final page size. For more information on the technologies and techniques involved in this kind of approach, see chapter 11.

STACKED SINGLE SHEETS

One of the most basic multiple-sheet formats is simply a stack of identically sized single sheets bound with some sort of simple mechanical binding, including paper clips, staples, three-ring binders, comb bindings, or spiral bindings. These bindings aren't fancy, but they are fast and cheap—sometimes fitting the constraints of a project perfectly.

Paper clips, string, and staples. Historian of technology Henry Petroski wrote in *The Evolution of Useful Things* (1994) that the staple and paper clip required significant technological development and went through many different forms before they were more or less standardized to the devices we know today. The styles we are most familiar with appeared only in the late nineteenth century; before that, people used straight pins, thread, or glue to hold together multiple sheets. Indeed, one of the oldest forms of binding still commonly practiced, especially by those starting in custom book binding, is stab binding (see figure 4.24), or *Japanese stab binding*, which originated in the Edo period of Japan (1603–1867) and was used extensively until the late 1800s. In the most common method, the book is bound by punching holes through the front and back covers and leaves, then sewing them together, rather than binding leaves along a spine.

Much less formal, of course, is the temporary binding provided by the ubiquity of paper clips and staples in offices today, which even in the computer age speaks to their success in meeting user needs by binding together a stack of paper quickly, cheaply, and conveniently. They also have the advantage of being removable (although paper clips are easier to remove than staples).

Figure 4.24. Stab binding. This book uses a simple form of stab binding to hold everything together. *Source: Rubaiyat of Omar Khayyam* (Edward Fitzgerald, Trans.). Barse and Hopkins. No date printed in the book, likely between 1899 and 1920. Photo by the author.

Naturally, paper clips and staples also have disadvantages. They have limited capacity—usually not more than twenty-five sheets, although some paper clips known as *binder clips* trade off a little convenience for a better grasp on thicker stacks. Because they are usually positioned at the top-left corner of the stack, paper clips and staples require the user to bend back the page over that corner rather than directly left, as in a more substantial codex binding that extends down the whole spine. So paper clips and staples are hard on paper, either by design (the stapler punches holes) or by accident (paper clips rust or their ends snag on the paper as they are removed). Finally, paper clips must be applied manually, and staples require a stapler, which often jams and mangles the staples themselves.

But for low-level designs—those that need or deserve little investment in time, money, or personnel—the familiarity and utility of these devices make them indispensable. For short-run and ephemeral work, you can staple or paperclip the sheets yourself, seeing your design through from inception to completion. For longer runs, some photocopiers include staplers, or you can ask a copy center or printer to staple a document for you. Printers and copy centers typically have machines that can staple pretty much anywhere on the sheet, and they can usually staple through three or four times more sheets than you can with a handheld stapler. However, they may balk at paperclipping any long document runs, since doing so would require significant (and expensive) manual labor.

We've mentioned zines elsewhere in this book—if you are looking to self-publish your work in a bound format, stab binding, stitching, or stapling is an inexpensive, and relatively fast, way to go. You could make a zine by stitching or stapling several stacked single sheets together, or, to look a bit ahead in this chapter to "Gathered Folios," a simple zine might include a few 8½" × 11" sheets of paper folded in half and stapled or stitched down the crease. This results in what is commonly known as a *saddle stitch binding.*

Binders. Binders are available with a variety of clips, grippers, and plastic sleeves to bind individual sheets into a codex. Many of them work well for small-run document design jobs, but they tend to be fussy, requiring lots of manual effort to get them on the sheets. Binders also tend to keep the document from opening flat because they grip about a half-inch of paper at the spine.

But one older binder technology still works well, even though it is just a step more elaborate than paper clips or staples: the ringed binder. You're probably very familiar with the standard three-ring binder, which comes in a variety of thicknesses and colors. As a technology, binders require a simple piece of office equipment to punch, or drill, holes through which the binder's rings extend to hold the sheets. These machines can drill either single holes or consistently spaced multiple holes. Or you can ask a copy center or printer to drill your document to your specifications.

Like staples and paper clips, ringed binders are a simple technology with many formatting advantages. Ringed binders also allow easy access to both sides of the sheet, they will lie flat, they're relatively sturdy, and users know how to use them. It's also easy to include tabbed dividers of cardboard or plastic to separate the sections of a document so users can access its contents easily. Ringed binders also work well for documents that must be updated occasionally. Reprinting the entire document would be too expensive, but if the document is in a ringed binder, outdated pages can easily be replaced. This time-honored technique requires careful version control; you must keep track not only of the correct version of the document but of each page as well. It's still a workable system, though, especially if the document is one that wouldn't translate well to a computer screen. Binders are also pretty tough, so they're a good choice for reference documents.

However, ringed binders are hard on paper. Most hold each sheet by only three points, making it easy for a user to rip out a sheet inadvertently. Drilled holes, particularly on the first or last sheets, also tend to rip if the document is very thick or if it's crammed into a binder that's too small. Manufacturers have devised many ways to make the holes stronger, including plastic inserts and adhesive hole reinforcement products, but it remains a basic weakness of the format.

Comb and spiral bindings. A further step up the ladder of technological complexity brings us to *comb bindings* and *spiral bindings* (figure 4.25).

Comb and spiral bindings extend the binder concept by using several holes instead of just three. Comb bindings insert a series of curved "fingers" through these holes, and as the name suggests, spiral bindings

Figure 4.25. Wire spiral binding and plastic spiral binding. Printed course materials can be inexpensively bound using plastic spiral binding, while many sketchbooks use wire spiral binding as it allows the user to fully open to the working page while lying flat. *Source*: Created by the author.

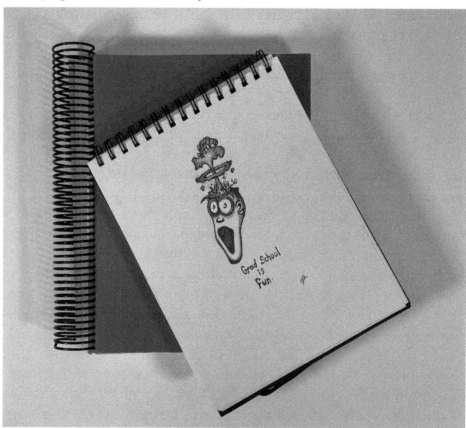

use a continuous spiral of wire or plastic inserted through the holes. Both types are available in wire or plastic, and both have the advantage of allowing a document to lie flat. But spiral bindings also allow users to bend the spine completely back.

Wire spiral bindings require an expensive machine to run the spiral wire through the row of drilled holes, so they're usually an option only if you are working with a professional printer. However, the equipment required for plastic and wire comb bindings is relatively inexpensive, and many offices and copy centers have those machines.

Perfect bindings. *Perfect binding* is the trade term for what are more commonly known as "paperback," "softcover," or "square-back" bindings. Perfect bindings use a hot-melt plastic glue to bind sheets together along the spine. Typically, the stacked sheets are fanned at an angle before the glue is applied to the spine. After they are straightened up again, placed in the cover, and clamped, a little bit of glue remains between each sheet, holding them all together and sticking them to the cover. Well-equipped print shops often have machines that can do all this in one pass.

Perfect bindings are durable and sharp-looking, but they're also more expensive than the other bindings we've discussed, primarily because they require special equipment and supplies. They also create a document that does not lie flat well, particularly if the cover stock is very stiff. One compromise is to score the cover stock about one-quarter inch out from the spine, which creates a hinge point that allows the binding to fold open a bit more easily.

GATHERED FOLIOS

One simple way to bind multiple-sheet documents is to format them as gathered folios (figure 4.26). As you'll recall, a folio is just a sheet folded in half. If we nest several folios together, they can be stitched with one

Figure 4.26. Gathered folios. *Source*: Created by the author.

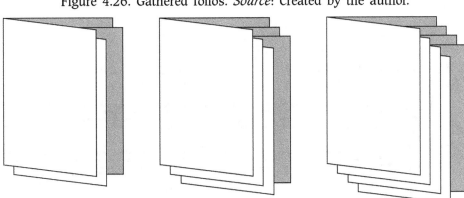

or more staples in the fold to create a quick and inexpensive binding. This technique is also known as saddle stitch binding or *saddle-stitching* because the nested folios are laid over an angled block that looks like a saddle for the staples to be inserted.

Gathered folios are a very usable format for mid-level design documents, including event programs, longer pamphlets or brochures, and mid-length instruction booklets (up to around eighty pages). The format lies open well, and it also allows us to use a cover stock of a somewhat thicker paper to increase the document's durability and raise its level of design. Because each folio is made from a different sheet, you can choose which sheets should be printed in color. It's common, for example, to use a color folio cover around an otherwise black-and-white document.

But gathered folios run into a couple of problems if the document gets very long. First, the more folios nested together, the more likely the document won't stay closed, particularly if the paper is relatively stiff. Second, because of the thickness of the paper, each nested sheet sticks out a little farther on the fore-edge than the previous sheet did, leading to a condition called *creep* (figure 4.27).

Depending on the thickness of the paper, creep can become evident in as few as four or five gathered folios. The creeping edge is often trimmed off at the very end of the production process, but this can cause problems with the fore-edge margin, which gets smaller and smaller for the folios sticking out furthest (i.e., those in the center of the gathering). Fortunately, most page layout programs can automatically adjust for creep to eliminate this problem.

STACKED SIGNATURES

For longer documents, such as books, nested folios won't work as well as multiple stacked *signatures*—essentially, stacks of gathered folios that

Figure 4.27. Creep at the fore-edge of gathered folios. Creep happens when several folios are gathered together; each nested sheet sticks out a little farther on the fore-edge than the previous sheet did. *Source*: Created by the author.

can be bound together. A signature is a single sheet with multiple pages printed on it—typically between four and sixty-four pages. After printing, the sheet is folded so that the pages line up in the correct order. Multiple signatures can then be stacked and bound before they're trimmed on the top edge, bottom edge, and fore-edge, releasing the individual leaves. After several signatures are stacked, bound, and trimmed, they are actually just a stacked collection of gathered folios, usually of two, four, six, or eight sheets each. (Chapter 11 discusses this production technique in more detail.)

A bound stack of signatures can also be glued into a square-back, perfect binding, just as in a perfect binding of single sheets. Or the bound stack can be inserted into a *case binding* (figure 4.28). Case bindings are formed by binding together the signatures first, by either sewing or gluing. Then the bound signatures are glued into a previously prepared "case" composed of the hardcover boards, the spine, and the cover fabric. Most hardcover books have case bindings.

Case bindings are extremely durable and friendly to the users' hands. When closed, they stay closed, and they lie open more easily than perfect bindings. They are impressive, giving the document an air of importance. But they're also among the most expensive of all the bindings used in document design, and they definitely require the services of a full-fledged commercial print shop.

Figure 4.28. Stacked signatures in a case binding, viewed from the top edge. The case is made of fiber boards covered in fabric or strong paper and joined together to cover the spine. *Source*: Created by the author.

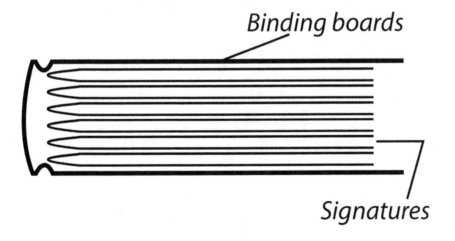

SPECIAL FEATURES

For projects that merit a higher level of design, a variety of special features can add interest or functionality to a document's format. Most of these special features require the services of a commercial print shop, and although they add considerably to the unit price of the document, sometimes rhetorical or practical situations call for spending a little extra money. In this section, we'll discuss three special format features often used in document design: trims, perforations, and composite formats.

SPECIAL TRIMS

Special trims are typically performed after a document is printed on an oversized sheet. For example, if you want part of the design to extend all the way to the edge of the finished document (called a *bleed*), the document must be printed on a slightly larger sheet, leaving a half-inch or so of paper all around for the printing press to grab (figure 4.29). The paper is then trimmed to its finished size to remove the unprinted edges. Naturally, this increases the cost of the printing because it requires larger, specially sized paper. To finish with a trimmed sheet of 8$\frac{1}{2}$" × 11", for example, you'd need to start with an original sheet size of about 9" × 11$\frac{1}{2}$".

Other special trims can be much more elaborate. For example, the leaflet in figure 4.30 is trimmed square on three edges but in a complex

Figure 4.29. Trimming a bleed. If part of a document's design extends to the very edge of the page, it must be printed on a larger sheet and then trimmed to its finished size. If designing for a trimmed document, keep all text within a safe area (often referred to as the "live area") set back from the trim line to account for any trimming inconsistencies. *Source*: Created by the author.

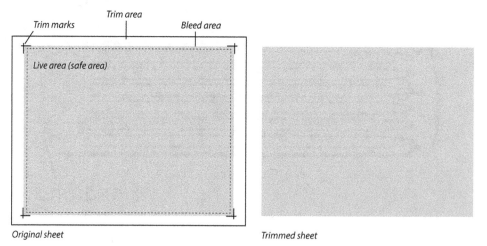

Trim area

Trim marks

Bleed area

Live area (safe area)

Original sheet

Trimmed sheet

Figure 4.30. Die-cutting. The special die-cut sheet, $3^1/_{16}"$ × 9", from a $15^1/_4"$ × 9" sheet, gives a 3D appearance to the die-cut images when the leaflet is folded. *Source*: Holiday Inn.

Side A

Cover, folded

Side B

die-cut shape on the fourth. When the leaflet is folded, this die-cut makes the different images stack up, giving a dramatic, 3D effect.

But before deciding to use a specialty trim or die-cut, make sure that some rhetorical, practical, or usability function is served. Specialty trims are quite expensive, and they tend to decrease the durability of the document by creating unusual edges and corners that inevitably get bent back or torn. To counter this problem, the leaflet in figure 4.29 is printed on heavily varnished paper.

PERFORATIONS

Perforations allow users to tear away a part of a document. For example, a leaflet from the Corcoran Gallery in Washington, DC, explains the benefits of membership in the gallery and includes a two-panel tear-off membership application (figure 4.31).

Figure 4.31. Perforations. This leaflet for an art museum (4" × 9¼" folded, from a 9¼" × 23½" sheet) includes a perforated tear-off membership application form. The form can be filled out, folded, glued or taped, and sent in the mail, with the postal information and postage already provided. *Source*: Corcoran Gallery of Art, "Join."

Perforation

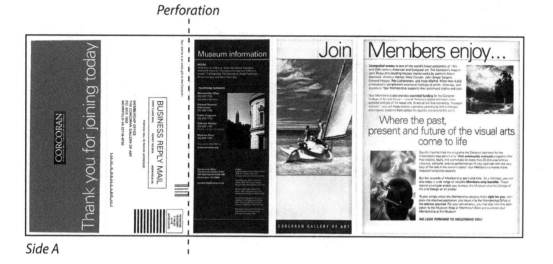
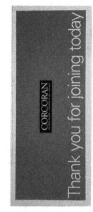

Side A   Cover

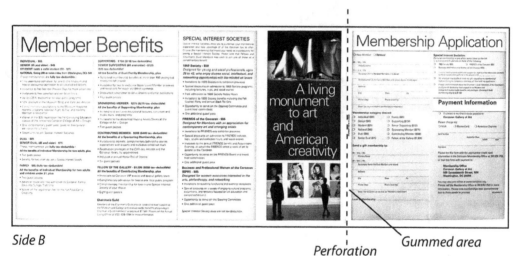

Side B

Perforation   Gummed area

After tearing off the membership form and filling it out, users can fold the torn-off panels in half, lick a gummed area to stick them together, and pop the application in the mail using the prepaid postage. This format encourages a pattern of use: read the pamphlet to understand the benefits and then tear off, fill out, and mail the application to join—easy, useful, and user-oriented!

COMPOSITE FORMATS

Finally, you can combine some of the formats we have discussed to create documents with composite formats. An example of a composite document is a brochure from the Denver Museum of Outdoor Arts in figure 4.32.

Figure 4.32. A composite format: two sizes of folio and a leaflet. This document set is 6" × 11" folded, from three gathered folios and an inserted four-panel leaflet. The outermost folio of this brochure is in a smaller format than the rest, adding some visual interest. The stitching holding the folios and the leaflet together is between the first two panels of the spread shown in the third view, with the other two panels folding out to show additional graphics and the map. *Source*: Denver Museum of Outdoor Arts, The Collection brochure. Cover image: Edward Hopper, *Ground Swell*, 1939, oil on canvas, 36½" × 50¼". In the Collection of the Corcoran Gallery of Art, Washington, DC, Museum Purchase, William A. Clark Fund.

First view and cover

Second view

Third view

   At first glance, it seems a simple gathered folio with a colored band printed on the cover. But on closer examination, the printed band is a much smaller folio stitched to the outside of the larger gathered folio. Turning to the inside, the user finds only two folio sheets, but within the folio sheets is nestled a four-panel leaflet that folds out to show a schematic of the public art a pedestrian would see following a planned viewing walk. The leaflet is stitched into the folio at one of its folds, and the whole package folds up to a convenient 6" × 11" size—just right for carrying with you as you go on your walk. In other words, this single document combines two formats of folios with a four-panel leaflet, making a flexible and interesting composite document.

Another example of composite formats is a booklet from the British Library advising patrons on how to order photocopies and other reproductions (figure 4.33). Overall, it's a simple folio booklet with a card stock cover measuring 9" × 12". But the designers have designed a pocket in the inside back cover, which holds the appropriate forms to order reproductions of library holdings.

EXERCISES

In this chapter you've thought a lot about how documents exist in the world. You'd considered document ecology and user perceptions, and thought about how perception, culture, and rhetoric shape a document's use and impact. You've learned a bit about how to make decisions about design based on factors like conventions, physical properties, what the document needs to do, and how much it might cost to get it to do those things. You've also thought a lot about how you might start folding, cutting, and generally shaping a document to your users' needs. These exercises are designed to help you think through these concepts and more by looking at how real humans use real documents.

Figure 4.33. Another composite format: a 9" × 12" booklet and order form. The back cover of the stitched booklet has a pocket to hold an order form. *Source*: British Library, "Reproductions."

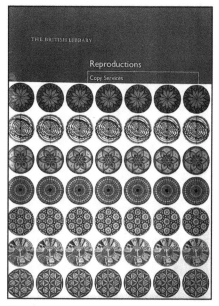

Cover

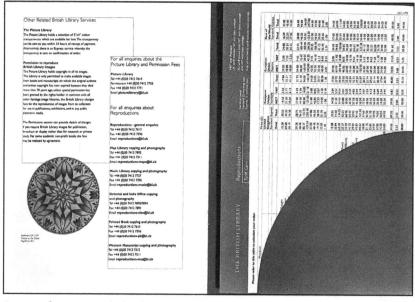

Request form in inside back cover pocket

1. Observing how users use documents in their own ecology can be fascinating. Go to a workplace and explore how documents are physically used there. Workplaces other than offices can be particularly enlightening: consider places like construction sites, factories, police stations, or restaurants. Ask if you can talk to some of the employees about the documents they use and observe them during work for an hour or two.

2. Ask them these questions during your visit:

   o What documents are essential to your work? (Remind interviewees that documents aren't just reports and memos but also signs, checklists, forms, websites, and so on.)

   o How often do you use these documents?

   o How and where do you use these documents?

   o How well do they stand up to use? What tends to mess them up?

   o Where do you keep them? Why?

3. Write a memo to your instructor reporting your findings. In the memo, reflect on the documents' physical ecology, how well the documents are designed to that ecology, and how you'd improve the documents to fit their ecology better.

4. Find a document with an unconventional, unusual, or specialized format and bring it to class. In groups, discuss the following about each of the documents your group members collected:

   o What needs was the document intended to meet?

   o Under what physical conditions would users likely use the document?

   o What factors led the designers to go to the extra effort and expense to create a document in this format? Was it worth the extra cost?

   o What has the document's physical life been like after it left the designer's hand and came to yours? Has it held up well?

5. Choose what you think are the most and least successful documents in terms of format. Prepare a presentation to the rest of the class to explain your choices.

o Take one of the documents from exercise 2 as a model and try to replicate its format in whatever page layout or word processing software you have access to. Don't forget to take into account page size, folds, trims, and bleed areas.

o In word processing programs, your options will naturally be limited. But try to press against the boundaries of your word processor's capabilities to set up a document as close as possible to your model. Don't be afraid to search for tips and tricks on how to get the program you are using to do what you want!

o After you've created a document format as close as possible to your model, discuss with your classmates what you learned. What were the software's limitations and capabilities?

REFERENCES AND FURTHER READING

Evans, P. (2004). *Forms, folds, and sizes: All the details graphic designers need to know but can never find*. Rockport.

Gaskell, P. (1995). *A new introduction to bibliography*. St. Paul's Bibliographies.

International Energy Agency. (2023). *World energy outlook 2023*. https://www.iea.org/reports/world-energy-outlook-2023

Johnson-Sheehan, R., & Baehr, C. (2001). Visual-spatial thinking in hypertexts. *Technical Communication*, *48*(1), 22–30.

Kostelnick, C. (1996). Supra-textual design: The visual rhetoric of whole documents. *Technical Communication Quarterly*, *5*(1), 9–33.

McKenzie, D. F. (2004). *Bibliography and the sociology of texts*. Cambridge University Press.

Monteiro, M. (2021). *Ruined by design*. https://www.ruinedby.design/zine

Petroski, H. (1994). *The evolution of useful things*. Vintage.

Perrin, A., & Atske, S. (2021). *7 percent of Americans don't use the internet. Who are they?*. Pew Research Center. https://www.pewresearchgrg/fact-tank/2021/04/02/7-of-americans-dont-use-the-internet-who-are-they/

Robinson, S. (2014). *The book in society: An introduction to print culture*. Broadview Press.

Rockley, A., Kostur, P., & Manning, S. (2002). *Managing enterprise content: A unified content strategy*. New Riders.

Spinuzzi, C., & Zachry, M. (2000). Genre ecologies: An open-system approach to understanding and constructing documentation. *ACM Journal of Computer Documentation (JCD)*, *24*(3), 169–181.White Flower Farm. 2004. *White Flower Farm, spring 2004*. Series 23, Box: HFLA 002.287, Folder: 0024. Seed Catalog Collection, HFLA 002. Denver Botanic Gardens, Helen Fowler Library Archives.

World Bank. (2023). *World development indicators*. https://data.worldbank.org/indicator/EG.ELC.ACCS.ZS

5

PAGES

LEARNING OBJECTIVES

Upon completing this chapter, you will be able to:

- Consider the page as a dynamic visual field

- Discuss pages through lenses of perspective, culture, and rhetoric

- Consider how your page designs can create meaning

- Actively think through the ways users' eyes address a page

- Discuss the way grids shape effective page design

～

One of the most challenging and dynamic tasks a document designer will undertake is page design. The page is the space where a document comes together in the user's field of vision—everything from content to context, from the visual marks on page or screen to the material framework that surrounds and delivers them. Designing pages requires close attention to the shape and dimensions of the visual field as well as to the characteristics of and relationships among the design objects we place there. It also must work dynamically within the medium and format you have chosen.

As a term, *page* has both general and specific meanings. Generally, a page is just a single coherent visual field in a document—whatever the user can see at once, without manipulating the document by turning a paper page or clicking to go to another web page (see figure 5.1). But both print and electronic media add some specific twists to that visual field. In print documents, a page is only one side of one sheet of paper, but when we open a bound document, the visual field we see is a *spread* of two pages connected at the spine. And as we saw in chapter 4, each *panel* in a folded document can be designed as a separate visual field or combined with other panels to create larger fields. Even electronic documents use the concept of page, such as a web page.

In this definition of page, *page design* is the process of placing design objects such as text, headings, and images consistently and effectively on the page, taking into account the actual visual field, the characteristics of the design objects, and the relationships implied among them by the principles of design.

Figure 5.1. Pages as visual fields. The concept of page as a visual field is useful across media and formats. *Source*: David Garrick (1717–1779), "A Theatrical Life," Folger Shakespeare Library; "Power in Numbers," Nucleus Gallery; "Countertops," Expo Design Center; "Natural Resource Management," NASA, http://www.nasa.gov.

Panels

Page

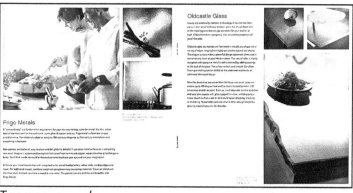

Two-page spread

Web page

To see what we mean, consider the design of a web page in figure 5.2, the 2021 Webby Award–winning website Black Disabled Creatives. The site is designed to showcase Black disabled creatives by creating a searchable space for employers to find highly qualified, but often extremely underrepresented and underappreciated, individuals. The page does this by creating a visual framework for the many design objects on the page, including navigational elements on the top and side, and interactive vignettes showcasing the many Black creatives who populate the site. The designers used

Figure 5.2. Pages on-screen. This web page frames content by using the principles of similarity, enclosure, alignment, contrast, order, and proximity to help users easily search for the type of professional and work they need done. When a user clicks the Menu button, contrast on the page changes to help choices stand out (as seen in the second view here). *Source*: Black Disabled Creatives, https://blackdisabledcreatives.com/.

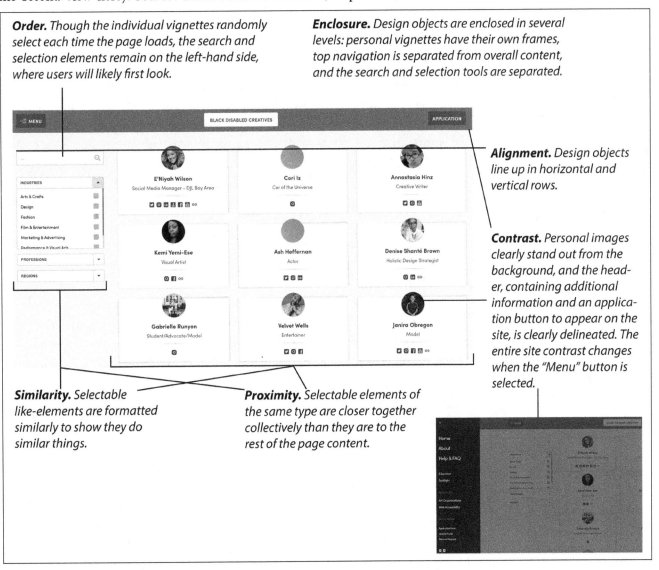

principles of similarity, enclosure, alignment, contrast, order, and proximity to organize the many design objects on the page into an understandable system, group together like information, and distinguish different types of information (e.g., searchable and selectable categories are enclosed at the left side of the page). Contrast is further employed when the user selects the menu hamburger at the upper left: when selected, a black bar featuring white text appears, which gives users insight into the page's intent, along with additional resources and information.

This is a good example of how page design can bring order to the chaos of information within the primary visual field in documents: the page. Though the site includes information for an increasingly large number of creatives across many different industries and professions, users are rapidly able to narrow down options to find the person they are looking for. Page design guides users through the many pieces of information in the visual field by using design principles to create clear relationships among design objects.

In this chapter, we'll discuss some of the many techniques for designing effective pages. First, we'll discuss pages through the three theoretical perspectives introduced in chapter 2: perception, culture, and rhetoric. Building on these perspectives, you'll learn how to assess content for its structure and set up a grid system to organize content into logical and visually distinct spaces. Finally, you'll learn when to break the consistent system you've created to emphasize important content.

THREE PERSPECTIVES ON PAGES

PERCEPTION

Take a minute to think about what you're seeing right now. On your desk, in your hands, or on your screen, you see a book. More specifically, because much of the book is obscured from view, you are looking at either a two-page spread in a book or possibly only part of a single page. Regardless, you know how to use the book—you'll continue reading, then turn the page or scroll your view. This view might seem very familiar. But think for a moment about *why* the pages are designed this way. Pages don't have to be rectangular, and they don't have to feature a nice symmetrical shape with strong edges and a frame of empty margins. So why do they? (Most of them, at least.) As we learned in chapter 2, Gestalt psychology suggests that one reason is our visual need to know where the page ends and the background begins. Gestalt theory posits that we perceive things with consistent edges as *figures* on a *ground*. If you are holding a physical book, it clearly illustrates that principle: the sharp and consistent edges identify the joined pages as an object within your field of vision, or a figure on a ground. If you are reading this on a device, you are likely still treating the device, visually, the same way you would a physical book. When we read documents, we so trust the figure-ground distinction that we allow the rest of the visual field to recede from our awareness—the pages of the document make up all of the conscious visual field. (Have you paid any attention to the background *around* or *behind* this book while you've been reading it?)

But this is just the beginning. Just as the pages before you seem like figures on a broader visual field, you can also see figures (*design objects*, as we discussed in chapter 1) within the page field. For example, there's a design object you recognize as where the text goes—the primary *text field*. Nested in the primary text field, you recognize smaller areas of coherent and separate text figures: *paragraphs* and *headings*. You can also see some areas in the text field that you take as coherent and separate *graphic elements*. And you see some other text fields at the top and bottom of the page: the *header* and the *footer*. Finally, you can tunnel down to even smaller design objects: sentences, words, and individual letters.

As an experienced reader, you're familiar with these design objects—so much so that you typically use them to organize your reading of the page without consciously thinking about them as figures on a ground, as we are here. But these objects are not random; they're the result of a document designer's careful work to assemble design objects into a coherent visual system. A great part of page design is creating a visual field filled with design objects and using users' perceptions to suggest clear relationships among those objects.

CULTURE

The designer's careful work, however, doesn't spring up from nothing. Part of design relies on cultural conventions—the long-standing but always changing expectations users have about where some kinds of design objects will be located on a page and how those objects will look. Understanding such conventions involves more than simplistically applying page design rules to documents. Instead, it requires designing while keeping in mind the many cultural factors surrounding documents in the user's own experience.

For example, most people are familiar with the design objects in a typical formal letter—letterhead, address block, salutation ("Dear . . ."), text block, closing ("Sincerely"), and signature. But conventional elements, like those of a letter's page design, developed over time due to specific cultural influences.

Consider what letters have looked like under *different* cultural conditions. Today, the sender of a letter must pay for it to be delivered, usually by purchasing a postage stamp and pasting it to the letter. Before postage stamps were introduced in the mid-nineteenth century, however, the *receiver* of the letter had to pay the postage—and the more sheets in the letter, the higher the postage due. Sending a longer-than-necessary letter was wasteful, expensive, and rude. So letter-writers found ways to conserve the amount of paper they used on letters. One technique was called *crossing* (figure 5.3).

As you can imagine, crossing made reading the letter quite difficult because it consisted of two competing page designs set at 90-degree angles, with each direction making the other less legible. Some readers were willing to accept this inconvenience in the name of economy, but

Figure 5.3. The page layout of a crossed letter. Different cultural conditions often lead to different conventions of page layout, as you can see from this crossed letter. Because of the expense of postage, the writer would complete a page of the letter normally, then turn the sheet 90 degrees and write a second page across the first set of text lines. *Source*: Kenneth Spencer Research Library.

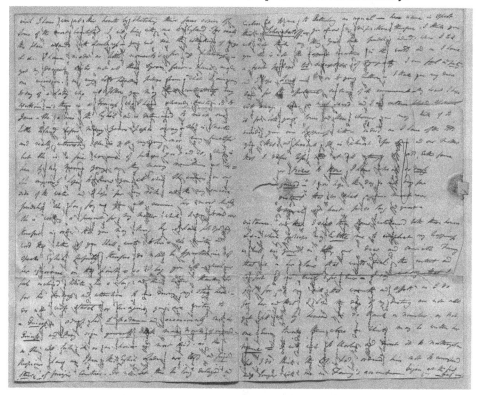

others were less understanding. The narrator of one of Benjamin Disraeli's novels remarked that he appreciated a woman correspondent "provided always that she does not *cross*" (Kimball, 2004, p. 112). At the time, however, cultural conditions made it somewhat acceptable to cross. Those social conditions have since changed, so crossing has fallen entirely out of use.

Page design conventions arise from the culture surrounding documents as much as from general perceptual principles. Sometimes those conventions can even overpower perceptual principles, producing document designs that are not ideal but that work in their cultural context. By the same token, sometimes breaking well-established conventions can draw more attention to a document, but doing so always involves some risk.

Moreover, there is never just *one* culture surrounding documents, but many different cultures. Some are extremely local—for example, a company or work group can have its own cultural expectations. Others are broader, such as an immigrant culture or a professional culture within a nation. Still others are extremely broad, such as ethnic cultures, national cultures, or even continental cultures (European, South American, etc.). Smart designers pay attention to the power of culture on users' page design expectations, habits, preferences, and conventions.

RHETORIC

From the perspective of visual rhetoric, page design is important because it can further your goals as a page designer and meet the reader's needs. For example, page design can promote an ethos that reflects well on the client, matches the purpose, sets a tone for the document, and encourages users to use it. Consider the page designs in figures 5.4 and 5.5. Both share the same format: the computer screen. And the topics of both are health oriented: Medicare and quitting smoking.

Despite their similar topics, these web pages differ remarkably in their use of page design to impart a different sense of ethos to different audiences. These page layouts reflect the designers' anticipations of two different sets of users: one more mature and more interested in familiar designs, making it easy to find news and health-related information, and the other younger and more interested in novelty. The page layouts also try to appeal to the interests and needs of each audience by building for the client organizations an ethos of responsibility and maturity for the web page on Medicare, and a more youthful persona for the web page on smoking. Both sites also use page design for an important ethical purpose: to get people interested in their own health.

Figure 5.4. A page design for a mature audience. This web page uses a relatively traditional design, with clear navigational areas on the top, over a content area, the focal center of the page. This central content area includes images and descriptive linked headlines, which suggests that its audience wants information quickly, clearly, and efficiently. *Source*: AARP, https://www.aarp.org/.

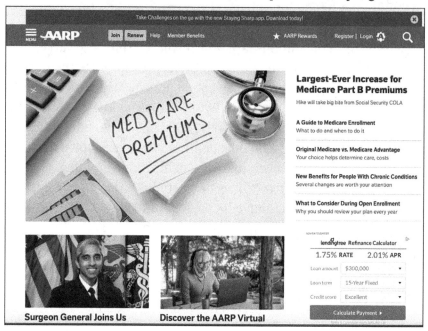

Figure 5.5. A page design intended for a young audience. The banner across the top uses the logo for Take Down Tobacco, with the cartooned logo for Kick Butts Day farther down the page. Both are parts of a youth-oriented smoking prevention program. The site uses a striking color scheme of black, red, and white. The focal point of the page is the banner running across the top, which features youth of various ages holding up anti-tobacco protest signs. *Source*: Take Down Tobacco, https://www.takedowntobacco.org/about.

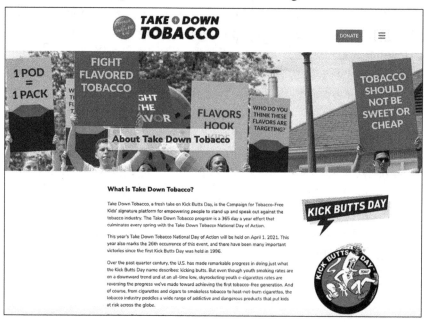

VIEWING PAGES

Before we discuss how to create a successful page design, we must address how and why users usually encounter pages as they do. Users typically engage in three modes of looking at documents:

- Sometimes they *skim* the document, looking for something that will catch their eye.

- Sometimes they *scan*, looking for particular information.

- Sometimes they settle down and *read* passages of text or images.

Users switch rapidly between these three modes, often within the same document. They might scan first, looking for some particular section that will help them solve a problem. When they find the appropriate section, they will likely slow down to read it carefully. Then they might skim the rest of the document to see if there's anything else interesting or helpful that they might want to check out. Or they might start with skimming, letting their eyes rest idly here and there until they find something that piques their interest enough to slow down and read. Or they might start by reading, find that the text is less useful than they thought, and skim through the rest of the document to see if anything looks more useful. It all depends on the users' needs and agendas when they encounter the document.

We should design pages that accommodate users in whatever mode makes sense for their situation—skimming, scanning, or reading. Doing so will foster users' interaction with the document, increase their comprehension, and help them complete the tasks that brought them to the document in the first place.

Design Tip: Anticipating Modes of Viewing

As a designer, you must anticipate what modes of viewing users will likely need and create a page design that helps them to read the way they want. For example:

- If the user will need to find specific information quickly, like a procedure to follow to solve a problem, you'll want to create a document that scans easily. For example, you would probably want to make navigational elements like headings very prominent.

- If the user will want to look at lots of images, such as in a catalog, you'll need to design the pages with the images and accompanying text arranged in a consistent, regular framework. This arrangement will help users quickly find the images they want to look at more carefully.

- If the user will want to read long sections, such as in a formal business report, you'll want to create a document with consistent and comfortable blocks of text.

- If users will just skim the document—such as a newsletter—you'll want to create a document with vivid contrast to make each story attractive and engaging.

CREATING MEANING WITH PAGE DESIGN

It's easy to dismiss page design, believing either that it doesn't matter as much as text in the transmission of information or that it is primarily aesthetic. On one hand, as Rudolf Arnheim (1997) pointed out, for a long time our culture has privileged lexical meaning (conveyed in words) over visual meaning (conveyed in images), teaching us all at some time in our childhood to put down the crayons and pick up the pen, or later, the keyboard. On the other hand, even those who create page designs often consider their work an aesthetic act—the process of creating a beautiful wrapper for "content"—which again they see primarily as words. In either case, these attitudes suggest that any meaning we gain from visual cues is a mere adjunct or shell to the words themselves.

But in documents, we cannot even experience words without seeing them. And the way words and other design objects are arranged on the page conveys a significant amount of meaning to users. This meaning is often *metatextual*—about the text. Page design uses visual cues to create order among the chaos of words, to guide users through the information held in the document, and to help users get what they need out of the document. In short, good page design creates *meaning*. Page design does this by implying a logical structure for documents, using the visual variables and principles of design to create design objects and define relationships among them (as we discussed in chapter 2). Page design is the craft of developing a consistent system in the document for implying those relationships.

The possible relationships among design objects on pages are numerous, but for the most part they boil down to *connection, hierarchy, sequence,* and *balance*. In practice, these relationships can be deeply interrelated; any document can exemplify all four simultaneously, often using similar techniques to imply these relationships. But in the following sections, we'll discuss each individually, examining how we can use design objects and apply principles of design to show these relationships and create visual systems of meaning on pages.

CONNECTION

The most basic relationship between page design objects is that of simple connection. Using visual cues, we can imply that one design object is somehow logically connected to other design objects, thereby reinforcing the logic of the page.

As we discussed in chapter 2, relationships of connection are mostly governed by the principles of design: *similarity, enclosure, alignment, contrast, order,* and *proximity.* We can use these principles to break up content into manageable chunks and then group those chunks to show

Figure 5.6. Principles of design create connection among design objects. Notice how the minimalist design of this page describing faculty achievements follows the principles of design to present clear information. *Source*: Auburn University.

Horizontal alignment and the *similarity* of page objects implies the equality of described faculty.

Vertical alignment connects the various page objects vertically—each column represents one faculty member's information.

Proximity implies that objects close together are connected. Here, the caption is close to the photo, creating a strong relationship between the candidate's photo and their name.

Order implies that what comes after is related to what came before—for example, that the text below the photo is connected to the photo.

Negative space implies that each of these objects is a separate page object. Each separate page object has a different task: headings introduce sections, paragraphs organize text, captions explain images.

FACULTY ACHIEVEMENT **HIGHLIGHTS**

CHAKRABORTY   COUNTERMAN   EL-SHEIKH   FANTLE-LEPCZYK   LAYNE

Dr. Imon Chakraborty, assistant professor in the Department of Aerospace Engineering and director of the Vehicle Systems, Dynamics, and Design Laboratory, was recently awarded a grant from the Federal Aviation Administration for his study, "Flight Simulation-Driven Research into Simplified Vehicle Operations for Urban Air Mobility."

The urban air mobility (UAM) concept of operations aims to field efficient electric vertical takeoff and landing (e-VTOL) aircraft to transport occupants quickly between intra-urban and possibly inter-urban locations. The FAA is interested in understanding the safety, certification and pilot-training impacts of UAM aircraft.

Dr. Brian Counterman, an assistant professor in the Department of Biological Sciences in the College of Sciences and Mathematics, conducts research on the environment, genetics and mating preferences impacting the Dogface Butterfly. He is the recipient of a National Science Foundation (NSF) CAREER Award for $1,565,641 for "Physiological Genomics of Sexually Dimorphic Developmental Plasticity on Butterfly Wings."

Dogface Butterflies are native to the Black Belt Prairie region and have areas on their wings with distinct pink, yellow and ultraviolet markings. Counterman's research focuses on the genetics of plastic responses or plasticity, which is how the environment can influence their development.

Dr. Mona El-Sheikh, the Leonard Peterson & Co. Inc. Professor in the Department of Human Development and Family Studies, received continued funding of $564,836 from the National Institutes of Health for her multi-year project on family aggression and young adult health. Since 2019, funding for the project has totaled over $2.4 million.

Early findings reveal bioregulatory (sleep) and behavioral (physical activity) protective factors for youth who experience socioeconomic adversity. Recent analysis show the effects of family aggression on mental health depend in part on the bioregulatory functions of sleep. The project aims to advance understanding of associations between adolescent sleep and mental and physical health, and cognitive and academic functioning.

Dr. Jean Fantle-Lepczyk, a research professor in Auburn's College of Forestry and Wildlife Sciences, has co-authored a study that found the economic costs of biological invasions in the U.S. have exceeded $1.2 trillion since 1960. The multinational study, published in the journal Science of the Total Environment, was conducted with a team of seven scientists from the U.S., Germany, France, the Czech Republic and the United Kingdom.

Fantle-Lepczyk described invasive species as those that were moved by humans to a new ecosystem and have begun breeding outside the area to which they were initially introduced.

Dr. Desmond Layne, head of the Department of Horticulture, has been honored by the Association of Public and Land-grant Universities for leading Auburn's Transformation Garden project and his completion of the association's Food Systems Leadership Institute.

Layne, a fellow in the institute, and 25 other fellows from across the country were recognized for contributions to their organizations, higher education and food systems during the association's annual meeting.

Layne has led the creation and planning of Auburn's 16-acre Transformation Garden. The project will encompass every aspect of plant-based agriculture, from fruits and vegetables to ornamentals to row crops and more.

|28| AUBURN RESEARCH

the relationships among them. For example, consider the layout in figure 5.6, a faculty achievement highlights page from a university publication. As the callouts describe, the page layout arranges content into design objects and uses similarity, enclosure, alignment, and proximity to show particular relationships between those design objects.

HIERARCHY

But what kind of connection should we show between design objects? Many kinds are possible, but the most common in document design is *hierarchy*. Smaller design objects build up into larger and larger *composite* design objects: letters make up words, words make up lines, lines make up paragraphs, paragraphs make up sections. Hierarchy even goes beyond the page, providing the most common organizational principle of all documents: subsections make up bigger sections, sections make up chapters, and chapters make up whole documents.

Showing these hierarchical relationships visually is a key part of page design, particularly in technical documents. But unfortunately, we often don't incorporate adequate visual cues to make the hierarchy obvious to users. Imagine that you decide to make a paper glider and look up instructions. You find the document shown in figure 5.7.

Not only is this document difficult to read, it's also hard to understand. Using only line breaks, it sets up a very basic set of sections: a title, introduction, materials, conditions, a warning, and the instructions themselves. Line breaks alone, however, don't distinguish the sections adequately, much less show a hierarchical relationship among the sections. All of the different kinds of content on the page—titles, headings, sentences, and steps—are consigned to a single bland, indistinct generic field.

For this document, we would need a visual system that shows at least four distinct levels of hierarchy:

- The document as a whole (title)

- Sections within the document (headings)

- Text within the sections (paragraphs)

- Specifics (items or steps)

Figure 5.7. Unclear visual hierarchy. This document is difficult to use because of the unclear hierarchy of page objects. All of the text looks the same, regardless of its level within the document hierarchy. *Source*: Created by the author.

How to Fold a Basic Paper Glider. These instructions guide you through the process of folding a basic paper glider. You will need a piece of paper and something to fold it on. Don't throw the glider at anyone's face. Start by folding the top-left corner of the paper all the way across to the far-right side. Then fold the top-right corner all the way across to the far left. Next you'll fold the paper in half down the center. You will then open the paper along the fold, then fold the point down until the point touches the bottom line of folded paper. Fold in half again, then fold each wing down so that roughly half of the wing extends beyond the body of the plane. And that's it! Your glider is now ready for flight!

We can use the visual variables and principles of design to show this hierarchy on the page by carefully managing typography and the positioning of elements, using diagrams and photographs where necessary (see figure 5.8). Clear visual cues help users distinguish the levels of hierarchy. They also enable users to scan the document to find the information they need, and they make skimming easy.

Figure 5.8. Clear visual hierarchy. In this version of the paper glider instructions, the visual hierarchy of page objects is much clearer, allowing users to scan through the document and see its structure more easily. *Source*: Created by the author.

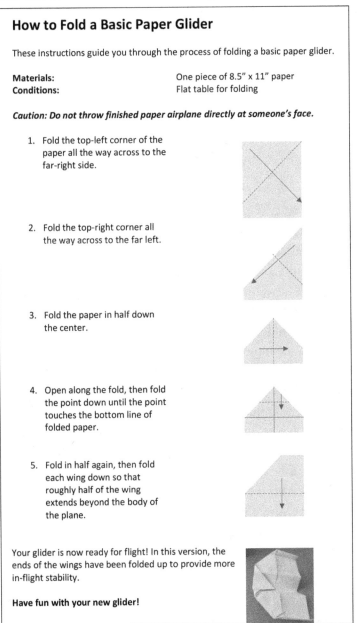

How to Fold a Basic Paper Glider

These instructions guide you through the process of folding a basic paper glider.

Materials: One piece of 8.5" x 11" paper
Conditions: Flat table for folding

Caution: Do not throw finished paper airplane directly at someone's face.

1. Fold the top-left corner of the paper all the way across to the far-right side.

2. Fold the top-right corner all the way across to the far left.

3. Fold the paper in half down the center.

4. Open along the fold, then fold the point down until the point touches the bottom line of folded paper.

5. Fold in half again, then fold each wing down so that roughly half of the wing extends beyond the body of the plane.

Your glider is now ready for flight! In this version, the ends of the wings have been folded up to provide more in-flight stability.

Have fun with your new glider!

SEQUENCE

One of the most important relationships when designing a page is *sequence*—the order in which users will encounter design objects as they skim or scan the page. Of course, as we noted in chapter 1, users will look at whatever they choose, and in whatever order they choose. But good page designs provide a clear and natural *flow* for users' eyes to follow from design object to design object.

A good flow gives users a clear entry point to the page—a clear sense of where to look first—and then visually implies a clear path through the rest of the objects on the page. The design principle that most comes into play in creating flow is *order*, but *similarity*, *alignment*, *contrast*, and *proximity* also help users pick out where to look first and where to look next.

From practice and convention, users employ two common strategies to guide them through the sequence of objects on the page: *patterning* and *focusing*. As designers we can use these strategies to inform our page designs.

PATTERNING

Most users look at a page in a series of glances that form more-or-less consistent patterns we apply to most new pages we see. We develop these reading patterns by reading many thousands of pages over our lifetime, which builds a strong habit in terms of where we look first at a new page and how our eyes flow through the page from that point. If your designs follow these patterns, users can usually read, skim, and scan more successfully.

Perhaps the strongest pattern in documents written in English and most other European languages is known as the *Z pattern* (figure 5.9). This pattern applies particularly to documents that are designed to be read, rather than scanned or skimmed.

Text in English is usually represented on pages in successive lines that readers look at from left to right and top to bottom. Because they must move from line to line down the page, readers' eyes actually travel in a zig-zag pattern. We can use this pattern to predict what zones of the page are more likely to receive a reader's attention. Three tendencies are particularly useful for designers:

- Readers tend to look first at the upper left corner when they encounter a page or text field.

- Readers then tend to look across the top of the page or text field on their first pass through the Z. They do so in quick

Figure 5.9. The Z pattern. A reader's gaze tends to skip to the right across a line of text in a series of saccades about seven to nine characters long. At the end of the line, gaze returns quickly to the left margin to find the next line. *Source*: Created by the author.

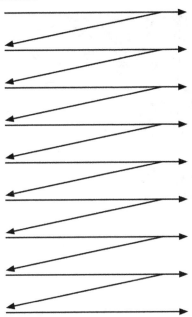

eye movements called *saccades*, followed by short *fixations* on particular points of the page. At the end of the line, their eyes leap quickly from the right side of the page back to the left margin to begin the next line, a movement known as the *return sweep* (Findlay and Gilchrist, 2003, p. 84).

- Readers' eyes hit the left margin repeatedly as they zig-zag across the text field.

This Z pattern occurs most often when a page uses plenty of non-textual visuals (see, e.g., Babich, 2017). In other cases, we read using an F pattern, where our eyes move most of the way across the first few lines or skim most of the way across the page closer to the top (eye tracking shows that our eyes stop, or fixate, more often closer to the top of the page) then our eyes return to the left, move down, read or skim part-way across, return to the left, then proceed rapidly to the bottom with increasingly limited across-the-page skimming or reading.

In either case, because of the way our eyes move across a page, we can identify *power zones* (figure 5.10). Any design objects placed in the power zones will likely get readers' attention before anything placed outside of these zones. And the more quickly readers are browsing or scanning the document, the more likely they will pay attention *only* to the power zones, looking for something that will capture their attention.

Figure 5.10. Power zones. Objects placed in power zones (marked here with shading) receive greater emphasis than objects placed lower down or further to the right. The strongest point of emphasis is at the top left of the page, where readers look first on most pages. *Source*: Created by the author.

Research has shown that users apply the Z pattern to web pages, which aren't usually intended to be read through (Outing & Ruel, 2004). These patterns apply as well to multiple-column layouts, which add a second level of Z patterns to the user's eye movements. In addition to scanning the entire page, readers use the Z pattern to scan individual columns from left to right. They also use the Z pattern to determine which column to read next.

Readers also apply the F pattern to web pages, though typically when they are considering content, not navigation, and this patterned response to content also carries over onto mobile devices. Keep in mind, of course, that though users may naturally employ an F (or Z) pattern when addressing content, that patterned response to content is not necessarily desirable. In both cases, what we need to remember as designers is that while eye-tracking studies like the ones that gave us the F and Z patterns (among others) show us where the eye goes, they also show us where users are *missing* information.

Therein lies the problem: a page design can present ambiguous flows that make a user unsure of where to look next, or present content without providing visual cues or landing places for the eye, allowing users to miss important material. One common mistake arises when columns of text are wrapped around images, as in figure 5.11. Columns inherently

Figure 5.11. Flow and reading order. Be sure to give users clear cues about where to look next as they work through a page. *Source*: Created by the author.

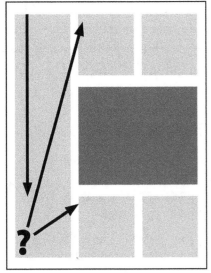
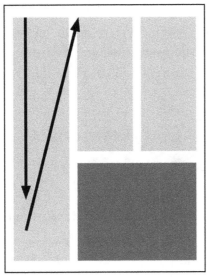
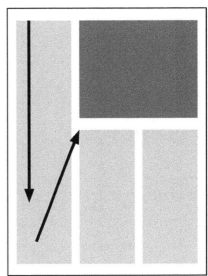

Ambiguous reading order   *Clear reading order*   *Clear reading order*

create a break in proximity between one line of text (the bottom of the first column) and the next line (the top of the second column). Ideally, users can adapt quickly to find out where to read next after finishing one column; the conventional Z pattern prompts them to look for the next block of text at the top of the page. But setting an image in the middle of the page gives users ambiguous cues about where to go next. Do they skip up to the top of the next column *above* the image or to the top of the next column *below* the image? Shifting the image to either the top or the bottom of the page eliminates this problem.

Giving users a clear path through the document will help them avoid any confusion. And if you keep in mind the well-practiced patterns that users bring to documents they read, you will have a better sense of where to position important design features so readers can find them easily.

Keep in mind, however, that a reader's culture and intent play an important role in how they address a document. The Z and F patterns we've described here apply primarily to documents read in English, or similar left-to-right languages. Readers of right-to-left languages, for example, such as Arabic and Hebrew, may follow an inverse Z or F pattern (as described by the Nielsen Norman group) or something else entirely. After all, there are many different patterns: Z; F; layer cake, where the eye quite literally moves in a pattern that looks like a cake; spotted, where the eye moves in chunks; marking, which involves locking the gaze as a page is scrolled; bypassing, where particular elements are consistently skipped; commitment, where *everything* is scanned; and likely many more (see Pernice, 2017).

If you're designing a document for another culture, study the conventions and habits of users in that culture and create designs that reflect those users' expectations. You can find out a lot about other cultures' reading habits by talking to and observing users in that culture. Analyzing successful documents from the target culture will also help you create responsive documents that cross the cultural divide successfully.

Design Tip: Using Power Zones

Here are some things to keep in mind when using power zones:

- Put the page entry point inside the power zones. Doing so will help users start their page journey on the right foot.

- Put design objects that are high in the page hierarchy inside the power zones. These objects include navigational cues such as titles and headings.

- Put mission-critical design objects inside power zones, where they'll get the most attention. Mission-critical objects might include cautions and warnings.

By corollary, avoid putting relatively unimportant information inside power zones. If you place objects that shouldn't receive a lot of attention inside the power zones, they'll get in the way of users skimming or scanning the document for important information. For example, it's usually best to avoid centering headings, which places these important visual cues outside of the power zones.

FOCUSING

Because patterning is a general ordering strategy users bring to documents from previous reading experiences, it works best in the conventional documents users find most familiar—particularly text-heavy documents such as books and reports. But in viewing documents that users will more likely skim or scan, such as posters, signs, catalogs, or flyers, patterning doesn't work quite as well.

In such documents, users rely less on established reading patterns than on the characteristics and arrangement of design objects to order their path through the page, *focusing* their attention dynamically. Like hunters, they latch onto the visual "scent" of something that appeals to them—because they're looking for something in particular, because it looks either familiar or unusual, because it has elements or features they find intriguing, or because it's the biggest, brightest, most striking object on the page. Once that first design object has grabbed their attention, users look for objects connected to it.

The key for designers, then, is to capture the user's attention, focusing it on a particular design object as a starting point, then leading the user's eye to related objects connected by the principles of design. Designers facilitate this process through page design by giving the user a clear starting point and making the flow from that point visually clear. Then they use principles such as *alignment*, *contrast*, *order*, and *proximity* to create a sense of flow.

Consider the flyer for a photography exhibition in figure 5.12. It uses visual cues to capture attention and guide users through the

Figure 5.12. How page layout guides the user's eye. The design objects on this flyer are arranged to lead the user's eye down and across the page in a predictable and logical order. *Source*: University of Virginia Art Museum, "Reverie and Reality: Photographs by Rodney Smith."

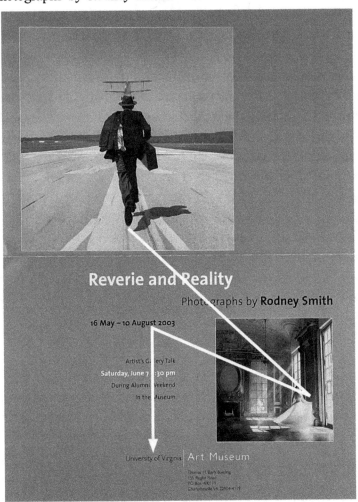

essential information on the page. The user's eyes might focus first on the compelling image of a man running down an aircraft runway, then move down and right through the title and subtitle to another image. With this arrangement, it's difficult for users to see the images without casting their eyes across the title—an essential piece of information. Now that the flyer has the user's attention, other essential information, such as the date, time, and location of the exhibition, can appear in smaller fonts to the left of the image. Finally, the user's eyes flow down to the sponsor of the exhibition (the University of Virginia Art Museum) at the bottom of the page.

The *alignment* of design objects also encourages this reading pattern. The title "Reverie and Reality" aligns with the left edge of the runway stripe on which the man is running. The photographer's name (Rodney Smith) aligns with the second image. The subtitle aligns with the word "Reality" in the title, as well as with the vertical rule between "University of Virginia" and "Art Museum." And the time, date, and location information right-aligns along an axis slightly displaced from that between "Reality" and the museum title block at the bottom of the page.

Even if the users only glance at the sheet without taking in every detail, the design's use of *contrasts* in color, value, and size in its typography helps emphasize the most important information. The contrast between the background color and the text causes text blocks in white to pop out more vibrantly than the text in black, which draws attention to the most essential information: the exhibit title, the date and time, and the location (the art museum). Finally, the use of white text and black text is appropriate for a flyer that is advertising a photographer's black-and-white photos.

Of course, users can look wherever they want, but the techniques in this page design draw users' eyes through the information in a logical and carefully orchestrated order, giving them the visual cues they need to hunt down information, first by emphasizing the most essential information (what, when, and where), then by filling in subsidiary information.

BALANCE

The fourth relationship among design objects is *balance*, which provides a sense of unity and coherence to a page. Ideally, design elements should be placed on a page so that they balance both horizontally and vertically. Designers determine balance by assessing the relative *visual weight* of different objects in the design. Visual weight depends on the subjective vibrancy or dynamism of individual objects in the design. Visual variables such as these can affect the weight of a design object:

- *Size:* Larger objects have greater weight than smaller objects.

- *Shape:* Unusual shapes have greater weight than conventional shapes.

- *Color:* A splash of any color typically has greater weight than a surrounding page of black and white or grays; a dramatically different or highly contrasting color has greater weight than a less-contrasting color.

- *Value:* The more visually dense the object in comparison to its surroundings, the greater its weight. Black on a white field (or white on a black field) carries the greatest visual weight because of the high contrast of value between white and black.

- *Position:* Objects in power zones have more weight than those not in power zones; objects clustered together can form a greater weight, as can objects set within large areas of negative space.

Design objects with a lot of visual weight tend to use several of these factors simultaneously. For example, a visually weighty object might be large, surprisingly shaped, brightly colored, and highly contrastive in value.

Keep in mind two caveats about balance:

- *Balance is relative.* As you might have guessed from the previous list, no object *really* has any particular visual weight, except in comparison to other objects that seem lighter or heavier or less or more emphatic, attractive, or surprising.

- *Balance comes from users as much as from the design objects or page design.* Users just skimming or browsing through a document might be attracted by a dynamic but merely decorative photo, giving this element a lot of weight. But users looking for specific information will ignore that same photo in favor of an information-filled heading that might lead them to their goal.

In other words, balance is not always a dependable indicator of good design. Sometimes *unbalanced* designs can seem more dynamic and interesting, and this can be used to good rhetorical effect. However, balance can create an overall coherence and unity between design objects, making it clear that they work and belong together.

Assessing balance can be difficult. As designers, we're often so close to our work—both literally and figuratively—that we may have trouble assessing our designs as a whole. One way that some designers do this is to literally take a step (or several) back. If you are working on-screen, for example, you might zoom your view out so that you can see the entirety of a design area, then step to the far side of the room to view your work. In so doing, you often lose focus on the minutiae of design and are finally able to see how the larger design elements interact. Another tried-and-true method is to print your design out and attach it, *upside down*, to a viewing surface, and then step as far back as possible to assess the work. Again, the unfamiliar view may allow you to see how the design elements interact and give you a better sense of the visual balance of the piece.

The simplest kind of balance is *symmetry*. *Symmetrical layouts* center everything on a single vertical (or, less often, horizontal) axis. Symmetry often implies a certain formality or traditionalism, so it is often used for conventional genres like invitations or designs that invoke history or nostalgia (figure 5.13).

Figure 5.13. Symmetrical layout. The symmetrical layout and the nineteenth-century typefaces mesh with the historical theme of this leaflet. *Source*: Art Institute of Chicago, "Honoring Heroes in History: Illustrations from Coretta Scott King Award Books, 2001–2005."

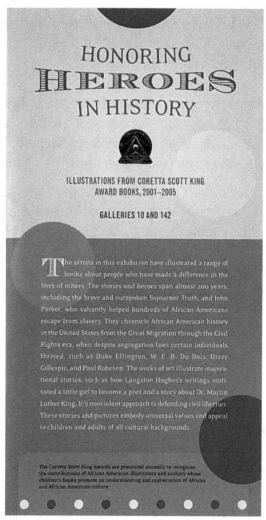

Asymmetrical layouts can also have balance, as we can see in figures 5.14 and 5.15. These examples still use a primary axis (either horizontal or vertical), but the axis is off-center. In asymmetrical layouts, the more dynamic or weighty objects are often balanced by larger areas of less dramatic objects, such as paragraphs.

USING GRIDS FOR PAGE DESIGN

So far, we've discussed page design primarily in terms of isolated pages or spreads, such as one might see in a single web page, flyer, or information sheet. Each of these documents might have a unique page design—and they probably should. But most documents have several or even many pages. These longer documents need a structured and consistent page

Figure 5.14. Asymmetrical layouts can also show balance. The asymmetrical layout of this membership brochure uses a broader right-hand area and large, high-value type to balance the dynamic image on the left. *Source*: Art Institute of Chicago, "Join Us."

Figure 5.15. Horizontal asymmetrical balance. The dark shading and images along the bottom of this cover exert a strong downward visual emphasis. But the negative space at the top, which incorporates the title ("Cityscapes Revealed"), half-rosette, and the museum's name, creates an equally strong weight to lift the balance back toward the top. The result is horizontal asymmetrical balance. *Source*: National Building Museum, "Cityscapes Revealed: Highlights from the Collection."

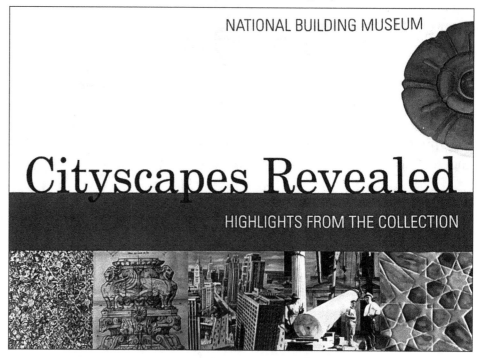

layout system that extends across the entire document, or at least significant parts of it. Users rely on that consistency to guide them through the document. And as users grow familiar with a consistent page design system, they read, scan, or skim each individual page more quickly and efficiently.

One effective method for designing a consistent page design system is to create a consistent *grid* on each page. A grid divides pages into a series of rectangular areas that can organize content—whether text or graphics—that can be dropped into these boxes.

Take a look at figure 5.16, a two-page spread from a visitors' guide to a tour of public art in downtown Denver, Colorado. This spread was clearly designed to accommodate lots of small pictures and captions—just what you'd expect in a brochure guiding users in seeing many pieces of art. The grid separates photographs and captions into nine rows and six columns. Some of the photographs are a consistently small size; others are sized precisely to equal several of the smaller photographs. For example, the banner area over the top of the spread extends across

Figure 5.16. Using a design grid. Notice how this brochure follows a consistent design grid so users can recognize the relationship between illustrations and their captions. *Source*: Denver Office of Cultural Affairs, "Public Art Guide to Downtown Denver."

six of the pictures below, unifying and connecting them. Two headings also fall into this pattern: the running title "Public Art Downtown" in the rectangle at the top left and the heading "Platte Valley," which aligns with the second column of photos.

The grid in this spread isn't explicit; you cannot actually see the lines arranging the content. But implicitly, the grid manages the page layout. The diagram in figure 5.17 shows the features of the grid used in figure 5.16.

Figure 5.17. The design grid for the page spread shown in figure 5.16. The gripper edge will be trimmed away in line with the trim marks after printing, leaving the printed area of the document extending to the edge of the finished sheet. The inset superimposes the grid over the original document. *Source*: Created by the author.

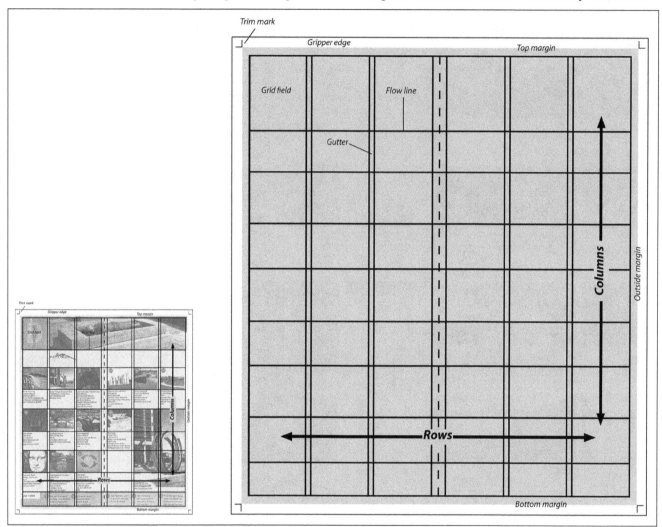

The positive areas of a grid, where content can be inserted, are formed into consistent rectangular areas arranged in rows and columns. Each of the rectangular areas in the grid is called a *grid field*. Collectively, the area covered by the grid fields is known as the *live area* of the page. Content can be made to fit within individual grid fields or to extend across multiple grid fields. Vertically, the grid fields are arranged into *columns*, and *flow lines* divide columns horizontally into *rows*.

Negative areas play a key role in defining the positive areas. The negative spaces between grid fields are called *gutters*. The negative space between the grid and the edge of the page or screen is called a *margin*.

For one-sheet formats, the margins are referred to as *top*, *bottom*, *left*, and *right*. For a spread formed by two pages, the margins are referred to as top, bottom, *inside* (next to the binding or fold), and *outside* (away from the binding or fold). These negative spaces play a central role in defining the positive areas of the page.

More about . . . Asymmetrical Layouts

Before the early twentieth century, nearly all page layout was symmetrical, regardless of the purpose of the document. But in the 1920s, a group of German designers began to challenge the symmetrical approach. This group was founded at Das Staatliches Bauhaus, a design school active from 1919 to 1924 in Weimar, Germany, and from 1924 to 1933 in Dessau, Germany. The leading lights of the Bauhaus movement were Walter Gropius (1883–1969), a founding member of the school; László Moholy-Nagy (1895–1946), a talented photographer and designer; and Jan Tschichold (1902–1974), an innovative and outspoken typographer.

In his book *Die Neue Typographie* [*The New Typography*] (1928), Tschichold promoted the vitality of asymmetrical layouts as better fitted to a modern industrial age than symmetrical layouts.

Asymmetrical layouts provide many options for displaying different kinds of content. These layouts allow designers to show useful relationships between pieces of information, without expecting users to read linearly through an entire document. Tschichold's ideas deeply influenced designers, and we can still see his impact today, when most designs are asymmetrical. For more information on the Bauhaus, see Phillip Meggs's book *A History of Graphic Design*.

A grid such as the one in figures 5.16 and 5.17 takes advantage of several principles of design at once:

- *Similarity:* In this example, the page grid keeps all of the tour images the same size. This similarity of size helps users recognize that other images, such as the sculptures across the top or at the bottom right of the page, aren't necessarily stops on the tour. Using such a consistent page design grid can help us create page layouts with the same arrangement of design objects on each page.

- *Enclosure:* By nature, a page grid encloses design objects, setting them apart from their surroundings. This enclosure might be positive, with elements such as lines, rules, borders, or shading distinguishing one grid field from the next, or it might be negative space that implies separation. In this example, the consistent horizontal and vertical gaps between different design objects provide a strong sense of enclosure.

- *Alignment:* Because page grids are clearly rectilinear, they encourage good alignment of design objects, both horizontally and vertically. In this case, the many rows and columns of the grid helped the designers align the small photographs and captions. Not only are these photographs a primary feature of the brochure, but they also serve a primary function: to show users what to look for among the many pieces of public art in downtown Denver.

- *Contrast:* The page grid emphasizes contrasts between design objects. The images (in full color) contrast well with the buff-colored background, which in turn provides good contrast for the text in the captions, and the green banners at the top and bottom contrast with the mostly blues and browns of the content between them. A consistent grid also gives us the opportunity to *break* consistency sometimes. For example, the circled numbers on the tour notes at the bottom of the spread straddle the grid boundary, suggesting to users that these are footnotes, not additional entries.

- *Order:* Page grids can encourage a particular order for users to navigate a page or spread. In this example, three factors encourage users to read in order across the rows rather than down the columns: the strong horizontality of the row of images, the visual separation created by the intervening rows of captions, and the proximity of the captions to the images.

- *Proximity:* The page grid creates a system of proximities that show the relationship of design objects. As we noted in chapter 1, the proximity of two objects doesn't mean much except in relation to their proximity with other objects; it's not a matter of absolute distance but of relative distance. Grids help us manage these relationships consistently. For example, the captions are made to "hang" from the bottoms of the photographs, which encourages users to connect the correct caption with the correct photograph.

The grid is a powerful convention, and most users will follow it—whether the designer wants them to or not. For example, in this case the grid suggests an order the designers probably didn't intend. The grid might lead users to scan an entire row of photographs all the way *across* the spread. But the numbers on the images suggest that the designer wanted users to read the left page first and then go back up to the top to read the right page. It's important to match the grid design with the users' expectations and needs, as well as with the content and your intentions as a designer. Despite this small flaw, however, this example shows how designers use grids to design a consistent, yet flexible reading system.

In the sections that follow, we'll discuss more fully the advantages and features of grids, what we can place in the grid system, and how we can design grids that meet users' needs while complementing the document's format and content.

DESIGNING GRID SYSTEMS

So how do we design grid systems that work? We need to balance two factors that complement each other to meet users' needs. *Content* applies pressure from *inside* the grid, determining the number and width of columns and rows. *Format* applies pressure from *outside* the grid, determining how big the grid can be. We must balance these pressures so that the design objects, the grid, and the format all combine to meet user needs.

Take a look at figure 5.18, a page spread from a product catalog for kitchen cabinets and accessories. Most users looking at catalogs either *skim* while browsing for something interesting or *scan* for a particular

Figure 5.18. Content, format, and user needs determine grid designs. Notice how the design grid in this spread (8½" × 11" pages forming a 17" × 11" spread) accommodates the content and the format to help users skim for something interesting or scan for a particular product. *Source*: Home Depot, "Home Organization Design Guide."

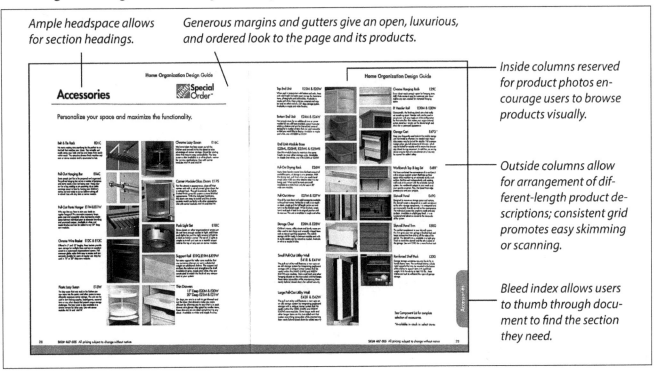

item. This grid design accommodates content within the format to help users accomplish those tasks. It is designed to match the design objects, and the objects are sized to fit the grid. The grid divides each page into four columns, the inner two holding the product images and the outer two holding the product descriptions. This arrangement helps users scan or skim down the central columns. In other words, the grid works from the inside out, arranging content so that the most important design objects and user activities are the focus of the design.

At the same time, the relatively large format (17" × 11" spread) supplies plenty of room for multiple images, allowing users to see many items at once and compare them. The format offers room for generous margins, and gutters provide an ample negative space, framing and supporting the content while bringing the focus to the images from the outside. The margins also provide an area for important metatextual page elements, such as headers, footers, and a bleed index.

DESIGNING GRIDS FOR CONTENT

Design the grid system for a document to accommodate every kind of content you need to place in it, *without exceptions*. When planning a grid, you must plan for all of the different design objects in the project and particularly the line length of the basal text.

ASSESSING CONTENT

An effective grid fits all of the different kinds of design objects the project calls for, as well as a wide range of sizes. A document might contain the following objects:

- Headings (levels 1–4)

- Basal text

- Images

- Captions

- Headers/footers

- Tables

- Footnotes

- Special text objects: notes, cautions, warnings, sidebars, pull quotes

The grid design should accommodate all of these design objects, allowing you to place objects so that they either fit within individual grid fields or extend all the way across multiple grid fields. No design object should extend only part-way into another grid field unless you have a specific reason for allowing it to do so.

Generally, the more complex the document and the greater the number of different kinds of design objects, the more columns you'll need in the basic layout grid. More columns give you greater flexibility in laying out various kinds of content, allowing some design objects to extend across multiple grid fields and others to stay within one. For example, you can create a grid with multiple narrow columns to accommodate small graphics and then allow the text to extend over several columns to create a readable line length.

> No design object should extend only part-way into another grid field unless you have a specific reason for allowing it to do so.

Design Tip: Creating Grids in Page Layout Programs

All page layout programs (including InDesign and Quark) include tools to set up page design grids. For example, with InDesign you can set up a grid either on each page individually or on a master page that governs the layout for the entire document.

Doing so requires setting up guides (Adobe's term for gridlines) in the Layout: Create Guides dialog. This dialog lets you specify how many rows and columns you want, whether they should fit within the margins or the edge of the page, and how wide the gutter should be. After setting these specifications, you'll see the guides as pale blue or pink lines on the page or spread. These lines won't print, but they will help you align objects to the grid you've created. By default, InDesign will even automatically "snap" objects to these guides, ensuring accurate alignment.

Creating grids in word processing software is more difficult, but it can be done for small-scale jobs. Microsoft Word allows you to set type into columns, or you can use linked text boxes to arrange design objects into something like a gridded design of boxes with text flowing between them. If you use the latter technique, set the text box borders to "none," and select the "Snap objects to grid" option in the Draw: Grid dialog to help you align the text boxes.

TEXT SIZE AND LINE LENGTH

The single most important factor determining the width and number of columns is the relationship between *text size* and *line length*. A century of researchers have tried to determine the absolutely most efficient line length for reading, with mixed results. But generally, most designers consider that users typically prefer to read lines of text about 7 to 12 words long, or between 50 and 75 characters per line. Users can read longer or shorter lines with little or no loss of efficiency, but they report that this average line length is most comfortable to read.

Of course, this guide doesn't lead us to an *actual* width for our columns in terms of inches, picas, or centimeters, because the length of 7 to 12 words, or 50 to 75 characters, depends on the typeface and font.

Text set in a small typeface might have a relatively short line length in absolute terms. This text in Tahoma 7-point generates 10-word lines with a line length of about $2^3/_4$.

However, text set in a larger typeface must have a relatively longer line length. This text in Tahoma 16-point generates lines of about 10 words with a line length of about 6.

And text set in a condensed typeface must have a shorter line length than one set in a broad typeface. This text, set in Arial Narrow 16-point, generates lines of about 10 words with a line length of about 4¾.

As a result, much of the question of the width and number of columns is determined by the typeface and font size of the text you intend to use in the document. In the zine in figure 5.19, for example, notice that the body text on the right-hand page of the spread is relatively small and divided between two columns. On the left-hand page, however, the introductory text runs across both columns and so must be set in an accordingly larger font size to yield about 7 to 12 words per line. (Also notice, as we suggested in the previous section, that the text on the left-hand page runs all the way across both columns—not just part-way—thus respecting the design grid.) This relationship between line length and font size can give you a good sense of the number and width of columns you'll need in your design.

There are some practical limits to line length, however. It's best not to create lines that are too wide or too narrow. Long lines of text can make users lose their place when their eyes move from the end of one line to the beginning of the next. Most users also feel comfortable with columns of text they can read without having to move their head from side to side. Given most people's field of view, this means that lines of text in a typical handheld business or technical document shouldn't be

Figure 5.19. Font size, line length, and grids. The text on the left-hand page of this spread extends across both columns of the grid, so the designers used a larger font size to get 7 to 12 words per line. On the right-hand page, the body text stays within the individual columns, so the font size must be smaller. *Source*: Dyno. (2021). "Security nihilism is the threat." *Earth First! Journal, 40*(2), 36–37.

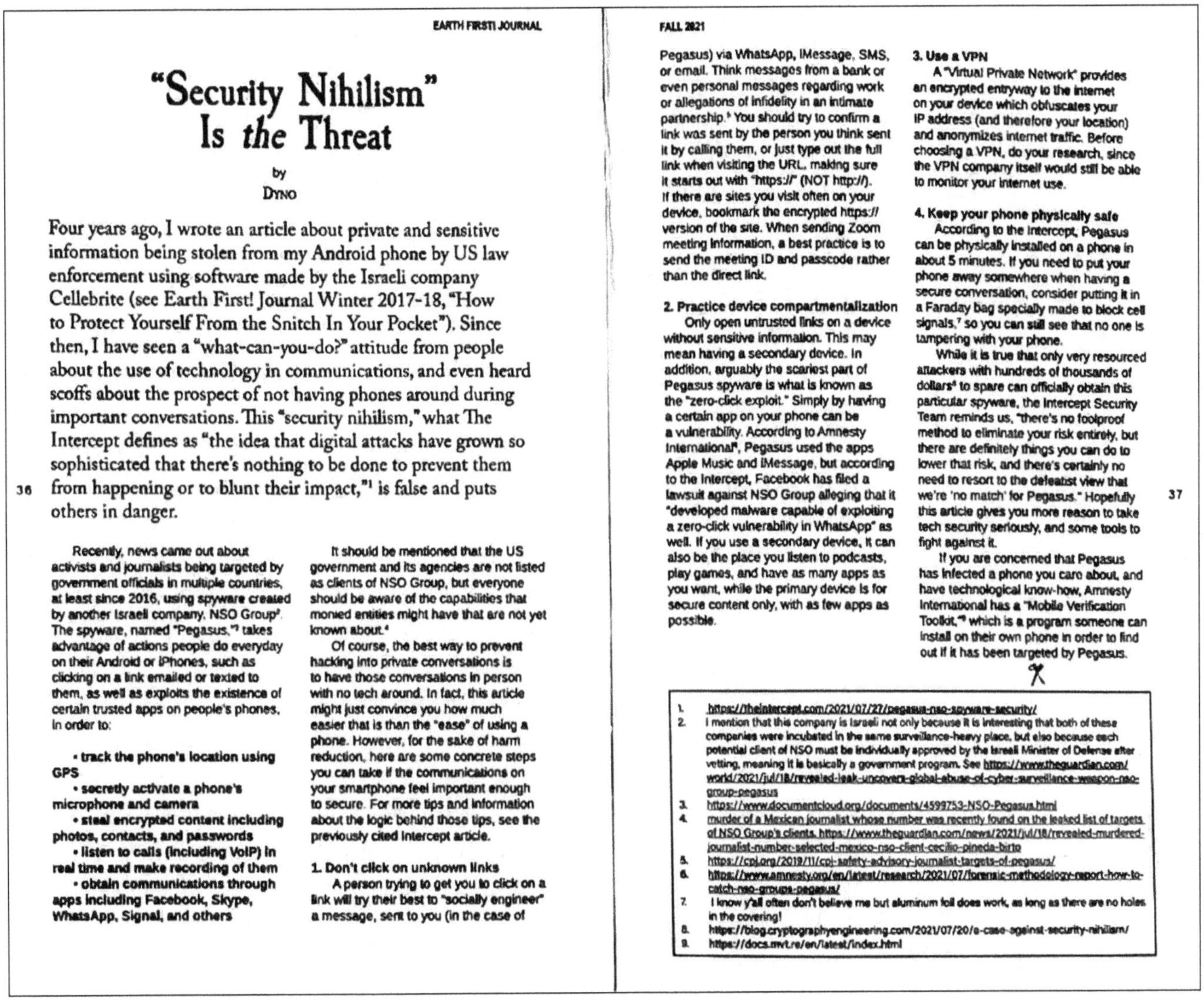

EARTH FIRST! JOURNAL

"Security Nihilism" Is *the* Threat

by
DYNO

Four years ago, I wrote an article about private and sensitive information being stolen from my Android phone by US law enforcement using software made by the Israeli company Cellebrite (see Earth First! Journal Winter 2017-18, "How to Protect Yourself From the Snitch In Your Pocket"). Since then, I have seen a "what-can-you-do?" attitude from people about the use of technology in communications, and even heard scoffs about the prospect of not having phones around during important conversations. This "security nihilism," what The Intercept defines as "the idea that digital attacks have grown so sophisticated that there's nothing to be done to prevent them from happening or to blunt their impact,"[1] is false and puts others in danger.

Recently, news came out about activists and journalists being targeted by government officials in multiple countries, at least since 2016, using spyware created by another Israeli company, NSO Group[2]. The spyware, named "Pegasus,"[3] takes advantage of actions people do everyday on their Android or iPhones, such as clicking on a link emailed or texted to them, as well as exploits the existence of certain trusted apps on people's phones, in order to:

• track the phone's location using GPS
• secretly activate a phone's microphone and camera
• steal encrypted content including photos, contacts, and passwords
• listen to calls (including VoIP) in real time and make recording of them
• obtain communications through apps including Facebook, Skype, WhatsApp, Signal, and others

It should be mentioned that the US government and its agencies are not listed as clients of NSO Group, but everyone should be aware of the capabilities that monied entities might have that are not yet known about.[4]

Of course, the best way to prevent hacking into private conversations is to have those conversations in person with no tech around. In fact, this article might just convince you how much easier that is than the "ease" of using a phone. However, for the sake of harm reduction, here are some concrete steps you can take if the communications on your smartphone feel important enough to secure. For more tips and information about the logic behind those tips, see the previously cited Intercept article.

1. Don't click on unknown links

A person trying to get you to click on a link will try their best to "socially engineer" a message, sent to you (in the case of

FALL 2021

Pegasus) via WhatsApp, iMessage, SMS, or email. Think messages from a bank or even personal messages regarding work or allegations of infidelity in an intimate partnership.[5] You should try to confirm a link was sent by the person you think sent it by calling them, or just type out the full link when visiting the URL, making sure it starts out with "https://" (NOT http://). If there are sites you visit often on your device, bookmark the encrypted https:// version of the site. When sending Zoom meeting information, a best practice is to send the meeting ID and passcode rather than the direct link.

2. Practice device compartmentalization

Only open untrusted links on a device without sensitive information. This may mean having a secondary device. In addition, arguably the scariest part of Pegasus spyware is what is known as the "zero-click exploit." Simply by having a certain app on your phone can be a vulnerability. According to Amnesty International[6], Pegasus used the apps Apple Music and iMessage, but according to the Intercept, Facebook has filed a lawsuit against NSO Group alleging that it "developed malware capable of exploiting a zero-click vulnerability in WhatsApp" as well. If you use a secondary device, it can also be the place you listen to podcasts, play games, and have as many apps as you want, while the primary device is for secure content only, with as few apps as possible.

3. Use a VPN

A "Virtual Private Network" provides an encrypted entryway to the internet on your device which obfuscates your IP address (and therefore your location) and anonymizes internet traffic. Before choosing a VPN, do your research, since the VPN company itself would still be able to monitor your internet use.

4. Keep your phone physically safe

According to the Intercept, Pegasus can be physically installed on a phone in about 5 minutes. If you need to put your phone away somewhere when having a secure conversation, consider putting it in a Faraday bag specially made to block cell signals,[7] so you can still see that no one is tampering with your phone.

While it is true that only very resourced attackers with hundreds of thousands of dollars[8] to spare can officially obtain this particular spyware, the Intercept Security Team reminds us, "there's no foolproof method to eliminate your risk entirely, but there are definitely things you can do to lower that risk, and there's certainly no need to resort to the defeatist view that we're 'no match' for Pegasus." Hopefully this article gives you more reason to take tech security seriously, and some tools to fight against it.

If you are concerned that Pegasus has infected a phone you care about, and have technological know-how, Amnesty International has a "Mobile Verification Toolkit,"[9] which is a program someone can install on their own phone in order to find out if it has been targeted by Pegasus.

1. https://theintercept.com/2021/07/27/pegasus-nso-spyware-security/
2. I mention that this company is Israeli not only because it is interesting that both of these companies were incubated in the same surveillance-heavy place, but also because each potential client of NSO must be individually approved by the Israeli Minister of Defense after vetting, meaning it is basically a government program. See https://www.theguardian.com/world/2021/jul/18/revealed-leak-uncovers-global-abuse-of-cyber-surveillance-weapon-nso-group-pegasus
3. https://www.documentcloud.org/documents/4599753-NSO-Pegasus.html
4. murder of a Mexican journalist whose number was recently found on the leaked list of targets of NSO Group's clients. https://www.theguardian.com/news/2021/jul/18/revealed-murdered-journalist-number-selected-mexico-nso-client-cecilio-pineda-birto
5. https://cpj.org/2019/11/cpj-safety-advisory-journalist-targets-of-pegasus/
6. https://www.amnesty.org/en/latest/research/2021/07/forensic-methodology-report-how-to-catch-nso-groups-pegasus/
7. I know y'all often don't believe me but aluminum foil does work, as long as there are no holes in the covering!
8. https://blog.cryptographyengineering.com/2021/07/20/a-case-against-security-nihilism/
9. https://docs.mvt.re/en/latest/index.html

more than 5 or 6 inches (30–36 picas). Of course, the greater the distance between the document and the user's eyes, the longer the line length can be without forcing readers to move their heads. That's why large signs often use much wider lines than smaller documents do.

Narrow lines (fewer than 7 words per line) can also decrease reading efficiency. In reading narrower lines, readers have a greater tendency to *regress*—that is, they flick their eyes back to go over text they've already read. They also tend to fixate for longer periods of time on individual words. Both of these tendencies cut down on reading efficiency.

NEGATIVE SPACE: MARGINS AND GUTTERS

We've discussed the positive elements in page designs—the objects we put in grids—but we also need to address the negative elements in page design. By *negative space*, we don't mean *empty space*, which doesn't have a particular function. On the contrary, negative space has specific functions, especially in regulating the proximity of design objects. Negative space can also help draw attention to important design objects, such as graphics, headings (which mark the beginning of a section), paragraph breaks (which mark the beginning of a new topic), bullets (which mark each item in a list), and navigation elements (e.g., headers and footers). In particular, you'll need to plan carefully for two kinds of negative spaces in every grid layout: *margins*, which separate the grid from the edges of the page or spread, and *gutters*, which separate the grid fields from each other.

When designing margins, take into account both usability and rhetorical effect. Margins are important usability features in documents. Users need margins to focus on the document as something separate from its background; margins enclose the document, giving it a clear frame and boundaries. Margins also give users somewhere to hold the document without covering the content with their fingers. The inside margins can also clearly distinguish between the two pages of a spread. Even on a screen document such as a web page, margins provide valuable space for the user interface (scroll bars and toolbars) and metatextual elements such as navigation bars.

But margins can also have a rhetorical effect. In general, document designers use wider margins to imply openness and luxury. As a result, wide margins are most likely to be employed in high-level designs that deserve the extra expense of paper. In figure 5.20, a glossy catalog from a university press, the designers have used ample margins to focus users' attention on the photographs. This design builds an ethos of importance for the press and the authors' works, making the products seem valuable and attractive. The result is also very usable, with plenty of open space for users to hold the catalog or write notes.

Narrow margins can imply a business-like air of efficiency and practicality, though at the expense of usability. Figure 5.21, a special report on COVID-19, uses narrow margins and a low level of design. This page design suggests that the people in charge of presenting the research will not waste money on fancy documents with lots of artistic negative space, but will provide them with important, unadorned facts. What the designers saved on paper and space, however, they lost on usability. The inside margins of the report are so tight that users would need to force the binding open to read it comfortably (if bound), and the narrow outside margins would not give users much space to hold the document.

Figure 5.20. Broad margins imply an ethos of importance. This brochure with a 16" × 8" spread has luxurious margins, giving each design object plenty of room to shine. However, it is generally more expensive to print a document with broad margins. *Source*: Texas Tech University Press, "America's 100th Meridian."

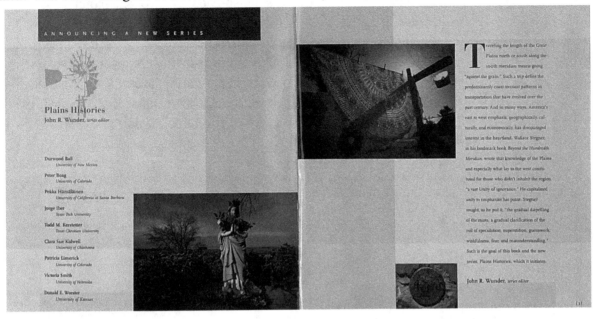

Figure 5.21. Narrow margins imply business-like efficiency. These pages from a Centers for Disease Control and Prevention (CDC) Special Report on COVID-19 waste little space on margins. This leads to a sense of utility, economy, and pragmatism. *Source*: CDC, "Covid-19: U.S. Impact on Antimicrobial Resistance, Special Report 2022."

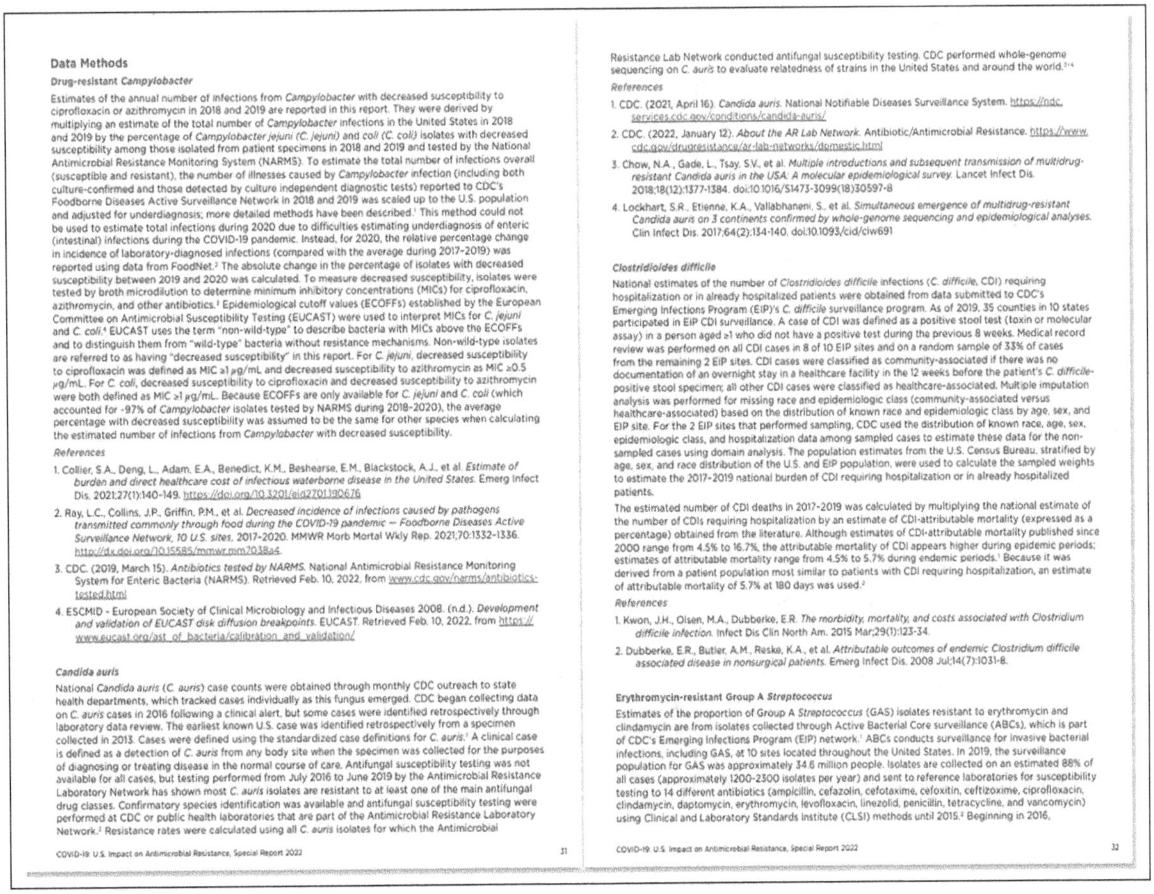

One common error in designing margins is forgetting that each page or panel needs consistent margins around *all* of its grid areas. For example, consider the inside spreads of two six-panel brochures. The leaflet in figure 5.22 is divided into three equal grid fields, and the text is aligned with the edge of each grid field. As a result, in an otherwise well-designed leaflet, the text margins lie uncomfortably close to the folds.

Figure 5.23 uses a consistent margin around each panel, in effect placing a double width at each fold. This preserves the negative space users need to distinguish between one panel and the next.

Figure 5.22. Using adequate negative space. In this 11" × 8¹/₂" leaflet's inside spread, adding more negative space to the inside margins between the three columns of the grid would keep the text out of the creases between the panels. *Source*: University of Virginia Career Center, "UVA Employee Career Services."

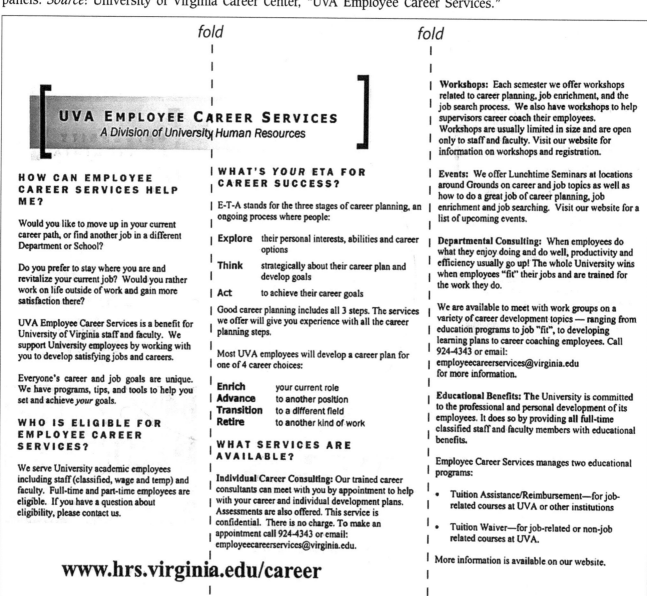

Figure 5.23. Creating effective inside margins. Inside margins must be big enough to accommodate the fold without obscuring or cramping content. This 11" × 8½" inside spread of a three-panel brochure has inside margins twice as big as the outside margins, creating adequate negative space surrounding each text field. *Source*: United Parcel Service (UPS), "International Shipping."

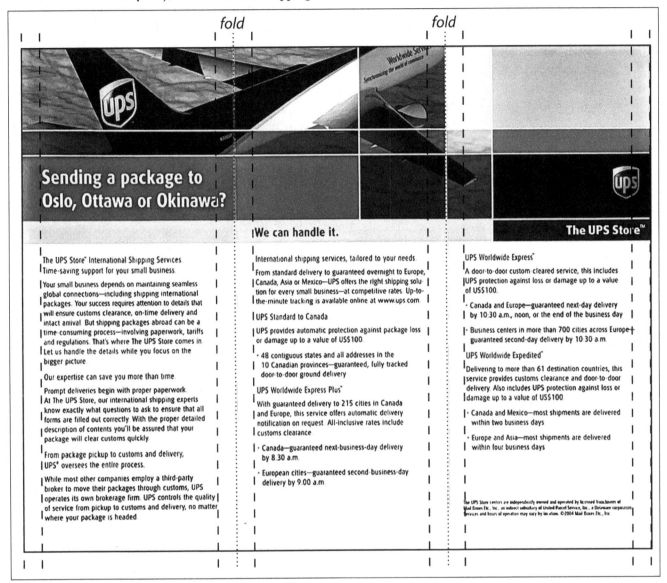

Gutters—the spaces between columns or grid fields on a single page or panel—also play an important role in enclosing content and distinguishing one column or grid field from the next. As a general rule, many designers use a gutter about half the width of the margin. But if the margins are particularly wide or narrow, you might need to make some adjustments. For example, you could modify the gutter width to fit the content, but that might make the line lengths too short. You could change the text to a smaller font size to maintain an adequate number of words per line, but that might make it too small to read. (Design requires compromise!)

DESIGNING GRIDS FOR FORMAT

As we discussed in chapter 4, choices about the format of the whole document determine the finer-grained decisions about page layout, as well as the type and graphics that we put into that format. In a sense, format exerts a pressure from the outside of the page, constraining what we can put on the page. Different elements of format to consider when creating a page design include the page size, the spread size, and their aspect ratio.

The page or spread size plays a significant role in the page design decisions you'll make. Smaller documents constrain us to simpler grid systems with fewer columns and rows, and with accordingly fewer options for arranging content within the grid. Larger documents, however, encourage more complex grid systems with more options for arrangement.

Size presents a particularly difficult issue in screen documents because you often do not know how big the user's screen might be or what its resolution might be. Screen documents must be designed to work equally well on multiple sizes of screens and at various resolutions. In practical terms, however, most designers aim at the most common screen sizes and resolutions, hoping that the majority of users will have a positive experience with the design. In fact, design programs often offer layout presets that help designers plan for common screen sizes.

More specifically, the width of the page plays a significant role in how we lay out the grid. As we just said, most designers aim for a line length of about 7 to 12 words, or 50 to 75 characters, per line of text. But the wider the page, the more words will fit onto it in a given font size. To maintain 7 to 12 words per line, you must either increase the size of the text or reduce the width of the column. So typically, the wider the paper, the more likely you'll need multiple columns of text or very large margins.

Finally, the aspect ratio of the format makes a big difference in page layout decisions because it changes the way we can arrange content on a page. A taller-than-wide format encourages a vertical arrangement of design objects—for example, vertical columns of text. But a wider-than-tall format, like typical computer screens and common two-page spreads, encourages a more horizontal arrangement of design objects as we try to encourage users to scan across as well as down the visual field. In addition, for bound or folded print documents, we need to create a grid that works both on each page individually and on spreads of two pages or multiple panels simultaneously.

One of the biggest distinctions between printed and screen documents is the aspect ratio of their formats (we mentioned this in chapter 4 as well). Most printed documents are taller than wide, whereas most screens are wider than tall (figure 5.24). Typical computer screens, for example, have aspect ratios of 16:9 (widescreen) or 4:3 (full-screen) (width:height), while a letter-sized sheet has an aspect ratio of 17:22. But printed documents can also be arranged in formats that fold out to show spreads of two pages at a time. A spread built on a tabloid-size sheet (11" × 17") shows two letter-sized pages but a whole spread with an aspect ratio of 17:11. This capability of pages complicates how we design grid systems because it changes the aspect ratio of the visual field, making it wider than tall when we consider the entire spread.

On the other hand, some formats can also make us accommodate both tall and narrow and short and wide visual fields at the same time. A common three-panel leaflet built on a letter-sized sheet (figure 5.25) means that we must design both for the single narrow panel users see first ($3^5/_8$" × $8^1/_2$", or an aspect ratio of about 15:34) and for the entire spread of three panels users see when they open the leaflet (11" × $8^1/_2$", or 22:17).

Figure 5.24. Common aspect ratios on page and screen. Aspect ratio has a significant effect on page grid decisions. Screen documents are typically designed for a wider aspect ratio than print documents. *Source*: Created by the author.

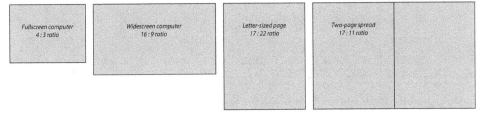

Figure 5.25. Aspect ratios in a three-panel leaflet. In this format, we must design both for the narrow panels and for the wide spread that's visible when users open the sheet fully. *Source*: Created by the author.

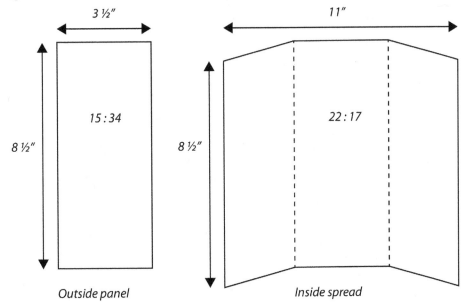

DIMENSIONS AND MEASUREMENTS

Even after we assess content and format, we still need to specify the width of columns, rows, gutters, and margins with precision and consistency. Fortunately, designers have developed a variety of ways to measure these layout elements.

Some designers (particularly in Europe) prefer to use measurements in centimeters and millimeters. But the most common measurement system in print page design uses English units of measurement based on the inch but with two units particular to document design: *picas* and *points* (see figure 5.26). As we discussed in chapter 1, 1 inch = 6 picas = 72 points. In other words,

- 1 pica = 12 points = $^1/_6$" = 0.167"

- 1 point = $^1/_{12}$ pica = $^1/_{72}$" = 0.0139"

The notation system for these units represents picas and points in this format:

[number of picas]p[number of points]

For example, 12p6 describes a measurement of 12 picas plus 6 points (or 12 and $^1/_2$ picas). This measurement would also be equal to about 2.08". If the text area of a book's page is 6.25", for example, then its

Figure 5.26. Inches, points, and picas. Six picas make an inch; 12 points make a pica. *Source*: Created by the author.

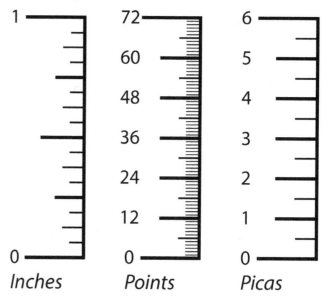

layout is set at 37p6 wide (37 picas plus 6 points = 37 and $\frac{1}{2}$ picas = 6.25"). Most page layout programs will allow you to specify dimensions of columns and graphics with this notation system.

As you might expect, this notation system is also used extensively in typography to determine font size, which is usually specified by points. It's also used in dimensioning design objects (e.g., columns of text and the dimensions of graphics).

Design Tip: Creating Grids for Web Pages

Creating grids for website design is more complicated than it is for print, in large part because the design should change depending on what device the user is viewing the site on, and because you want to avoid using layout tables as they create problems for mobile devices and accessibility in general.

A website needs to perform well on a computer with a large resolution monitor as well as on a mobile device. The changes that need to take place when designing for multiple screens are not just about making sure the webpage is *viewable* on different size screens, but also about making sure that the site is *usable* on different size screens. For example, the size a button needs to be when a user is selecting it using a mouse is much different than the size it needs to be when a user is selecting it using their finger. Images need to scale well from the small size of a handheld device to a widescreen desktop display, and you want to try to keep your text from wrapping as it changes across multiple devices in ways that can render a page difficult to use.

To simplify the design process, many experts (e.g., Marcotte, 2017) recommend starting with mobile design and scaling up. Designers use CSS (Cascading Style Sheets) to control how and when the page design changes. Among the more common things the CSS might check are the device viewport (the browser width) and the device orientation—checking to see if the user is holding their device in portrait or landscape mode.

For pages on screens, designers typically use measurements based in pixels rather than in inches or centimeters. This choice makes sense because it allows the size of the layout and design objects to match the output technology (a computer monitor) very precisely. Keep in mind, however, that pixels differ in size from monitor to monitor. On one monitor, 100 pixels might be about an inch; on a monitor of higher resolution with smaller pixels, 100 pixels might be more like a $\frac{1}{4}$ inch. When designing documents for the screen, take into account this potential difference in size of pixels.

**Design Tip:
Choose a Dimensioning System and Stick With It**

Whatever dimensioning system you choose, use it consistently throughout the entire document design, including the layout and all design objects. Inconsistencies will inevitably creep in if you measure some aspects of the design in centimeters, others in picas, and still others in pixels.

BREAKING THE GRID

After you've established a strong grid system, you can choose to *break* that system for emphasis or effect. For example, the magazine page in figure 5.27 uses the common technique of breaking the grid with a *pull quote*—a quotation from the story set off and enlarged to attract interest. Here, the pull quote is presented in reverse text on a green field that is centered between two columns of the grid. It differs from elements like the basal text, which stays within individual grid fields, or the subtitle, which extends across both grid fields.

A more vibrant example is figure 5.28, a layout from a glossy university library newsletter. The design has a strongly established grid with three columns per page (six per spread). The four inside columns on each spread are used for a story on songwriters; the two outside columns on each spread are used for sidebars, photographs, or other ancillary information, most of which stays strictly within the grid.

Notice, though, how two striking elements break out of the grid: photographs of a rare volume of Shakespeare and of a guitar. These design objects are placed so they extend across the grid borders. The Gestalt law of continuation makes the grid border seem to extend beneath these objects, making them appear to pop out in a third dimension, as if they were laid *on top* of the grid. The Shakespeare volume pokes a corner out of the gold-colored field enclosing the sidebar. The guitar extends across the spine gutter and bleeds completely off the page while itself enclosing a text object: the lyrics to a song. The normally rectangular columns of text are wrapped around the irregularly shaped guitar, allowing a bit of negative space between text and image. The effect is to make the design objects vibrant and interesting—but it depends on the fact that they break the convention of a well-established grid.

Breaking the grid should always be a strategic choice, used sparingly for the most interesting or important information on the page. The drama

Breaking the grid should always be a strategic choice, used sparingly for the most interesting or important information on the page.

Figure 5.27. Breaking the grid with a pull quote. This 8¹/₂" × 11" page layout uses a strong, two-column grid and then breaks it with a pull quote that straddles the two columns. *Source*: Donna Seaman, "The People of the Book: Riding the Third Wave," *American Libraries*, May 2005.

COLLECTIONS | 53

The People of the Book: Riding the Third Wave

Two new library-based programs examine the immigrant experience that permeates Jewish-American literature   by Donna Seaman

I entered the world of book reviewing, criticism, and literary journalism during the height of the multicultural movement, when academics, critics, and librarians were belatedly recognizing the significance of works by writers from groups left out of the official canon of American literature. During this awakening, books by African Americans, Latinos, Native Americans, and Asian Americans, as well as gay and lesbian writers, were read with fresh eyes, accorded serious critical attention, and embraced by the public. This enrichment of American letters has had lasting and profound effects, and librarians were in the vanguard of this movement, not only by making books available to readers, but also by sponsoring book groups and participating in book discussion events, including the American Library Association's latest "Let's Talk About It" (LTAI) program, "Jewish Literature: Identity and Imagination"(*AL*, Mar., p. 7–8).

> I thought about all that has changed (and all that has not) since my Russian-Jewish grandfather arrived in New York City.

I've given a great deal of thought to the relationship between ethnicity and literature, and as a Jew I've contemplated my heritage. Yet I did not expect to see Jewish literature recognized as a subset of American letters because Jewish writers have long been in the mainstream. Think Saul Bellow, Philip Roth, E. L. Doctorow, and Cynthia Ozick. But a confluence of events has convinced me that there is such a thing as Jewish literature that is recognized as unique and avidly read by Jewish and non-Jewish readers alike.

The gestalt of exile

I was inspired to reconsider the state of Jewish literature after I realized that I was coming across a surprising number of provocative works of fiction about what it means to be Jewish in the 21st century. I was quite taken with Julie Orringer's fiercely beautiful short story collection, *How to Breathe Underwater* (Knopf, 2003), in which she deftly portrays an array of young Jews, includ-

DONNA SEAMAN *is an associate editor for* Booklist, *editor of the anthology* In Our Nature: Stories about Wildness *(Univ. of Georgia, 2002), and host of the Chicago radio station WLUW program "Open Books." Her latest book is* Writers on the Air: Conversations about Books *(forthcoming from Paul Dry Books).*

ing Hasidic teens. As I read Russian-Jewish immigrant Lara Vapnyar's *There Are Jews in My House* (Pantheon, 2003), a stunning short story collection set in Moscow and Brooklyn, I found myself thinking about all that has changed (and all that has remained the same) since my Russian-Jewish grandfather arrived in New York City as a child. Another debut author, David Bezmozgis, who also writes about the Russian-Jewish immigrant experience in *Natasha and Other Stories* (Farrar, 2004), caused me to consider how, from the Exodus forward, exile has been at the core of Jewish existence, and how the Diaspora caused Jews to become known as the people of the book. As Jonathan Rosen, author most recently of a novel featuring a woman rabbi, *Joy Comes in the Morning* (Farrar, 2004), writes so succinctly in his brilliant interpretative treatise, *The Talmud and the Internet: A Journey between Worlds* (Farrar, 2000), the Talmud, the great gathering of Jewish tradition and thought, "offered a virtual home for an uprooted culture, and grew out of the Jewish need to pack civilization into words and wander out into the world."

When ALA's Public Programs Office asked me to help evaluate the materials for its new LTAI program, I had the opportunity to further clarify my perception of Jewish literature. A 22-year-old reading and book discussion series conducted at libraries nationwide, LTAI programs are led by local scholars and supported by exceptionally thoughtful and enlightening materials. Themes have included family and work, books for children, women's autobiography, Latino literature, Native-American writing, the African-American migration, and Japanese literature. ☞

Let's Talk About It in Chicago

There will be an informational meeting about the LTAI "Jewish Literature" program at ALA Annual Conference on Sunday, June 26 at 10:30 a.m. Libraries interested in being one of the hosts of the second round of programs may apply for a $1,500 grant by September 30. Details about this and other ALA Public Programs Office opportunities for libraries can be found at www.ala.org/publicprograms, or contact publicprograms@ala.org

MAY 2005 | AMERICAN LIBRARIES

---

of breaking the grid depends on the strength of the grid itself. If you break the grid too frequently, you'll undercut that strength, making the effect less emphatic. You might risk looking like the writer who ends all his sentences with exclamation points: after a while, the reader becomes

Figure 5.28. Breaking the grid boldly. Images of a rare volume of Shakespeare and of a guitar break across the grid in this design. The text wraps around the objects, making the objects seem more three-dimensional and realistic. *Source*: Texas Tech University Libraries, "The Portal."

oblivious to them. By the same token, if lots of elements break the grid, none of them will get any particular emphasis.

EXERCISES

In this chapter you've thought quite a lot about what layout actually looks like from a designer's point of view: the page as a dynamic visual field; how perspective, culture, and rhetoric impact page design and how those designs make meaning; how a user's eyes address a page; and the way grids work. You've thought about how objects create weight and balance, and how you might think about size. Now it's time to practice the things you've just read about. These exercises should help you better internalize everything that you just read.

1. Collect three documents that follow the conventional page designs discussed under "Using Grids for Page Design." Then look for documents that seem to flout these conventions. In class, form small groups and share your documents with each other. Discuss these questions:

    o What are the purposes of the conventions you see in these page designs?

    o What seem to be the motivations behind breaking page design conventions?

    o What perceptual, cultural, and rhetoric effects does following or breaking conventions seem to involve?

    o How do we decide where to strike a balance between conventions and breaking conventions?

2. Sometimes looking at another person's page design can give you a better idea of the kinds of details you must consider when you design your own pages. For this exercise, choose a document you explored in exercise 1 that uses a clearly defined grid. Draw out a schematic of the grid—much like the diagrams in this chapter—noting the dimensions of margins, grid fields, columns, rows, and gutters in inches, centimeters, or picas. (For the last one, you'll need a printer's rule.) Be as specific and accurate as possible.

    o What do you think were the purposes behind setting up the grid as the designer did? How consistent was the designer in using the grid they established?

    o What were the designer's motivations for breaking the grid?

3. Take the worst, most inconsistent page design from those you just explored and sketch out a grid system that will tame its inconsistencies. Then follow these steps:

    o Analyze what you think are the most serious problems of the page design.

    o Create several paper sketches to try out several options for new designs.

    o Choose your best design and recreate a page in page layout software. Use placeholder text ("lorem ipsum") for the text objects, and place frames or basic geometric shapes for the images.

o Discuss in what ways you feel your design improves the original document.

REFERENCES AND FURTHER READING

Arnheim, R. (1997). *Visual thinking.* University of California Press.

Babich, N. (2017). *Z-shaped pattern for reading web content.* https://uxplanet.org/z-shaped-pattern-for-reading-web-content-ce1135f92f1c

Elam, K. (2003). *Grid systems.* Princeton Architectural Press.

Findlay, J. M., & Gilchrist, I. D. (2003). *Active vision: The psychology of looking and seeing.* Oxford University Press.

Kimball, M. (Ed.). (2004). *The early novels of Benjamin Disraeli: "The young duke."* Pickering and Chatto.

Marcotte, E. (2017). *Responsive web design.* A Book Apart.

Meggs, P. (1992). *A history of graphic design* (2nd edition). Van Nostrand Reinhold.

Müller-Brockmann, J. (1981). *Grid systems in graphic design: A visual communication manual for graphic designers, typographers, and three dimensional designers.* Hastings House.

Outing, S., & Ruel, L. (2004). *The best of Eyetrack III: What we saw when we looked through their eyes.* Poynter Institute. http://www.poynterextra.org/eyetrack2004/main.htm

Pernice, K. (2017). *F-shaped pattern of reading on the web: Misunderstood, but still relevant (even on mobile).* Nielsen Norman Group. https://www.nngroup.com/articles/f-shaped-pattern-reading-web-content/

Samara, T. (2002). *Making and breaking the grid: A graphic design layout workshop.* Rockport.

Tschichold, J. (1995). *The new typography* (R. McLean, Trans.). University of California Press.

6

TYPE

LEARNING OBJECTIVES

Upon completing this chapter, you will be able to:

- Appropriately use typographic terms

- Discuss a general history of typeface design

- Articulate how culture shapes perception of type

- Recognize how type shape enables use

- Deploy different typographic systems for screen use vs. page use

~

Most people don't pay much attention to typography when they create documents. They simply open a word processor and type, accepting whatever defaults the program gives them. Sometimes that approach works for designers, too. You wouldn't want to spend an hour agonizing over what typeface to use for an email to a coworker.

But for more complex or important documents, paying attention to typography is an absolute must. Good typographic design doesn't just make documents look good—it gives users important clues about the structure of the document, the purpose of design objects, and the ethos of the organization that created the document.

Consider the three-panel leaflet in figure 6.1, which has a variety of text design objects, including a product logo (ABREVE), a corporate logo

(ArchiText), several designs of headings, a pull quote, captions, and two kinds of basal text. By our count, it uses six typefaces in twelve sizes and modifications (e.g., bold, italic, reverse text, and different colors) on the outside of the leaflet and one typeface in six sizes and modifications on the inside.

Despite its typographic complexity, the leaflet looks simple and clear. Users can easily understand the role of each text design object because they are easy to distinguish from one another. On the inside panels, the type used in the quotation is different from the caption that identifies the speaker; we can also tell the difference between the headings, basal text, and bulleted lists. On the outside middle panel, we can easily distinguish between the continents, cities, and street addresses.

These visual clues of typeface, color, and size help us *see* a logical structure, with ABREVE at the top of the hierarchy in the largest and boldest type and office addresses at the bottom in the smallest and lightest type. Finally, the clarity of the typographic design contributes to an ethos of clarity for the client organization—a company that specializes in translating technical documents. Obviously, the designers spent a lot of time thinking about the typography of this document.

Creating typographically successful documents requires a similar attention to detail. To understand how to use type, you'll need to learn minute distinctions—some that may seem too insignificant to worry about. But learning the finer characteristics of type will help you create more successful and usable documents.

In this chapter, we begin by considering how typography relates to the three perspectives discussed in chapter 2: perception, culture, and rhetoric. We'll help you build a vocabulary that will help you recognize

Figure 6.1. Typographic design in a leaflet. Careful typographic design creates a clear structure, distinguishes between different kinds of text design objects, and implies a positive ethos. *Source*: ABREVE, ArchiText.

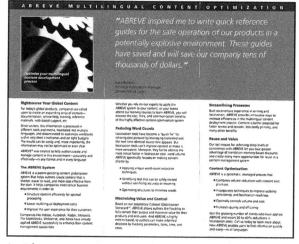

Outside   *Inside*

and analyze type more accurately and precisely. Then we'll discuss some of the common distinctions between different kinds of typefaces, as well as some practical advice about how to use type in print and on screen.

More about . . . Typographic Terms

These basic typographic terms will help you throughout this chapter:

- **Type:** letter shapes used for printing or on-screen display.

- **Typeface** (or just **face**): one consistent design of type. Times New Roman is one typeface; Arial is another.

- **Font:** one size of a typeface, including all of the lowercase letters, uppercase letters, numerals, and punctuation. Times New Roman 12-point is one font. Times New Roman 14-point is a different font. (*Font* is also a term used to indicate the digital file for a typeface, so it's often used synonymously with *typeface*.)

- **Font family** or **type family**: a collection of all of the fonts in a typeface, including specialized fonts such as light, medium, demi (between medium and bold), **bold**, *italic*, SMALL CAPS, condensed, and expanded.

- **Typography:** the practice of designing type—or more broadly, the practice of designing documents with type.

THREE PERSPECTIVES ON TYPE

PERCEPTION

Although people often think of reading text as an intellectual activity, it's actually a *visual* activity—a process of scanning shapes that we decipher as signifying something. In most written languages, these shapes (called *letters*, *characters*, or *glyphs*) signify phonetic sounds that combine into words and have their own distinctive shapes, known as *boumas* (pronounced BOH-məs). In other written languages, such as Chinese, individual shapes (*ideograms* and *pictograms*) can signify or represent entire concepts.

But if reading is a visual process, how does it happen? Researchers are still trying to find answers to this question, but some theories are generally accepted.

You first learned these shapes by rote, probably as your ABCs, laboriously practicing how to recognize and draw them. Later, you became so good at deciphering letters that you could recognize whole words merely by their shapes. Eventually, as you grew more practiced at reading and writing, you stopped thinking consciously about the shapes and skipped straight to their meanings.

However, the shapes themselves remain important to our ease of reading, which requires text to be created with clearly defined, consistent, and familiar type. From our long experience as both readers and writers, each of us holds archetypal models of what's acceptable for each letter of the alphabet. Of course, people's archetypal models might differ, depending on how and where they learned the alphabet. But as type design diverges from our archetype, it grows more and more difficult to read. Consider figure 6.2, which shows glyphs of the same letter represented in different typefaces. They're all *V*'s. But you can probably clearly recognize the letters on the left as such, while the letters further to the right probably seem less and less familiar. The *V* in Old Man Eloquent might look to some people like a lowercase *r*, and in Bickham Script Pro, the *V* could even masquerade as a *D*. In Saffran, the *V* almost loses its identity entirely. In other words, the letters on the left better exemplify "*V*-ness": two diverging diagonal lines joined at a bottom vertex. The farther away we get from this archetypal shape, the more difficult it is to connect the shapes and their corresponding significations.

Despite this tendency, users can still recognize remarkably distorted type. Our perceptions allow for a considerable amount of difference in letter forms. In general, however, the further away a letter shape is from our archetypal sense of the letter form, the more effort it requires for recognition; reading slows and comprehension drops as a result. Many organizations now draw upon the ease humans have in deciphering distorted letter forms as a way to defend against automated hacking attempts (figure 6.3). This technology is known as CAPTCHA, short for Completely Automated Public Turing Test to tell Computers and Humans Apart, and it uses distorted, segmented, and otherwise-altered text to prevent automated systems from hacking into accounts. As technology evolves, however, even distorted letter forms can be deciphered electronically as computers are taught how to link broken segments and extract letter features from the data (see, e.g., Wang et al., 2019; Chen et al., 2014).

Figure 6.2. How letter shapes affect reading. All of these shapes represent the letter *V*, but the less *V*-like the shape, the harder it is to recognize. Note also how each typeface sits differently within the same design space. These *V*s are all 72 points, and each was shifted to its typeface from the first 72-point *V* in Times New Roman without altering the design space. *Source*: Created by the author.

Times New Roman   *ATF Franklin Gothic*   *Papyrus*   *Bodoni URW*   *Bickham Script Pro*   *Old Man Eloquent*   *Saffran*

Figure 6.3. Users can recognize remarkably distorted type. Here, CAPTCHA technology is used to defeat hacking attempts on a website. People can recognize this distorted type, but it is more difficult for computers to do so, though machine learning algorithms are rapidly making this security measure obsolete. *Source*: Cloudflare (https://www.cloudflare.com/learning/bots/how-captchas-work/).

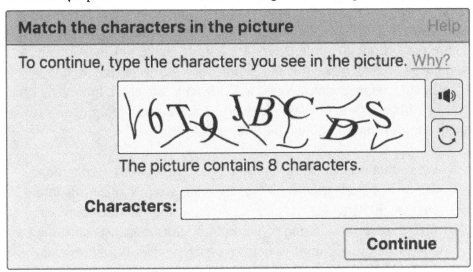

Figure 6.4. Reading blurred type. Blurred type is still readable. Some argue that this is because of boumas, even though individual words might be hard to make out; others argue that this identification is the result of recognition of individual component letters. *Source*: Created by the author.

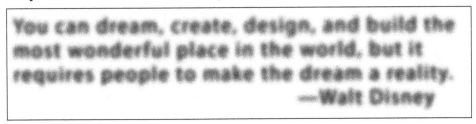

Figure 6.5. Reading transposed letters. Although some letters in this quotation are transposed, we can still make out the words because of their general shape. *Source*: Created by the author.

> Yuo cna deram, craete, disegn, nad biuld teh msot wdrnufeol plcae in hte wlord, btu it ruqiesre poplee to mkae hte darem a raleity.
> —Wlat Dsneiy

The significance of individual letters builds up to that of the general word shapes, and many typographers suggest that readers do not read one letter at a time but instead use the combined shapes of letters to perceive words. As we mentioned in chapter 5, readers' eyes move across a page (physical or digital) of text in small, rapid jumps called *saccades*.

Between saccades, they fixate on a group of three to four letters, and then jump on to another group—although not necessarily the closest or the next group. Readers often skip ahead and back as they read. They also don't typically fixate on all of the letters in a line of text. Instead, some argue, they read by deciphering the hazy boumas that surround each fixation. You can see how this works by looking at the sentence in figure 6.4. Although some letters are hard to make out individually, you should be able to make out the words successfully, if not easily.

Kevin Larson, a psychologist working with Microsoft, however, argues (along with many other psychologists), that we don't recognize words by their boumas but by individual letter forms. He writes:

> Word shape is no longer a viable model of word recognition. The bulk of scientific evidence says that we recognize a word's component letters, then use that visual information to recognize a word. In addition to perceptual information, we also use contextual information to help recognize words during ordinary reading, but that has no bearing on the word shape versus parallel letter recognition debate. It is hopefully clear that the readability and legibility of a typeface should not be evaluated on its ability to generate a good bouma shape. (2004, p. 12)

This argument is based on research into parallel letter recognition, which shows that "the letters within a word are recognized simultaneously, and the letter information is used to recognize the words" (2004, p. 3).

Does letter shape matter at all? Chances are, if you use culturally conventional letter forms, your reader will be able to read what you write based on shapes and context. Even if boumas do not actually allow us to distinguish words, they are still important from a design perspective. Design is ultimately about placing design objects onto a visual field and assessing how those objects interact. From this perspective, boumas matter even just at the shape level of design objects.

CULTURE

Type has an important role in culture, primarily because of the reciprocal development of printing technologies and cultures. As new printing technologies were developed, printed documents became increasingly common and cheap, deeply affecting culture. At the same time, cultures affected the growth of printing technologies as styles and ideologies changed.

In previous centuries, type design was an expensive, highly skilled, and labor-intensive process. Movable type, developed in China in the eleventh century CE by Bi Shēng, first took the form of a porcelain-like amalgamation of clay and glue, with one piece shaped for each symbol to be printed. But these blocks were fragile and wore out quickly. Carved wooden movable typefaces, which were more durable, were developed by Wang Zhen in the late 1200s, and cast metal movable type was developed in Korea in the early 1200s. The most enduring movable type process, however, is credited to Johannes Gutenberg (1398–1468), considered the originator of European letterpress printing, European type designers devised a solution to the problem of readily replicating immense numbers of the identical letters. They carved each letter of a font onto the end of steel punches and then used these punches to strike each letter into a brass matrix. The matrix was placed in a mold, and molten type metal—an alloy of lead, antimony, and tin—was poured into the mold to form each piece of type. Each matrix could thereby mold many identical pieces of type.

Each size of a typeface, however, required an entire set of punches and matrices. The collected type for each size became known as a font, which included multiple pieces of type for all letters, numerals, capitals, and punctuation in that size. Because type was heavy, awkward to store, and expensive, most letterpress print shops had only a few fonts of three or four typefaces, and they would often advertise their choices with typeface specimen sheets (figure 6.6). As a result, most early documents used relatively simple and restrained (although sometimes carefully crafted) typographic designs. European type design went through several stages related to the cultural values of the times. Early typefaces, known as *old style* faces, echoed common handwriting practices from the fifteenth through seventeenth centuries (figure 6.7). This made sense in that printing was a newer and cheaper technology than handwriting, and printers needed to convince audiences of the value of this new way of duplicating texts. Old style typefaces are typically known by their graceful curves and a moderate contrast between thick and thin strokes.

With the rise of the Enlightenment in the eighteenth century, type designers such as Giambattista Bodoni (1740–1813) and Firmin Didot (1764–1836) designed typefaces that incorporated regular curves and lines. In doing so, they sought to separate typography from handwriting and relate it to the universal, rational principles of abstract geometry. These faces are still called *modern*—and they were modern, for their time (figure 6.8). Modern typefaces are recognizable by their geometric consistency and high contrast between thick and thin strokes.

In the nineteenth century, designers extended Bodoni and Didot's separation of type from handwriting but left behind their notions of ideal type geometry. This produced typefaces that were decorative, brash, and visually interesting—but not always easy to read. This development

Figure 6.6. Example of a specimen sheet by John Baskerville, designer of the Baskerville typeface. *Source*: Photo by D. G. Ross from materials provided by Rare Book School, University of Virginia.

Figure 6.7. Old style typefaces. Garamond, an old style typeface, echoes Renaissance handwriting. *Source*: Created by the author.

Garamond

Figure 6.8. Modern typefaces. Bodoni, a modern typeface, was developed in the eighteenth century to emphasize ideal geometry rather than the common handwriting practices reflected by old style faces. *Source*: Created by the author.

Bodoni

matched the growth of commercial advertising and the rise of printed material as a commodity that needed to attract attention as well as convey information. As a result, many typefaces developed during this period echoed current styles and fads. For example, a fad for all things medieval, the Gothic Revival, led to the redrawing of *blackletter* typefaces (figure 6.9).

The end of the nineteenth century and the beginning of the twentieth century saw a return to old style typefaces. These were often redrawn for Linotype and Monotype, new printing technologies that used large machines to cast type on the fly; the type was melted down at the end of each print job and reused. These machines required the development of new matrices, which gave designers the opportunity to design new typefaces. For example, during this period Stanley Morison designed Times New Roman for *The Times* of London, and Edward Johnston designed Johnston Sans Serif for the London Underground signs (figure 6.10). At the same time, the eighteenth-century desire for a rational, geometric type arose again with modernism, with its interest in minimalism, abstraction, and universalism. For example, designers associated with the German Bauhaus movement boiled down letter forms to their minimal characteristics, designing abstract faces intended to be effective in all settings.

Later in the twentieth century, two technological developments changed typography dramatically, increasing its flexibility in the hands of document designers. First, in the 1950s, printing began to be performed by photography and offset presses (see chapter 11). Photography allowed much greater flexibility in typesetting than hot metal.

Figure 6.9. Blackletter typefaces. Blackletter typefaces like Amador echo northern European forms of handwriting. Many blackletter faces were drawn during the Gothic Revival of the nineteenth century. *Source*: Created by the author.

Amador

Figure 6.10. Changes to type in the early twentieth century. Stanley Morison's Times New Roman and Edward Johnston's Johnston Sans Serif (repackaged by P22 as Underground) were developed in the early part of the twentieth century to meet new needs and to work with new technologies. *Source*: Created by the author.

Times New Roman
Underground
(Johnston Sans Serif)

Second, in the 1980s, designers began to create documents with computers, using word processing and page layout programs. The digital type used in these programs led to a renaissance in type design. Nearly anyone with an interest in typography could become a type designer using a desktop computer and relatively inexpensive software. Echoing the development of postmodernism and the web, graphic designers in particular used digital typography to experiment with layered, off-kilter layouts and difficult-to-read typefaces, attempting to draw readers' attention to the physical text as well as to whatever its words were trying to convey. Borrowing a term from popular music, this style was often called *grunge* or *postmodern* typography.

Postmodern type designs tended to grab attention but at the cost of usability. So in recent years, information design has brought the emphasis back to focus on the user's experience of type, favoring designs that users find attractive, useful, and easy to read. Graphic designer and former editor of *Ray Gun* magazine David Carson, however, cautions designers about sacrificing a designer's perspective when thinking about type, arguing that we shouldn't "confuse legibility with communication" (2003). As a designer, it's up to you to determine your audience's needs and expectations.

RHETORIC

The perceptual and cultural perspectives lay the groundwork for the visual rhetoric of typography—that is, the use of typography to influence, affect, or guide readers toward their own goals or the goals of the client sponsoring the document.

This assertion is still in some ways controversial, primarily because of the continued popularity of the "crystal goblet" theory of typography. According to this theory, first propounded by typographer Beatrice Warde (1956), typography should be transparent, holding content like a crystal goblet holds a fine wine. When designed and set well, in other words, type shouldn't even be noticed by users, who could focus instead on the content the type held.

However, this theory underestimates the rhetorical significance of visual forms, assuming that there can be content separate from form. Typography can have important effects on users, particularly in terms of conveying structural and rhetorical meta-information about the document. For example, type reinforces the commonly hierarchical structure of documents, showing users how the different parts of a document fit together. Document designers use type to break down a document into sections and subsections, designing headings of different sizes to show the logical hierarchy of a document.

Applied consistently, these typographic contrasts guide readers through a document. In fact, headings serve as an integral index readers can skim visually, looking for whatever information they need most. Headings and paragraph breaks also make the text looks less like a dense gray field. Many readers are filled with a dogged despair at the prospect of working through long, undistinguished paragraphs; a more open typographic design with multiple paragraphs and section headings encourages readers to continue by giving them many resting places along the way. Think back to chapter 3, for example, where we noted how constructivist theories of perception indicate the importance of providing readers with clear visual cues. More open typographic designs also give readers multiple places to enter and exit the text—especially important if the document provides reference or procedural information, such as a manual. So your choice of typeface and typographic system you apply directly influences your audience's perception of a document's usability and shapes the type of experience (positive, negative, pleasant, disagreeable, etc.) they will have with the document.

The Value of Adding Typographical Contrast

Consider the three iterations of lorem ipsum shown here. In each case, the same text is shown. As more *visual contrast* is added however (headings, bullets, and spacing in this case), your eye is more able to find entry points (see e.g., Keyes, 1993). This visual contrast includes *typographical contrast*, which is typically recognized as the way the size, weight, form, texture, direction, structure, and color of letterforms differ from each other (see e.g., Williams, 2015; Wyatt & DeVoss, 2017). Try reading each section in lorem ipsum. As contrast and spacing increases, so too does your ability to "read" more text. As you see here, it doesn't take much to change the way your eye approaches text on the page.

1. Lorem ipsum dolor sit amet, consectetur adipiscing elit, sed do eiusmod tempor incididunt ut labore et dolore magna aliqua. Eros in cursus turpis massa. Odio morbi quis commodo odio aenean sed. Interdum velit laoreet id donec ultrices tincidunt arcu non sodales. Nisi vitae suscipit tellus mauris a diam maecenas sed. Vestibulum lectus mauris ultrices eros. Rutrum quisque non tellus orci ac auctor augue mauris. Convallis tellus id interdum velit laoreet id. At erat pellentesque adipiscing commodo elit at imperdiet. Gravida arcu ac tortor dignissim. Ultrices sagittis orci a scelerisque. Tellus in hac habitasse platea. Elementum integer enim neque volutpat. Mattis nunc sed blandit libero volutpat sed cras. Iaculis urna id volutpat lacus. Malesuada fames ac turpis egestas integer eget aliquet nibh praesent. Eget mi proin sed libero enim. Egestas egestas fringilla phasellus

faucibus scelerisque eleifend. Dictum fusce ut placerat orci nulla pellentesque dignissim enim sit. Nisl tincidunt eget nullam non nisi est. Tristique nulla aliquet enim tortor at. Quis commodo odio aenean sed adipiscing diam donec adipiscing. Ac felis donec et odio pellentesque diam volutpat. Erat imperdiet sed euismod nisi. Pulvinar mattis nunc sed blandit libero volutpat sed. Rhoncus est pellentesque elit ullamcorper dignissim cras tincidunt lobortis feugiat. Vestibulum rhoncus est pellentesque elit ullamcorper. Lectus mauris ultrices eros in cursus turpis massa tincidunt. Enim lobortis scelerisque fermentum dui faucibus in. Eros donec ac odio tempor orci. Sit amet tellus cras adipiscing enim eu turpis.

2. Lorem ipsum dolor sit amet, consectetur adipiscing elit, sed do eiusmod tempor incididunt ut labore et dolore magna aliqua. Eros in cursus turpis massa. Odio morbi quis commodo odio aenean sed. Interdum velit laoreet id donec ultrices tincidunt arcu non sodales. Nisi vitae suscipit tellus mauris a diam maecenas sed. Vestibulum lectus mauris ultrices eros. Rutrum quisque non tellus orci ac auctor augue mauris. Convallis tellus id interdum velit laoreet id. At erat pellentesque adipiscing commodo elit at imperdiet. Gravida arcu ac tortor dignissim. Ultrices sagittis orci a scelerisque.

Tellus in hac habitasse platea. Elementum integer enim neque volutpat. Mattis nunc sed blandit libero volutpat sed cras. Iaculis urna id volutpat lacus. Malesuada fames ac turpis egestas integer eget aliquet nibh praesent. Eget mi proin sed libero enim. Egestas egestas fringilla phasellus faucibus scelerisque eleifend. Dictum fusce ut placerat orci nulla pellentesque dignissim enim sit. Nisl tincidunt eget nullam non nisi est. Tristique nulla aliquet enim tortor at. Quis commodo odio aenean sed adipiscing diam donec adipiscing. Ac felis donec et odio pellentesque diam volutpat.

Erat imperdiet sed euismod nisi. Pulvinar mattis nunc sed blandit libero volutpat sed. Rhoncus est pellentesque elit ullamcorper dignissim cras tincidunt lobortis feugiat. Vestibulum rhoncus est pellentesque elit ullamcorper. Lectus mauris ultrices eros in cursus turpis massa tincidunt. Enim lobortis scelerisque fermentum dui faucibus in. Eros donec ac odio tempor orci. Sit amet tellus cras adipiscing enim eu turpis.

3. Lorem ipsum

Dolor sit amet, consectetur adipiscing elit, sed do eiusmod tempor incididunt ut labore et dolore magna aliqua. Eros in cursus turpis massa.

a. Odio morbi quis commodo odio aenean sed. Interdum velit laoreet id donec ultrices tincidunt arcu non sodales.

b. Nisi vitae suscipit tellus mauris a diam maecenas sed. Vestibulum lectus mauris ultrices eros.

c. Rutrum quisque non tellus orci ac auctor augue mauris. Convallis tellus id interdum velit laoreet id. At erat pellentesque adipiscing commodo elit at imperdiet. Gravida arcu ac tortor dignissim. Ultrices sagittis orci a scelerisque.

Tellus in hac habitasse platea. Elementum integer enim neque volutpat. Mattis nunc sed blandit libero volutpat sed cras. Iaculis urna id volutpat lacus. Malesuada fames ac turpis egestas integer eget aliquet nibh praesent. Eget mi proin sed libero enim. Egestas egestas fringilla phasellus faucibus scelerisque eleifend. Dictum fusce ut placerat orci nulla pellentesque dignissim enim sit. Nisl tincidunt eget nullam non nisi est. Tristique nulla aliquet enim tortor at. Quis commodo odio aenean sed adipiscing diam donec adipiscing. Ac felis donec et odio pellentesque diam volutpat.

Erat Imperdiet
Sed euismod nisi. Pulvinar mattis nunc sed blandit libero volutpat sed. Rhoncus est pellentesque elit ullamcorper dignissim cras tincidunt lobortis feugiat. Vestibulum rhoncus est pellentesque elit ullamcorper. Lectus mauris ultrices eros in cursus turpis massa tincidunt. Enim lobortis scelerisque fermentum dui faucibus in. Eros donec ac odio tempor orci. Sit amet tellus cras adipiscing enim eu turpis.

More broadly, typefaces can suggest a specific tone or ethos to readers; typography creates what Katie Salen has called a "voice-over" that influences how people see the document. Some typefaces are very conservative. Times New Roman, for example, until 2007 the default face of Microsoft Word and the most common typeface used in the business world, suggests a tone and ethos of business. Other common typefaces are more informal:

Comic Sans, for example, looks as if it were written with a magic marker.

Curlz looks playful and childish.

Design Tip: Matching Visual Tone

Match the visual tone of the document you are designing with the rhetorical situation for which it's intended. You probably wouldn't write an important business memo to your supervisor in Comic Sans because you want to be taken seriously as a professional. By the same token, you probably wouldn't write a love letter in Times New Roman, lest your friend think they are being dumped (or worse yet, sued).

But context is everything. If you work at a day care facility or your relationship is very formal, the reverse might be more rhetorically appropriate.

Users recognize the visual tone of type and use it as a clue to the nature and purpose of the document, as well as to the character of the document's author. Each typeface has what Eva Brumberger has described as a "persona": some typefaces are commonly perceived as "playful," for example, while others might be seen as "sloppy." Similarly, typefaces may be read as "professional" or "friendly," among other attributes, a conclusion that she argues suggests that "visual language is analogous to verbal language in carrying connotations" (2003, p. 221). These attributes may be influenced by everything from cultural expectations to perceived gendering, and so it remains quite distinctly up to the designer to consider the audience, culture, and context of use when it comes to choosing and implementing type choices.

Vincent Connare, for example, the designer who developed Comic Sans, famously came up with the typeface as a replacement for Times New Roman's use in a talking dog's speech bubble in Microsoft Bob (released in 1995). He based the face off the text in Alan Moore's *Watchmen* and Frank Miller's *Batman: The Dark Knight Returns* comics. As he has noted in interviews, he had to develop the typeface because "dogs don't talk in Times New Roman! Conceptually, it made no sense." The dog needed to "speak" in a more comical tone, and Times New Roman just doesn't have that persona (Connare, 2017).

Similarly, typeface is used in packaging and marketing to influence audience expectations of a product. Research suggests that there are particular "shape-taste associations" (see Velasco & Spence, 2018) that lead audiences to perceive certain sodas as more or less sweet based on the curvatures of the typeface used in their logos or certain products as more or less sour based on the letter forms applied. Typefaces can quite literally prime an audience's expectations, leading them to believe certain things about a product just through letter forms.

Typography has an undeniable effect on how readers respond to documents. As a visual rhetorician, you'll want to take that effect into account and even use it to the advantage of your clients and users.

LOOKING AT TYPE

You've been using type since you first began to read and write. But if you're like most people, you might not have thought very carefully or precisely about type. Before you can manipulate type with confidence, you'll need to sharpen your senses to recognize the many small but significant differences between typefaces.

Knowing the details of letter forms is a key part of recognizing different typefaces, so we'll start there. Then we'll discuss the various categories of typefaces to help you make good typographic decisions in your own document designs. With your typographic senses sharpened, you will also be able to identify and classify typefaces more accurately.

LETTER FORMS

Typographers design typefaces so all the letters have consistent features. To do so, they need a comprehensive terminology that describes the shapes and parts of letters. This terminology can be complex, but knowing a little of it can help you recognize differences between typefaces and make good choices for your typographic designs. In general, glyphs in English employ the characteristics shown in figure 6.11.

STROKES

Type designers use a basic vocabulary of *strokes* to make most of the letters in a typeface, giving the characters visual consistency with each other as well as a distinction from other typefaces.

Figure 6.11. Characteristics of letter forms. Knowing this terminology will help you identify and distinguish different typefaces. *Source*: Created by the author.

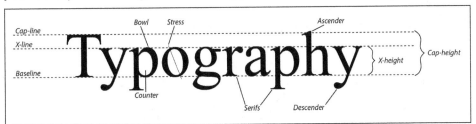

Strokes derive their name and their shapes from the fact that most typefaces are based on the stroke marks of a broad-nibbed calligraphic pen. Marks made by the pen are typically broader on vertical strokes and narrower on horizontal strokes. They also transition smoothly from broad to narrow as the pen forms a curve, such as in the letter *O*. These variations in stroke width survive even in many digital typefaces used today, and they can be used to identify typefaces. Some faces have significant contrast between the thick and thin areas of the strokes; others have a minimal contrast. Many typefaces also break with this tradition entirely, using geometric, irregular, or purposefully inconsistent strokes.

Regardless of their specific shape, strokes are the basic building blocks of most letters in the English alphabet. Figure 6.12 shows the basic strokes that combine to make English characters. For example, the uppercase letter *A* is made by two diagonal lines and one horizontal line, the lowercase letters *b* and *p* are made by a vertical line and an arch, and the lowercase letter *n* is made by two vertical lines and an arch.

SERIFS

Serifs are another tradition derived from the strokes of a calligraphic pen. The pen often leaves a slight horizontal mark at the beginning or ending of each vertical stroke; in time, medieval scribes emphasized such marks as deliberate decorations. These marks survive today as *serifs*—the small "feet" at the beginning or more commonly the end of vertical strokes. Typefaces with serifs are known as *serif typefaces*; typefaces without serifs are known as *sans serif* typefaces, because *sans* means "without" in French (figure 6.13).

Type designers continue to use serifs not just because of tradition or convention but because serifs can increase the readability of text in

Figure 6.12. Strokes from which all English characters are made. Strokes are the basic building blocks of letter shapes. *Source*: Created by the author.

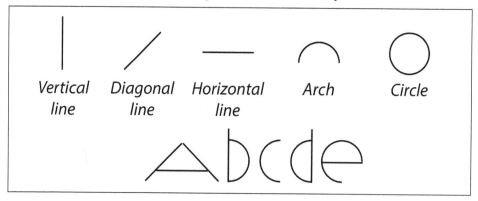

Figure 6.13. Serif and sans serif typefaces. Serifs are the little horizontal marks on the bottoms of vertical strokes. Sans serif ("without serif") typefaces lack these marks. *Source*: Created by the author.

The quick brown fox jumps over the lazy dog.
Serif typeface:
Garamond

The quick brown fox jumps over the lazy dog.
Sans serif typeface:
Franklin Gothic

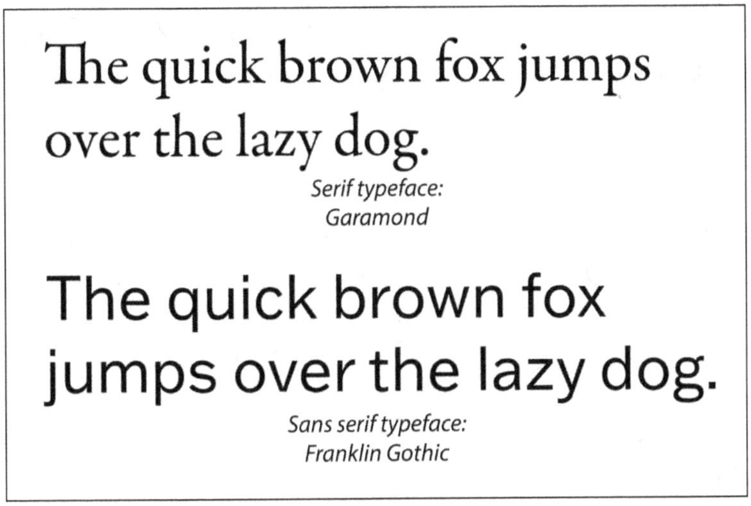

certain circumstances. When the serifs on many letters in a line of type printed physically are arranged consistently on the baseline, for example, they imply a subtle horizontal line themselves—almost like a dashed line that leads the user's eye across the page.

As figure 6.14 shows, serifs can also come in many different shapes, adding visual distinction to the typeface. (For advice on choosing between serif and sans serif typefaces, see "Serif vs. Sans Serif Typefaces" in this chapter.)

UPPERCASE AND LOWERCASE

Historically, when books were printed with individual pieces of movable type, printers kept the capital letters in one drawer or *case* (the "upper case") and small letters in another case (the "lower case"). We still use this terminology to describe capital (uppercase) and lowercase letters.

Figure 6.14. Differently shaped serifs. From left to right: Times New Roman, Fairplex Narrow OT, Bodoni 72 Oldstyle, Rockwell, Trajan Pro, and Bely. *Source*: Created by the author.

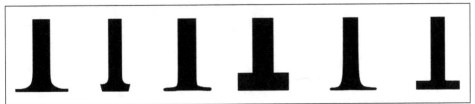

Most typefaces include a full set of upper- and lowercase letters, but some are formed entirely of uppercase letters. These typefaces are most often designed for titles or signs, where capital letters can sometimes stand out more emphatically than lowercase letters. However, it's important to avoid overusing capitals, as they cut down on readability (see "All Caps and Small Caps" in this chapter).

COUNTERS AND STRESS

An enclosed or mostly enclosed space in a letter form is known as a *counter*; the stroke that encloses a counter is called a *bowl*. Because of the shape of the counter itself and the varying width of the bowl stroke, counters can give a very distinct look to a typeface. This distinction applies particularly with the lowercase letters *a* and *g* because both letters can be designed as one-counter or two-counter forms (figure 6.15).

The shape of the bowl and counter can also make the letters appear to have a vertical alignment, known as *stress*. Stress is usually determined by drawing an imaginary line between the narrowest parts of the bowl that form a lowercase letter *o*. Most typefaces usually have a stress either *vertical* or *oblique* (slanted to the left; see figure 6.16).

Figure 6.15. One-counter and two-counter letters. These distinctive letters are often a good way to distinguish between typefaces. On the left, Comic Sans. On the right, Times New Roman. *Source*: Created by the author.

One-counter letters

Two-counter letters

Figure 6.16. Stress in letter forms. Stress is determined by drawing an imaginary line across the two narrowest points of the bowl stroke that forms the letter *O*. From left to right: Baskerville, Times New Roman, and Franklin Gothic ATF. *Source*: Created by the author.

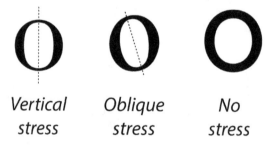

Vertical stress

Oblique stress

No stress

Old style faces usually have an oblique stress, while modern faces use a vertical stress. Sans serif typefaces almost always use strokes that vary little in width, so they may have little or no stress.

BASELINE, X-LINE, AND CAP-LINE

Although we need visual distinctions between letter forms to read efficiently, we also need consistency. To help create this consistency, typographers in Western languages use three horizontal lines (see figure 6.11):

- The *baseline*, the line on which all letters sit

- The *x-line*, the line at the tops of a lowercase letter *x*

- The *cap-line*, the line at the tops of uppercase (capital) letters

You may already be familiar with these lines because you practiced writing letters in copybooks that provided them as guides, and notebook paper still has baselines on which to write. For most documents, however, these lines are merely *implied* by the typeface, formed by the mostly consistent heights and alignment of the letters on a line of type. Serif faces make these implied lines somewhat stronger by including serifs at the baseline.

Typographers use these lines to make each letter approximately the same height, known as the *x-height* for lowercase letters and the *cap-height* for uppercase letters, measured from the baseline to the x-line or to the cap-line, respectively. Some letters break this pattern, giving them greater visual distinction. In many typefaces, a number of letters, such as *n*, *r*, and *h*, might rise slightly above the x-line. The letters *i* and *j* include diacritical marks—the dots—that sit above the x-line. And most dramatically, letters with *descenders* (*g j p q y*) dip below the baseline, and letters with *ascenders* (*b d f h k l t*) rise above the x-line. But these letters still include a relationship to the x-line; for example, the lowercase *f* has a horizontal line at or near the x-line.

MEASURING FONTS

Baselines also serve as a convenient way to measure fonts. The size of a font is measured in points from the baseline of one line of type to the baseline of the next line of type, set with no extra space between lines (see "Vertical Spacing" in this chapter). As we discussed in chapter 5, a point is $1/72$ of an inch (or $1/12$ of a pica).

However, some typefaces are designed to "sit" differently within this vertical distance. Typefaces can have a wide variability of actual sizes even within the same point size—for example, 12-point Antique Olive is actually a much bigger font than 12-point Centaur. When mixing typefaces in a design, compare fonts carefully by eye as well as by point size.

Fonts can also be specified in pixels for on-screen designs, but the size of pixels differs from monitor to monitor: some monitors have higher resolution and smaller pixels, others have lower resolution and larger pixels, and users themselves have a large amount of control over the resolution of their screens. So measuring font sizes by pixels is pretty imprecise. With designs meant to be used exclusively on-screen, however, dimensioning fonts with pixels can lead to greater legibility across a variety of monitors (see "Type on the Screen" in this chapter).

> When mixing typefaces in a design, compare fonts carefully by eye as well as by point size.

TYPEFACE CATEGORIES

Now that you have a better sense of the characteristics of letter forms, we can discuss typefaces in more detail. Many thousands of typefaces have been designed since Bi Shēng first shaped moveable type in the eleventh century—most of them since the development of digital type design in the 1980s. Some come standard with computer systems, and others can be purchased from type design bureaus like Linotype or Adobe or from individual type designers for a relatively small fee. At the time of this writing, for example, Adobe Fonts offers over 25,000 options for use with its Creative Cloud suite of programs—a number that is truly mind boggling when it comes to making choices. Google Fonts, whose slogan is "Making the web more beautiful, fast, and open through great typography and iconography," offers over 1,609 free font families in more than 135 languages. As AI use increases in all aspects of design, you can expect those numbers to continue to rise.

It's a luxury to have so many typefaces to choose from, but choices can also lead to confusion. In this section, we'll discuss four overlapping ways to distinguish between typefaces:

- Serif vs. sans serif typefaces
- Roman vs. italic typefaces
- Text vs. display typefaces
- Monospace vs. proportional typefaces

These distinctions will help you make good choices about what typeface to use in different situations. Keep in mind, however, that these are overlapping categories—for example, some typefaces are simultaneously sans serif, italic, *and* display faces. Also remember that many typefaces purposefully blur the boundaries between categories; type designers love to get creative and break the mold!

SERIF VS. SANS SERIF TYPEFACES

Serifs are one of the most common distinctions between typefaces: some typefaces have them, some don't. For the most part, use serif typefaces for paragraphs that are meant to be read through, such as those in a report or a book.

Because they don't have serifs, sans serif typefaces typically look more vertical than serif faces do. So use sans serifs where you want to direct the reader's eye down rather than across a line of type—for example, in lists and tables. Although some designers use sans serifs for paragraph text, most designers reserve them for shorter text objects like headings, captions, and titles.

Sans serif faces are sometimes called *gothic* typefaces because they typically lack the amount of ornamentation of serif faces. This simplicity also makes sans serifs work well for signs and other large-scale work because their shapes are easily legible from a distance.

ROMAN VS. ITALIC TYPEFACES

When printing first became common in the sixteenth and seventeenth centuries, two styles of type became common: *roman* and *italic* (neither is usually capitalized). Most people think of roman typefaces as regular, plain, upright, "normal" text. Most text meant to be read in sentences and paragraphs—such as the text of this book—is set in roman type-faces. Italic typefaces slant to the right, echoing the slant of Renaissance handwriting. As a result, italics use thinner strokes and thus have a less dense value than roman text (see figure 6.17).

You've probably used word processors to make roman text into italic (just press Ctrl+I, right?). But strictly speaking, the romans and italics you would be using are different fonts within the same type family. Earlier

Figure 6.17. Comparing roman and italic typefaces. Roman type tends to be more dense than italic type in the same point size. *Source*: Created by the author.

roman: more dense

italics: less dense

in the development of digital type, word processors made roman type into italic by artificially leaning it over at an angle, which often created awkward-looking italics. If you wanted a truly well-designed italic, you had to specify an entirely different font, which was tedious if you just wanted to italicize a word here or there. But newer digital type technologies can include specifically designed italic and roman fonts within the same font file, so when you use a word processor's italic command, you're actually invoking a specially designed italic font. You can see this more clearly if you look at type families in Adobe InDesign, for example, where you have the option to download separate fonts within a typeface.

Most of the time, use roman typefaces in your designs; they're easier to read, clearer, and usually denser than italics, creating a better figure-ground contrast (remember those Gestalt concepts from chapter 3?). When you do use italics, however, avoid using them for long stretches of text, which can fatigue readers quickly. (For information on using bold or italic for emphasis, see "Modifying Type for Emphasis" in this chapter.)

Some typefaces are designed to be set in text; they are intended to help readers make their way through many lines of prose in sentences and paragraphs. Common text faces are Times New Roman, Book Antiqua, and Bookman Old Style, but there are many others.

> For most situations, use text typefaces for text and **display typefaces** for display.

Display faces, on the other hand, are designed to make a bold visual statement. They work best in small doses—for example, in titles or headings. As you can see in the table on display faces in this chapter, these letterforms are best used in moderation because they require readers to slow down and pay attention to each letter. Sometimes that's the desired rhetorical effect, but readers may also lose patience and skip over something that looks too hard to read. So for most situations, use text typefaces for text and display typefaces for display.

MONOSPACE VS. PROPORTIONAL TYPEFACES

One of the great tragedies of type design started as a technological compromise. The letters in most typefaces naturally have different widths—for example, an *i* is narrower than a *w*. These typefaces are called *proportional* because each letter gets only its fair share of horizontal space. But when mechanical typewriters were first introduced in the late nineteenth century, they required that each letter be as wide as the next, since the typewriter platen or print head shifted over a consistent space with every key pressed. To solve this problem, typewriter companies designed *monospace* typefaces such as Courier, which made each letter the same width. They arbitrarily narrowed the wide letters and widened the narrow letters—*i* became about the same width as *w*, mostly by awkwardly extending the serifs. With the success of the manual typewriter, monospace faces became a familiar fixture on typed business reports and school essays.

Figure 6.18. Display typefaces. This table shows a few types of display typefaces. Script, blackletter, inscribed, engraved, grunge, and dingbats styles (and more) all work if you need to make a statement. *Source*: Created by the author.

Kind of Typeface	Description	Samples	
Script	Echoing handwriting or calligraphy	Mistral	*AaBbCcDdEeFf12345*
		Comic Sans	AaBbCcDdEeFf12345
Blackletter	Echoing fraktur or early modern script	Old English	𝔄𝔞𝔅𝔟ℭ𝔠𝔇𝔡𝔈𝔢𝔉𝔣12345
		Parchment	⚜⚜⚜⚜⚜⚜12345
Inscribed	Echoing stone inscriptions	Imprint MT	AaBbCcDdEeFf12345
		Castellar	AABBCCDDEEFF12345
Engraved	Echoing texts engraved on metal	Edwardian Script	*AaBbCcDdEeFf12345*
		Kunstler Script	*AaBbCcDdEeFf12345*
Grunge	Drawing attention to themselves as art	Chiller	AaBbCcDdEeFf12345
		Jokerman	AaBbCcDdEeFf12345
Dingbats	Providing glyphs as images	Wingdings	♍☺♋♌♎♏♐♑♒♓♔♕
		Webdings	♈♉♊♋♌♍♎♏♐♑♒♓

Although we rarely use manual typewriters anymore, monospace faces still show up in documents to this day. In general, avoid monospace typefaces; use proportional typefaces instead.

Avoid `monospace typefaces`; use proportional typefaces instead.

However, a few justifiable uses do remain for monospace faces. They're sometimes used in tables that require a precise vertical alignment of numbers and letters. They're also often used to show computer code in a document such as a manual.

DESIGNING DOCUMENTS WITH TYPE

Now that you have a better sense of how to categorize typefaces, you still need to make smart choices about what typefaces to use in your designs. In this section, we'll discuss some guidelines for choosing and combining typefaces.

CHOOSING TYPE FOR LEGIBILITY, READABILITY, AND USABILITY

When choosing a typeface, pay attention to three complementary features of type—legibility, readability, and usability:

- *Legibility* refers to the extent to which readers can make out the individual letters of a typeface easily and efficiently.

- Good legibility leads to good *readability*, a general measurement of the ease of reading, scanning, or skimming a document.

- Both legibility and readability lead to *usability*, a measure of how well readers can use a document to complete a task.

Usability is the ultimate goal; we want to create document designs that allow users to do useful things in their lives. When thinking about usability, though, you should also consider *user experience*. Just because someone *can* use a document, doesn't mean they *want* to use a document or even *enjoy* using that document. If you can design so that a document is usable and the user has a positive experience, you increase the chances that they will return to the document, share the document, or otherwise engage with the document in desirable ways. Ultimately, the usability of type—and the user's experience with that type—depends on legibility and readability, which rely on three important factors: typeface design, font, and x-height to cap-height ratio.

TYPEFACE DESIGN

It may sound simple, but users find some typefaces easier to read than others. The problem is that not everyone agrees on *which* typefaces are easier to read. As we discussed earlier in this chapter, users have both a conventional sense of what makes letters look normal to them and an ability to recognize remarkably distorted type. But users can lose interest when they have to read type that is very different from what they consider "normal."

For the most part, the most legible and readable typefaces are those with relatively simple, unadorned designs—those closest to the archetypal forms most people carry in their heads. That would suggest that for many purposes, a sans serif face would be the most legible. And, in fact, that's exactly what you see most often on large-scale work, such as signs or posters.

The only ornamentation that some users seem to prefer, especially when reading paragraphs of text, is the serif. But this preference depends on each user's experience, training, and culture. If a user is accustomed to reading sans serif faces, they will prefer reading sans serif faces.

Studies on typeface preference suggest that designers have many options. Jo Mackiewicz's (2006) article on audience perceptions of fonts in PowerPoint, for example, suggests that there are several options for usable, comfortable design with personalities relevant to a given rhetorical situation, and that those typefaces do not have to be what some might label as "boring." In Mackiewicz's study, Gill Sans and Souvenir Lt were rated as comfortable-to-read, professional, and interesting as well. So rather than declare a simple rule about whether to use typefaces of one category or another, it's best simply to ask users what they prefer or to observe what typefaces seem to work best with the users—for example, by comparing documents designed for a similar purpose and audience.

FONT

Legibility and readability also depend on the size of type. Obviously, too small a font can be difficult to see (which is why legal documents put the details in the "fine print," where they're hard to make out). By the same token, using too large a font in a typical print or screen document can make it hard for the eye to take in enough of the shape to discern meaning—much like looking at a billboard from a few feet away. (Of course, if you really are designing a billboard, you should use as large a font as possible!)

X-HEIGHT TO CAP-HEIGHT RATIO

You'll recall from earlier in the chapter that the distance between the baseline and the x-line is the *x-height*, while the distance between the baseline and the cap-line is the *cap-height*. But some typefaces place the x-line closer to the cap-line than others, creating different ratios between the x-height and the cap-height. Typefaces with a relatively tall x-height are known as *high-ratio typefaces*; those with a relatively short x-height are known as *low-ratio typefaces* (figure 6.19).

High-ratio typefaces, such as Times New Roman, Lucida Bright, and Georgia, have an x-height relatively nearer the cap-height than low-ratio typefaces, like Centaur, Perpetua, or Garamond, which have an x-height closer to halfway between the cap-line and the baseline. A high-ratio typeface is often more legible than a low-ratio typeface in the same general size. So you can often use a high-ratio typeface in a smaller size to conserve space while retaining legibility.

> You can often use a high-ratio typeface in a smaller size to conserve space while retaining legibility.

MATCHING TYPE TO VISUAL RHETORIC

In addition to usability, typography can play a role in the rhetorical impact of a document. Even in the blandest of documents, some element of rhetorical persuasion is at work—if nothing else, the need to persuade users that the document holds useful information pertinent to their lives. Other documents, such as annual reports, brochures, and flyers, are more obviously highly persuasive.

As we discussed earlier in this chapter, typography creates a voice-over for the content; it gives a tone to whatever the document is conveying. So type should mesh with the client's rhetorical purposes and users' expectations. As we learned from examining typography's cultural background, typefaces also have cultural and ideological connections that still influence

Figure 6.19. X-height:cap-height ratios in different typefaces. High-ratio typefaces, with their relatively tall x-height, tend to be more legible in smaller font sizes than low-ratio typefaces. *Source*: Created by the author.

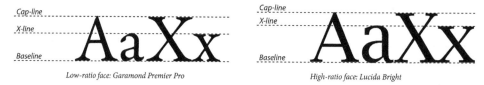

Low-ratio face: Garamond Premier Pro   High-ratio face: Lucida Bright

readers today. Bauhaus faces—with their uniform strokes rooted in modernism—still imply clarity, universality, and minimalism. Many grunge faces imply a validation of difference, otherness, and novelty. Choose typefaces and fonts according to their rhetorical purpose—particularly, according to what tone or sense of ethos you want to convey to users.

Choose typefaces and fonts according to their rhetorical purpose—particularly, according to what tone or sense of ethos you want to convey to users.

CREATING A TYPOGRAPHIC SYSTEM

Ideally, a document's typographic design should work as a system that shows users how to use the document. For example, typography should accomplish all of these tasks as a coherent system:

- Show users the document's structure

- Help users find where to look first on a page

- Tell users what's most important on the page

- Indicate how each textual element relates to other elements on the page, in terms of the document hierarchy

So when choosing typefaces for a document, it's essential to think holistically about all of the different kinds of text your document will include, as well as how all of those kinds of text are related to each other in the document hierarchy. As we discussed in chapter 5, the most common way to show a document hierarchy is by thinking about the various kinds of textual design objects and their relationships. For example, documents might include different kinds of type, like those listed in figure 6.20.

Choose type so that each of these different kinds of text is visually distinct from the others as design objects. But also make sure that each kind of text has a clear visual relationship in a hierarchy with other kinds of text. In other words, balance the design principles of *similarity* and *contrast* to create a unified, systematic design.

Most designers approach this problem for technical documents by choosing one or two type families for each document. Often, they'll use one type family for the basal text (for sentences and paragraphs) and another for the many metatextual design objects users need to navigate the document, such as the title, headings, headers, and footers in a paper document or site titles, page titles, footers, and navigational links in a screen document. This technique creates a good sense of unity and consistency, while retaining an important distinction between basal text and navigational/structural text.

Figure 6.20. Kinds of text in a typical technical document and a typical website. It's important to assess what different kinds of text will be included in your document design so you can create a system of typography that distinguishes between content types visually. *Source*: Created by the author.

Technical document	Website
Document title	Site title
Heading level one	Page title
Heading level two	Footer
Heading level three	Heading level one
Heading level four	Heading level two
Header and footer	Heading level three
Basal text (paragraphs)	Basal text (paragraphs)
Numbered lists	Numbered lists
Bulleted lists	Bulleted lists
Tables	Tables
Captions	Hyperlinks within paragraph text
Callouts	Hyperlinks in the navigation bar

Within this relatively small palette of typefaces, we can use *type variations* to make hierarchical distinctions between different kinds of text, visually creating the different levels of the document's hierarchy. Readers must be able to recognize a first-level heading, and they should be able to recognize which second-level headings belong under that first-level heading. The key is to match the typeface or font to its logical importance in the document hierarchy and use adequate contrast so users can tell the difference between one level and the next.

Don't be afraid to use contrast when choosing typefaces for textual and metatextual elements. This contrast can apply within a single typeface, using a variety of fonts and styles, or it can work by combining different typefaces with contrasting appearances. Figure 6.21, for example, uses typographic contrast to create hierarchy within the same textual elements shown in figure 6.20. This hierarchy tells users how to distinguish visually between the levels of the hierarchy; it also helps them distinguish between the metatextual elements (all in a sans serif typeface) and the textual elements (all in a serif typeface).

> Don't be afraid to use contrast when choosing typefaces for textual and metatextual elements.

We can use the visual variables of design objects we discussed in chapter 2 to make sure our designs have adequate typographic contrast. In particular, four kinds of contrast are particularly useful in providing typographic distinction between design elements and levels of the hierarchy: contrasts of size, shape, value, and color.

CONTRASTS OF SIZE

Perhaps the easiest way to create contrast in a document is by using a range of *sizes* of type for design objects at different levels of the document hierarchy (see figure 6.21). Readers usually assume that large fonts imply more importance than smaller fonts.

For example, figure 6.22 shows three versions of a flyer for a speaker at a Society for Technical Communication meeting. Although Version A gets the basic ideas across—who's speaking, what the topic will be, and when and where to show up—it gives equal weight to all pieces of information. The only visual cue that tells the reader how to read the flyer is the design principle of order: the convention of starting at the top and going to the bottom.

This isn't necessarily a problem, but adding size contrast would improve the flyer's visual impact and help readers understand the relative importance of the information being presented. Version B prints the title in a different font, giving greater prominence to the speaker's topic. Of course, if the speaker is a big celebrity—someone a lot of people would want to see—then the speaker's name might be more important than

Figure 6.21. Typographic contrast. Creating a typographic hierarchy requires using a system of contrasts in typefaces and fonts. *Source*: Created by the author.

Title (Franklin Gothic ATF 28 pt)
Heading level one (Franklin Gothic ATF 22 pt)
Heading level two (Franklin Gothic ATF 18 pt)
Heading level three (Franklin Gothic ATF 16 pt)
Heading level four (Franklin Gothic ATF Italic, 14 pt, indented)
Header and footer (Franklin Gothic ATF 9 pt)

Paragraph text (Lucida Bright 10pt)
1. Numbered lists (Lucida Bright 10pt)
- Bulleted lists (Lucida Bright 10pt)

Tables (Franklin Gothic ATF 9pt)
Captions (Franklin Gothic ATF 9pt)
Callouts (Franklin Gothic ATF 8pt)

Figure 6.22. Using contrasts of size to create hierarchy. Contrasts of size can help create typographic hierarchy. Version A offers no visual distinction between the different kinds of textual content. Version B uses font size to make the title appear most prominent, and Version C emphasizes the speaker's name. *Source*: Created by the author.

Managing Complicated Design Projects	Managing Complicated Design Projects	Managing Complicated Design Projects
Rachel Hawkins Project Manager DesignRight Solutions	Rachel Hawkins Project Manager DesignRight Solutions	**Rachel Hawkins** Project Manager DesignRight Solutions
Marriott Hotel Ballroom May 17, 2025	Marriott Hotel Ballroom May 17, 2025	Marriott Hotel Ballroom May 17, 2025

Version A   *Version B*   *Version C*

the topic. Version C places the name of the speaker at the top of the typographic hierarchy. Each of these versions gives users a remarkably different sense of the information being presented, but the only difference is in the size of the various text design objects.

To set up a reasonable typographic hierarchy in a document, start with the size you want to use for basal text, and then size the various hierarchical levels up or down from this standard. For typical business and technical documents, start with something between 9 points and 11 points for text. Then step up to larger sizes (up to around 36 or even 48 points) for headings and titles and down to smaller sizes (sometimes down to 6- or 7-point fonts) for captions and callouts.

> For typical business and technical documents, start with something between 9 points and 11 points for text. Then step up to larger sizes for headings and titles and down to smaller sizes for captions and callouts.

A 9-point font for text might seem small to you, particularly if you work primarily in Microsoft Word, which uses 11-point type as a default. But if you look at a wide variety of business and technical documents, you'll notice that most use fonts smaller than 12-point for basal text. Again, be sure to check font sizes visually, as well as by point number; 9 points in some typefaces is smaller than in others.

Of course, there are exceptions. If the document itself is large or intended to be viewed from a distance, start with a larger size of basal text—in fact, as large as possible. Nobody will read a banner, sign, or flyer in 9-point type. Also, you may want to use a slightly larger font for screen documents so they can be read easily on low-resolution screens (see "Type on the Screen" in this chapter). In CSS, this can be accomplished by setting font size specifically with relation to absolute size (x-small, small, large, x-large, and so on), or relative ("larger" or "smaller" as keywords) to the parent elements font size.

CONTRASTS OF SHAPE

Typefaces come in a variety of contrasting shapes. Typically, the more contrast in the shape, the easier it will be for users to tell the difference between one level of the typographic hierarchy and another. Figure 6.23 shows how shape can make a big difference in the amount of contrast between two typefaces.

Of course, we can get different shapes by using italics within the same typeface, but italics alone don't always supply adequate contrast between design objects such as basal text and headings, where we want an especially clear differentiation.

CONTRASTS OF VALUE

We can also use contrasts of value to show typographic contrast. *Value*, you'll recall from chapter 2, refers to the contrast between the brightness or darkness of a design object and its background. Contrasts of value can make a very clear distinction between levels of the hierarchy. The more dense or high-value-contrast the type, the more likely the user's eyes will be drawn to that type—particularly when the surrounding type has a lighter or lower-value contrast. Value increases with size, simply because there's more ink on the page (or pixels on the screen) devoted to the type:

as fonts get larger, they look more dense

than smaller fonts in the same typeface.

Figure 6.23. Similarity and contrast of shape. Bookman JF Pro and Palatino have relatively similar letter shapes, but both differ considerably from Myriad Pro. *Source*: Created by the author.

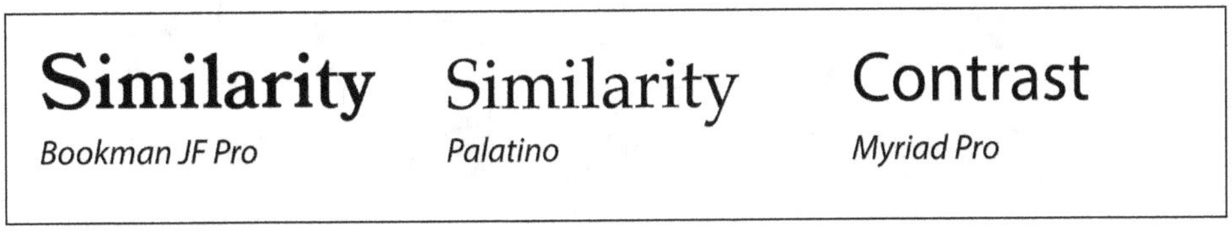

Similarity   Similarity   Contrast
Bookman JF Pro   *Palatino*   *Myriad Pro*

Figure 6.24. Density of typefaces. Some typefaces are designed to look more dense than others. *Source*: Created by the author.

Low-value typefaces	High-value typefaces
Trebuchet MS 12 pt	**Myriad Pro Black 12 pt**
Palatino Linotype 12 pt	*Magneto 12 pt*
Futura PT 12 pt	**Impact 12 pt**
Century Old Style Std 12 pt	**Krungthep 12 pt**

But some faces are designed to look denser than others, even at the same font size (figure 6.24).

Many type families also come in a variety of values, also known as *weights* (figure 6.25). You can use these weights to great effect, keeping a general typographic coherence within your design while taking advantage of their natural value contrasts.

CONTRASTS OF COLOR

Finally, you can use contrasts of *color* to distinguish different textual elements from each other. Many document designs take a step up from

Figure 6.25. Using a variety of weights in the same type family. Weights of fonts in the same type family allow designers to use a consistent type design but vary the value for different purposes within the document. *Source*: Created by the author.

Transducer Hairline
Transducer Regular
Transducer Medium
Transducer Black

one-color to two-color design by using colored headings. Doing so can definitely attract a user's eye to the structure of the document. And for on-screen designs, colored type is no more expensive than black and white.

However, one disadvantage of colored type is that any color contrast other than black on white (or vice versa) will decrease the value contrast. In other words, type of any color other than black usually looks less dense than black type on a white background. So make sure you choose relatively dark colors for type on a white background or relatively light colors for type on a dark background. Avoid low-value-contrast color combinations, such as bright yellow type on a white background or dark gray type on a black background. If you are designing for online use, you can check your URL for contrast errors (and many other errors as well) using a tool like the WAVE Web Accessibility Evaluation Tool (https://wave.webaim.org/). For print documents, something like WebAIM (Web Accessibility in Mind) allows you to put in the hex codes of your color choices and check their contrast (https://webaim.org/resources/contrastchecker/).

> Avoid low-value-contrast color combinations, such as bright yellow type on a white background or dark gray type on a black background.

SETTING TYPE

Setting type successfully requires paying careful attention to the spaces between, above, below, and beside pieces and groups of type. Word processors and page layout programs give considerable control over spacing, both horizontally and vertically.

VERTICAL SPACING (LEADING)

In the age of letterpress printing with movable type, the distance between each line of type was set by placing strips of lead between the lines. Accordingly, the distance between lines of type is still known as *leading* (rhymes with "heading"). Most typefaces are already designed to provide a little extra space between one line and the next, even when they are set with no leading (i.e., single-spaced). For example, a 10-point font will likely be designed so there are about 12 points from baseline to baseline, and a 9-point font will be set on an 11-point *body* (the more traditional way of saying the same thing). This practice keeps descenders and ascenders from running into each other from one line to the next. Leading is specified with a notation such as 10/12 or "10 on 12," which means a 10-point font leaded with baselines 12 points apart.

With digital type we can specify more or less leading with great precision. Most word processing and page layout programs also allow you to adjust leading decimally or by percentages, usually with 1.0 or 100 percent representing the "normal" leading (such as 10/12), 1.1 or 110 percent as a little more than normal (10/13), and 0.9 or 90 percent as a little less (10/11). Most designers set the leading initially by eye and then note the exact leading so they can set other paragraphs the same way.

The vertical space created by leading is essential for readers to recognize each horizontal line of type. In Gestalt terms, the individual letters—though not actually connected to each other in most faces—must line up in such a way as to invoke visual continuation. One way to do so is to make sure that the height of the *vertical* space between lines exceeds the width of a typical *horizontal* space between pieces of type or words. Otherwise, readers may have trouble recognizing the line of type as a strong figure (see chapter 3).

For example, figure 6.26 shows paragraphs set with different leadings. The 10/8 leading makes the text look like columns of letters rather than rows of type. The ascenders and descenders overlap, causing awkward

Figure 6.26. Paragraphs with different leadings. Lines leaded too closely together can make ascenders and descenders run into each other, creating problems with continuation. Lines leaded too far apart can impede readers from making the return sweep to the next line of text successfully. *Source*: Created by the author.

10/12 leading *(single spaced)* *Century Old Style Std*	Lorem ipsum dolor sit amet, consectetur adipiscing elit. Mauris dui arcu, venenatis a rutrum vel, pulvinar suscipit nisl. Maecenas dignissim semper lacus, in scelerisque mauris dapibus nec. Morbi eu venenatis augue. Ut eu velit dolor. Vivamus suscipit quis urna et egestas.
10/8 leading *Century Old Style Std*	Lorem ipsum dolor sit amet, consectetur adipiscing elit. Mauris dui arcu, venenatis a rutrum vel, pulvinar suscipit nisl. Maecenas dignissim semper lacus, in scelerisque mauris dapibus nec. Morbi eu venenatis augue. Ut eu velit dolor. Vivamus suscipit quis urna et egestas.
10/25 leading *Century Old Style Std*	Lorem ipsum dolor sit amet, consectetur adipiscing elit. Mauris dui arcu, venenatis a rutrum vel, pulvinar suscipit nisl. Maecenas dignissim semper lacus, in scelerisque mauris dapibus nec. Morbi eu venenatis augue. Ut eu velit dolor. Vivamus suscipit quis urna et egestas.

interferences. On the other hand, as you can see from the 10/25 paragraph, if the lines of type are set too far apart, readers might have a hard time making the visual leap from one line to the next.

LEADING FOR PARAGRAPHS

You can specify the leading both *within* and *between* paragraphs of text. Within paragraphs, you can often rely on the default leading (such as 10/12) built into the computer font, especially for low-level designs such as letters, memos, and internal reports that must be completed quickly and cheaply and don't justify much of a design effort. For higher-level designs, however, consider adding a bit more leading than the default—10/13 or 10/14, for example. This amount of leading is significantly less than double-spacing but more than single-spacing, and it makes reading easier.

In addition, set paragraphs with long lines for the particular font to greater leading than paragraphs with shorter lines for the font. For example, if you have a long line with small type, you'll need to increase the leading slightly to help users keep their place as they read from one line to the next.

Set paragraphs in sans serif with slightly greater leading than those with serif faces (figure 6.27). Sans serif typefaces lack the horizontal cues of serifs, so users need a bit more separation between lines to make each line of text a strong figure.

Figure 6.27. Leading for serif and sans serif typefaces. Set sans serif typefaces with a slightly larger leading than serif typefaces. Sans serifs need the extra space to make up for the lack of the horizontal cues provided by serifs. *Source*: Created by the author.

Lorem ipsum dolor sit amet, consectetur adipiscing elit. Mauris dui arcu, venenatis a rutrum vel, pulvinar suscipit nisl. Maecenas dignissim semper lacus, in scelerisque mauris dapibus nec. Morbi eu venenatis augue. Ut eu velit dolor. Ut pellentesque, sapien ac vehicula bibendum, enim dolor porta leo, quis imperdiet lectus arcu in arcu. Suspendisse faucibus felis condimentum felis rutrum ultrices. Donec vestibulum ligula ac ipsum efficitur, quis iaculis lacus maximus. Mauris sagittis leo sit amet dolor maximus mollis. Curabitur leo augue, cursus vitae orci ac, iaculis fringilla arcu. Integer nisl tellus, blandit ut neque faucibus, maximus ultrices arcu. Pellentesque eu facilisis sapien. Praesent id lectus sagittis, rutrum elit et, molestie sapien. Morbi sed efficitur arcu. Nulla sed justo vel lectus faucibus iaculis sit amet non arcu. Vivamus suscipit quis urna et egestas.

10/12 leading
Century Old Style Std

Lorem ipsum dolor sit amet, consectetur adipiscing elit. Mauris dui arcu, venenatis a rutrum vel, pulvinar suscipit nisl. Maecenas dignissim semper lacus, in scelerisque mauris dapibus nec. Morbi eu venenatis augue. Ut eu velit dolor. Ut pellentesque, sapien ac vehicula bibendum, enim dolor porta leo, quis imperdiet lectus arcu in arcu. Suspendisse faucibus felis condimentum felis rutrum ultrices. Donec vestibulum ligula ac ipsum efficitur, quis iaculis lacus maximus. Mauris sagittis leo sit amet dolor maximus mollis. Curabitur leo augue, cursus vitae orci ac, iaculis fringilla arcu. Integer nisl tellus, blandit ut neque faucibus, maximus ultrices arcu. Pellentesque eu facilisis sapien. Praesent id lectus sagittis, rutrum elit et, molestie sapien. Morbi sed efficitur arcu. Nulla sed justo vel lectus faucibus iaculis sit amet non arcu. Vivamus suscipit quis urna et egestas.

10/14 leading
Tahoma

Between paragraphs, you can specify a certain amount of space before or after each paragraph, usually in points or decimal inches (e.g., "6 points before and 3 points after" each paragraph). This technique streamlines setting type for single-spaced paragraphs in business documents; rather than double-spacing after each paragraph, you can just set each paragraph to include a certain amount of space "after." Typically, set the leading between single-spaced paragraphs larger than the leading between lines within paragraphs so readers can recognize each paragraph as a strong figure, as well. Above all, avoid double-spacing, which slows down reading by providing too much space between lines of type. Teachers and editors might request double-spacing because it gives them space to write comments and mark changes between the lines of type on a printed manuscript or school report. Most users don't need double-spacing, and, with modern electronic commenting features, many teachers and editors don't either!

> Avoid double-spacing, which slows down reading by providing too much space between lines of type.

LEADING FOR HEADINGS

The ability to set leading between paragraphs is particularly useful for headings, which in word processing and page layout programs are essentially specially formatted paragraphs. Headings should usually rest a little closer to the text they introduce than to the text they follow. Look at the headings in this chapter. You can see that they're leaded to give more space before than after each heading. As we discussed in chapters 2 and 3, this practice implements the Gestalt principle of proximity to help readers associate the heading with the section it belongs to rather than with the section that precedes it.

HORIZONTAL SPACING

In this section, we'll discuss the various aspects of the horizontal dimension of type, including the space between letters and words, line length, paragraph justification, and indentation.

LINE LENGTH

As we discussed in chapter 5, one of the biggest factors in setting easily readable type is the length of each line of type. Although users can read text in many line lengths, most feel comfortable with a line length that corresponds to about 7 to 12 words. Although this is basically a typographic issue, it's also a page layout issue because line length is established by the page layout grid within which the type is set. This grid determines how type can be arranged, forming a frame to constrain the length of lines. (See chapter 5, "Using Grids for Page Design.")

KERNING, LETTER SPACING, AND LIGATURES

Adjusting the spacing between *individual* letter pairs is generally called *kerning*; adjusting the spacing between *all* letters in a word, line, sentence, or paragraph is called *letter spacing*. There's some variation in terminology here, however: letter spacing is also known in some word processing and page layout programs as *character spacing*, *tracking*, or (even more confusingly) *track-kerning*. Let's just say that you have a lot of control over the horizontal space between letters.

Strictly defined, kerning shifts a letter horizontally so it extends into the negative space of a neighboring letter (figure 6.28). For example, in the word "for," you might want to move the *o* slightly under the ascender of the *f* to avoid a too-large negative space between the letters. In some digital fonts, such as Georgia, the *f* is designed to do this automatically. Other digital fonts, such as Comic Sans, require manual kerning. However, the automatic adjustment to the *f* in Georgia causes problems in other combinations: in the letter pair *fl*, the ascenders overlap each other, looking especially awkward in large fonts. In this case, you may need to add extra space between the letters or, better yet, use a ligature.

Ligatures are special combined letters such as *fi* and *fl*; they echo the natural connections that people make between letters in handwriting (figure 6.29). Ligatures tend to look formal and elegant, but ligatures are not often set as the default in word processing programs. In Microsoft Word, for example, you must select the text in which you want ligatures to appear, then enable them from the Format -> Font -> Advanced menu,

Figure 6.28. Automatic kerning. Georgia automatically kerns the *o* under the curl of the *f*, but Comic Sans does not. *Source*: Created by the author.

Georgia   Comic Sans

or use Alt and a four-digit code (search online for Alt codes or Unicode characters). On a Mac, you can simply hold down a letter key (such as A) and any associated ligatures will appear (i.e., æ). So reserve ligatures primarily for high-level design projects and large font sizes, such as those used in titles and signage.

For text intended to be read (rather than scanned or skimmed), it's usually best to trust the digital typeface itself for appropriate space between letters. In a well-designed typeface, the type designer will have built in a much better sensitivity to kerning than you're likely to have time to tinker with. As fonts get larger, however, the spaces between different letter combinations might create awkward negative spaces. In particular, you may need to adjust kerning between individual letters in large-format work. The higher the level of design, the more attention you'll want to dedicate to kerning.

Letter spacing is a rougher technique than kerning; it's typically used to make text fit in the space the layout provides—a technique also known as *copyfitting*. In most programs, you can specify letter spacing with great flexibility. You can set the type to a looser (expanded) letter spacing or to a tighter (condensed) letter spacing. The resulting inconsistency can make the text more difficult to read, so avoid using letter spacing for paragraph text except as a last resort. Again, the exception is large-format work. It's common to set the letter spacing tighter the larger the font, as otherwise many typefaces begin to look too spaced out.

WORD SPACING, JUSTIFICATION, AND HYPHENATION

Spaces between words are what readers use to distinguish one bouma from the next, and as such, these humble bits of negative space are

Figure 6.29. Ligatures. Because Georgia uses automatic kerning, the *f* runs into the *l* and comes too close to the dot of the *i*. Ligatures solve the problem by connecting letter pairs into one glyph. *Source*: Created by the author.

fl fi   fl fi

Georgia letter pairs with 0.01 em kerning

Georgia ligatures

often underestimated in importance. For easy read-ability, use consistent spaces between words. One of the most common ways designers mess up consistent word spacing is by using fully justified text. Typically, paragraphs can be aligned in three ways:

For easy readability, use consistent spaces between words.

Left-aligned: All lines of text are lined up to the left-hand edge of the column. Use left-aligned text for most designs. It helps readers get through the text quickly because it gives them a consistent vertical line to return to as their eyes seek the next line of text on the left side of the paragraph. This alignment is often called *ragged right* because the variety of line lengths leaves a ragged edge on the right-hand side of the column.

Right-aligned: All lines of text are lined up to the right-hand edge of the column. Use this alignment for some callouts, labels, and columns of tables.

Fully justified (or just *justified*): Lines of text are stretched out to meet both the left and right edges of the column.

Designers have debated extensively about the merits and problems of left-aligned and justified text. Research suggests that most readers find either alignment read-able, but justified text can slow down reading slightly.

Use left-aligned text for most designs.

Fully justified paragraphs often look good from the designer's per-spective—everything lined up squarely and neatly. But from the reader's perspective, full justification often creates awkward word spacing, particu-larly in narrow columns. If the word processing or layout program does not have enough space to fit in another word in the line, it often stretches the spaces between words to make the line meet the right edge. The resulting inconsistent word spacing can make reading very difficult, and the resulting wide spaces between words can also lead to *rivers* and *lakes*: bodies of empty space that distract the reader's eyes from the text. A careful designer can mitigate or avoid this inconsistent word spacing with careful layout and typography, but doing so requires extra time and work. In particular, designers often use *hyphenation* to break up words at the end of a line, making smaller chunks that are easier to align to both sides.

But hyphenation causes its own problems. Hyphenation breaks up the natural shapes of words and makes readers complete the return sweep to the next line before reading the whole word. Multiple consecutive lines

In general, don't allow more than two or three consecutive hyphenated lines.

ending in hyphens can also create a distraction, since the hyphens will line up vertically at the right side of the column. Most word processing and page layout programs allow you either to disallow hyphenation entirely or to limit how many hyphenated lines can appear in a series. In general, don't allow more than two or three consecutive hyphenated lines.

INDENTATION AND CENTERING

Indentation usually refers to extra horizontal space added at the beginning or end of lines of type, measured from the margin or column edge. Three kinds of paragraph indentations are commonly used in document design: first line, hanging, and whole paragraph.

First-line indentation, which you can see in this paragraph, is a traditional way to show the beginning of a paragraph. Typically, designers indent the first line of a paragraph by somewhere between 0.25" (0p3) and 0.5" (0p6), depending on the typeface and font of the paragraph (for more about inches, picas, and points, see chapter 5). Another common practice is to forgo first-line indents on the first paragraph after a heading but to use them in subsequent paragraphs. This helps readers move from the heading to the first paragraph more easily.

Hanging indentation, exemplified by this paragraph, makes the first line of the paragraph start at the left margin or column edge but indents all subsequent lines.

Hanging indentation is also the technique of choice for bulleted or numbered lists, since the bullets or numbers stick out and mark each new item:

- This is the first item in the list of three bullets. The second line of text in this item will wrap to line up with the text of the first line, not with the bullet.
- Second item
- Third item

You can also use hanging indents for alphabetized lists, like indexes or bibliographies, since it makes the alphabetized word stick out from the rest of the paragraph, increasing the reader's ease of scanning.

To emphasize a single paragraph, you can indent the whole paragraph on one or both sides. This practice creates additional negative space around the text, drawing attention to it and making it a stronger figure. It's also a common way to indicate a long quotation, usually of three or more lines.

Paragraph indentations are usually specified in picas or inches in the United States, in centimeters in Europe, or in pixels for screen design. Some designers might also use older dimensional units such as *quads* (four spaces, also known as an *em*) or *half-quads* (two spaces, or an *en*), but with digital typesetting, these are becoming less common because they aren't consistent or universally standardized.

Centering paragraphs is rarely a good idea, primarily because it creates an inconsistent boundary on the left side of the paragraph—as you can see here. This inconsistency makes readers' return sweep very difficult as their eyes saccade across the paragraph from right to left, looking for the next line. The next line never begins at a consistent place, so it's hard to find. Centered paragraphs do have some traditional uses, such as in formal invitations, but in most situations, centering paragraphs causes too many problems.

PUNCTUATION AND SPECIAL FEATURES

One of the most overlooked features of typography is punctuation. We can't hear a period or a dash, but these and other signs are essential to the meaning of texts. In this section, we'll discuss some of the details about using these quiet heroes of typography.

In word processing and page layout programs, you can access some of the specialized punctuation marks discussed here by using key combinations (system, program, or Unicode) or the Insert: Symbol dialog box. Check your software documentation for instructions on how to do so.

We'll concentrate on the punctuation marks themselves and the common typographical practices for using them in English. Punctuation as a grammatical practice differs from culture to culture, language to language, and writer to writer, so refer to grammar handbooks to find out how to use punctuation marks correctly.

More about . . . Spacing and Punctuation

The legacy of the typewriter has caused a particularly bad habit: double-spacing after periods and colons. On a typewriter that uses a monospace typeface, one space after a period or colon does not create enough visual separation between one sentence and the next, so adding an extra space increases legibility. But periods and colons in proportional typefaces are already designed with appropriate surrounding negative space. Accordingly, use only one space after a period or colon in a proportional typeface.

If you can't break the habit of double-tapping the space bar after every period, then make a new habit. After entering text in a document, do a global search, replacing every instance of two spaces with one. Most programs will let you do this with a quick and painless Replace All command. And if you receive text from other authors to include in your design, this step will ensure consistent single-spacing after all punctuation marks.

PERIODS, COMMAS, SEMICOLONS, AND COLONS

Readers use these common punctuation marks and the space after them to see the beginnings and endings of syntactical units such as phrases, clauses, and sentences. To see a sentence as a visual figure, readers need a certain amount of negative space between one sentence and the next. That space has to be more than the typical space between words in the sentence, or readers won't recognize the space as meaning "a new sentence begins here" rather than just "a new word begins here." The period contributes to that space because it's usually set at or near the baseline, leaving plenty of negative space above it. That negative space is almost as important as the period itself in visually distinguishing one sentence from the next.

HYPHENS AND DASHES

Hyphens and dashes look similar, but this class of marks includes three completely separate punctuation marks, each with its own specific purposes.

Table 6.1. Hyphens, En Dashes, and Em Dashes

Mark	Name	Description	Use
-	Hyphen	These short, relatively thick, horizontal marks indicate a tight connection between what comes before and what comes after. In some faces (such as Californian FB or Goudy Old Style), hyphens are diagonal (-, -).	Use hyphens to join some compound words (e.g., "forget-me-not") or to break a single word at the end of a line between syllables.
–	En dash	An en dash gets its name from being about as wide as a capital letter N. It's longer and usually thinner than a hyphen in the same font.	Use en dashes between numbers: 25–35 books 8:00–9:00 a.m. 1996–2009 Some designers (particularly in Europe) use an en dash for parenthetical or appositive expressions, adding a thin (¼ em) space before and after the dash – as you see in this sentence.
—	Em dash	Em dashes are about as wide as a capital M.	Use em dashes to set off parenthetical or appositive expressions in sentences—like this.

QUOTATION MARKS, APOSTROPHES, AND PRIMES

Quotation marks, apostrophes, and primes also look alike, but they have significantly different purposes and meanings (figure 6.30).

Primes, also sometimes known as hash marks or tick marks, are used mostly for showing dimensions in feet and inches (5' 11") or degrees and minutes of an arc (270' 30"). Most typefaces include vertical primes, but you may need to use the Symbol font (available on almost all computer systems) for the more sophisticated-looking angled ones. Quotation marks and apostrophes, however, are almost always curved or slanted.

Figure 6.30. Quotation marks, apostrophes, and primes. Primes are upright or slightly angled, and they usually taper from top to bottom. Quotation marks curve around the quotation with the "tail" pointing up on the beginning mark and down on the ending mark. Apostrophes look like a single ending quotation mark, with the tail pointing down. *Source*: Created by the author.

ELLIPSIS MARKS

Most digital typefaces include a special glyph to indicate ellipsis—not just three periods but a single, combined *ellipsis mark* (. . .). Use this glyph if it's available in the digital font you're using, as it's usually spaced differently—usually with more space between dots—than three periods in a row would be (. . . instead of ...). Most word processing programs now automatically insert an ellipsis mark when you type three periods in a row. If the ellipsis occurs at the end of a sentence, don't forget to include the period as well.

The result may look like four dots, but it's really a period and an ellipsis mark, followed by a space before the beginning of the next sentence.

Design Tip: Converting Primes to Quotation Marks and Apostrophes

Many word processing programs automatically change primes to quotation marks and apostrophes as you type (the "smartquotes" setting in Microsoft Word), but page layout programs don't. If you place copy from a TXT file into a word processing or page layout program, you'll run into another problem. Basic text files include only primes, so you'll need to replace the primes with quotation marks. Just use your program's Replace function to make this replacement. (You can do it at the same time you replace the double spaces after periods with single spaces!)

SPECIAL TECHNIQUES

In this section, we'll discuss some guidelines for setting diacritical marks, numerals, fractions, drop caps, lists, and reverse type.

DIACRITICAL MARKS

Diacritical marks—small dots, curls, or lines added to the tops or bottoms of letter forms, such as á, ç, ñ, d-, ü—are important elements of many written languages. For example, some languages include only consonants, relying on the diacriticals to indicate the intervening vowel sounds. If you are designing a document for readers in a language that uses diacriticals, take particular care to use them correctly. Many digital typefaces can now create a full range of diacritical marks. Most can be accessed by Unicode key combinations (also known as Alt codes) or, again, by holding down letters on a Mac (see "Digital Type" in this chapter).

ARABIC NUMERALS

Typefaces typically include either *inline* Arabic numerals, which are a little shorter than a capital letter in the font, or *oldstyle* Arabic numerals, which have a basic shape closer to the size of a lowercase letter and dip above or below the baseline and x-line, depending on the numeral (figure 6.31). Oldstyle numerals provide more distinction between the forms of the glyphs, increasing legibility when they're used in a line of text. However, the inconsistency of oldstyle numerals can cause problems in other settings where alignment is especially important. Tables and forms, for example, should use inline numerals.

Figure 6.31. Using inline and oldstyle Arabic numerals. Use inline Arabic numerals in tables and forms. Use oldstyle Arabics for numbers within paragraphs. *Source*: Created by the author.

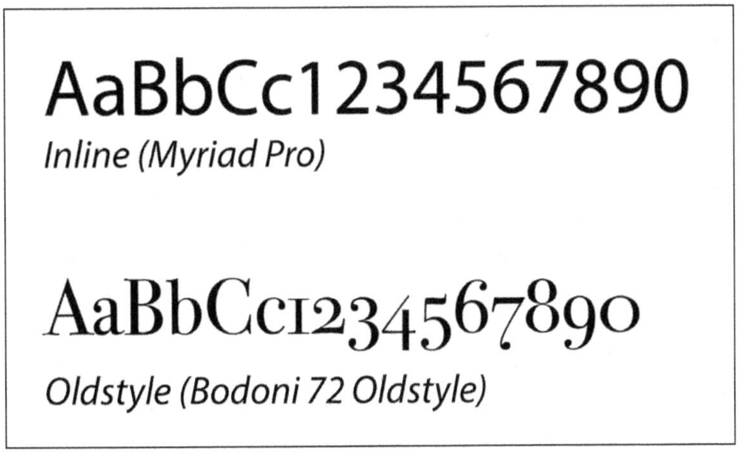

AaBbCc1234567890

Inline (Myriad Pro)

AaBbCc1234567890

Oldstyle (Bodoni 72 Oldstyle)

FRACTIONS

In many computer fonts, common fractions are designed as one combined glyph, $\frac{1}{2}$, as opposed to the inline 1/2, which is formed of three glyphs (1, /, 2). In the one-glyph fraction, the numerator is slightly raised (superscripted), the denominator is slightly lowered (subscripted), and both are kerned tightly to the slash and reduced in size.

Setting the common fractions ($\frac{1}{8}$, $\frac{1}{4}$, $\frac{3}{8}$, $\frac{1}{2}$, $\frac{5}{8}$, $\frac{3}{4}$, $\frac{7}{8}$) is easier now than it used to be, since most word processing programs automatically replace typed fractions with one-glyph fractions. You can set less common fractions manually or just allow them to stay as inline glyphs.

Drop caps. For greater distinction at the beginning of a chapter or section, designers sometimes use a drop cap. This technique increases the font of the initial letter of the text, making it as large as two, three, or even four lines of the text font (see figures 6.32a–b).

Using a drop cap can be a good technique to tell readers where to start reading; the density of the large initial letter naturally draws the reader's eyes to that point. For example, you might consider using drop caps to begin chapters or to begin the text on a single-page design such as a flyer. However, it's easy to overuse this technique, so employ it with caution.

REVERSE TYPE

Thus far we've focused primarily on black text on a white field, but designers often use white or light-colored text in a colored field for greater emphasis. This technique is known as *reverse type*. It's not a good idea to use reverse type for long passages of text; we've likely all struggled through web pages with black backgrounds and white text. But reverse type is useful for emphasizing the column heads in a table or creating a strong visual band across the page for a major heading. It can also mark a header or footer emphatically.

Figure 6.32a. Example of drop caps. The size of the first letter depends on the impression you want to make and how you want to place the accompanying text. Here, the text wraps around the first letter. *Source*: Created by the author.

Figure 6.32b. Another example of drop caps. In this case, the text is placed adjacent to the design, creating a more linear appearance and leaving the drop cap with more surrounding negative space. *Source*: Created by the author.

Some drop caps are set into the column, with the text wrapping around them (like this paragraph)

Other drop caps are set into the margin beside the paragraph (like this paragraph).

Figure 6.33. Visual density and reverse type. Because the visual density of the dark background can overpower the lighter-colored type, it's a good idea to bold the reverse text. *Source*: Created by the author.

Column Head 1   Column Head 1

Myriad Pro Regular   Myriad Pro Bold

When using reverse type, keep two things in mind. First, the visual density of the colored field can overpower the lighter-colored type. You'll need to increase the value or density of the typeface slightly so the text will show up adequately against its background (figure 6.33). To increase the visual density of the white area enclosed by the text, use a larger font or a demibold or bold font in the same typeface. Doing so will make the type slightly broader and more visually dense.

Second, make sure that the field gives the type adequate breathing room, particularly above and below the type. Leave plenty of space between the baseline and the bottom of the field especially, to accommodate descenders (see "Vertical Alignment" in chapter 9).

MODIFYING TYPE FOR EMPHASIS

With the development of digital type technologies used on word processors, it's now possible to apply type modifications such as bold, italic, underlining, caps, small caps, outline, shadow, and WordArt. Designers often use these modifications as a quick way to add emphasis to typography, and each has its own advantages and disadvantages. In all situations, however, it's best to use these modifications sparingly. If you bold everything, the bold text loses its meaning. If you bold only one word in a paragraph, however, that word is sure to stand out due to the contrast with its surroundings.

BOLD AND ITALIC

In most situations, use bold rather than italic for emphasis. Bold increases the value (density) of type, while italic decreases value. Increasing the value makes a lot more sense if you really want a bit of text to pop.

As we mentioned previously, bold and italic type modifications are not the same as bold or italic fonts, although the border between the two has grown increasingly hazy. Word processors often make bold and italic modifications by applying algorithms to type, stretching (for italic) or widening (for bold) the strokes, usually with some distortion. Boldface and italic fonts, however, are separate fonts in a typeface, supplied in a variety of sizes and designed individually to work well both internally and within their font family. For quick, low-design-level work, it's usually fine to use the bold and italic commands for bold and italic. But for high-design-level work, use the specialized bold and italic fonts when they are available.

However, faces designed to the OpenType standard include the roman, bold, and italic fonts within the same digital file (see "Digital Type" in this chapter). With these faces, it's not necessary to use a separate bold or italic font; just apply bold or italic in the word processor or page layout program, and the appropriate font will be used automatically.

UNDERLINING

Avoid using underlining unless you want to indicate an active hyperlink. Like double-spacing after a period, using underlining for emphasis came from the typewriter age. Since most typewriters can print in only one font and typeface, the only option for typists who wanted to emphasize a word was to type the word, move the platen back, and then underline the word. So underlining was a compromise—the quickest (but not the best) way to emphasize text within a typewritten document.

Underlining has some significant typographic disadvantages. It cuts across descenders, changing the letter forms people need to read efficiently. Note how underlining cuts across the descenders in this phrase: ugly typography. Underlining also obscures the essential space *between* words, which readers need in order to distinguish one bouma from the next. And underlining distracts readers when it's used on more than one consecutive line of text (figure 6.34).

About the only justifiable use of underlining today is on websites and hyperlinked documents. This practice is still typographically unfortunate for the perceptual difficulties just noted, but people have gotten used to

Figure 6.34. Underlining text. Underlining obscures descenders and the spaces between words. *Source*: Created by the author.

You can dream, create, design, and build the most wonderful place in the world, but it requires people to make the dream a reality.
—Walt Disney

being able to click on underlined text. That means it's especially important to use underlining on websites *only* for links and not for emphasis. And, despite the sometimes visual clunkiness of always using underlined text to indicate an active hyperlink, it's good web design. Underlining all links makes a site more accessible. Putting links in a different color can help, but color should not be the only means of conveying information, and, beyond simple convention, underlining ensures that your links will be obvious when they are presented in complex blocks of text (see, e.g., Hinds, 2023). Think about your design in several different dimensions—audience, purpose, and context. If someone is online and needs to be able to rapidly find information, is it better to have really cool, unique visual design, or to have design that clearly indicates use? The two aren't necessarily exclusive, but make sure you are thinking of your audience as actively *doing*, not just *looking*.

ALL CAPS AND SMALL CAPS

Writers often use text in all caps for emphasis, but doing so often doesn't work well. As we said earlier in this chapter, users tend to read by whole word shapes rather than letter by letter. This suggests that readers find words with distinctive shapes easier to read than those with fewer differences. ALL CAPS TEND TO MAKE EACH WORD LOOK PRETTY MUCH THE SAME: A VAGUELY RECTANGULAR SHAPE. Words in lowercase, however, have all sorts of distinctive bumps and ridges, particularly along the tops of the words. These distinctions make them easier to read than words in all caps.

This problem becomes more significant when more than one line is set in all caps. Ironically, some designers use all caps for caution and warning statements, which defeats the point: making sure people read them (figure 6.35).

Figure 6.35. Using all caps. Setting warnings in all caps can make them more difficult to read. *Source*: Created by the author.

WARNING: USING THIS PRODUCT IN WET CONDITIONS CAN LEAD TO ELECTROCUTION. WATER CAN CAUSE A SHORT BETWEEN THE SWITCH AND THE OPERATOR'S HANDS. DO NOT OPERATE WHILE STANDING IN WATER.

One compromise designers often use for headings is SMALL CAPS. Small caps retain some of the shape of lowercase words, while giving some extra emphasis. But like all caps, small caps are still mostly rectangular. Use them in moderation.

USING TYPOGRAPHIC STYLES

As you can see, typography requires managing many small details—so many that it's hard to keep to a consistent typographic system within each document. You want every paragraph to be set in the same type with the same leading, justification, before- and after-spacing, and so on. You want every second-level heading to look like all of the other second-level headings and every callout to use the same typographic design. Maintaining this kind of consistency manually can be a nightmare.

An even bigger problem is if your client decides that Gill Sans wasn't really the look they wanted for the headings in this document and asks you to change them all to Eras Demi on the Friday before the document goes to press. If you set the type for each of the headings as Gill Sans manually, you'll have to change each one of them to Eras Demi manually as well. There goes your weekend.

But with a little foresight and some computer skills, you can change each of these headings in a matter of seconds with one simple command. This technique is known as using *styles*. A style is simply a set of characteristics you can apply with one click to a paragraph or a string of characters. You can use this technique when creating documents for paper or for electronic media.

> A style is simply a set of characteristics you can apply with one click to a paragraph or a string of characters.

Design Tip: Outline, Shadow, and WordArt

If you get the urge to use outline, shadow, or WordArt—*stop*, put down the mouse, and step away from the computer. Outline is a product of the same kind of computer algorithms that fake italic and bold type. It removes visual density from the text, giving it inadequate figure-ground contrast. Avoid using outline for almost any purpose.

Shadow is almost as bad as outline, creating a gray step in density between figure and ground and thus decreasing contrast. However, on colored backgrounds, shadows can increase figure-ground contrast and create the illusion of the type popping out in three dimensions. Shadow is often used on large signs for that reason. Even so, use shadow cautiously.

WordArt is tempting, colorful, and fun to play with—but it can easily lead to ill-considered and cliché designs. Most of its templates use awkward color combinations and distortion to draw attention to the type, which unfortunately can decrease legibility. And because everyone using Microsoft Word has access to the same templates, most people recognize WordArt as a cheap design shortcut.

STYLES ON PAPER

Most word processors and all page layout programs allow you to create many different styles. A style designed for a whole paragraph is called a *paragraph style*; a style designed for a string of words or letters within a paragraph is known as a *character style*. Paragraph styles include styles set for your headings (1, 2, etc.), styles set for quotes, styles for lists, and even styles for intense emphasis. If you manage all these from a style menu, you never need to move from word to word or paragraph to paragraph to make adjustments in a long document. You'll be able to control it all from a simple menu.

PARAGRAPH STYLES

Imagine that you've decided to make your first-level headings Gill Sans Bold 18-point, leaded with 12 points before and 3 points after. Setting each heading this way manually would be a pain. But if you design a style that carries all of these characteristics, all you have to do is apply the style to each heading—a one-click operation.

Even more conveniently, if you decide you want to change all the first-level headings to Eras Demi, you don't have to change every Heading 1; you just change the style. The software then applies the new characteristics to all of the paragraphs you've identified as using that style.

Typically, page layout and word processing software lets you set up styles three ways:

- *Modify a default style.* Most programs include a variety of ready-made styles for paragraphs, such as several levels of headings, captions, and footnotes. You can apply these ready-made styles or modify them to suit your design.

- *Format a sample first, then define the style from the sample.* In this method, you format one paragraph to have the design you want, and then create a new style from that paragraph.

- *Define the style from scratch.* Most programs have a dialog box for defining a new style from scratch by specifying the name of the style and all of the characteristics one at a time. You can then apply this style to as many paragraphs as you like.

Most programs include a menu or dialog that lists all of the available styles for you to choose from. Most programs also let you export styles from one document to another, so you can create many documents with the same formatting—for example, if you need to follow an organization-wide style sheet.

A final advantage of using paragraph styles is that most programs can auto-generate a table of contents from headings styles. The procedure usually goes something like this:

1. Set up the document so that each heading is styled consistently as a Heading 1, a Heading 2, and so on.

2. When you're finished writing the document, place the cursor on the page where you want the table of contents to appear and choose the command to create a table of contents.

Upon receiving this command, the software goes through the document looking for paragraphs styled as Heading 1 or Heading 2 (or whatever styles you have specified for the table of contents to include). It then copies the text from those paragraphs and assembles them in order, along with the page numbers on which the headings appear. Voilà—a table of contents! The best part of this technique is that if you need to change the order, contents, or headings in the document, all you have to do is regenerate or update the table of contents to create an accurate navigational guide for users.

CHARACTER STYLES

One limitation of paragraph styles is that the entire paragraph must use the same style. That's where *character styles* come in. Character styles work very similarly to paragraph styles, but they're applied to a string of characters instead of to an entire paragraph. They're also a bit simpler because they don't include settings for paragraph-level features such as leading.

Character styles are very useful for applying special formatting to words or phrases that might appear in a paragraph, such as glossary terms, proper names, or trade names. You can apply a paragraph style to the entire paragraph and then apply one or more character styles to individual words or phrases within the paragraph. The program will apply the character style first, so even if you change the paragraph style, the words to which you applied the character style will stay the same.

STYLES ON WEBSITES: CSS

In web development you can also apply a similar technique, known as *Cascading Style Sheets (CSS)*. CSS is a powerful tool, allowing the same kind of style functionalities of page layout and word processing programs, plus many other features like layering and positioning design objects.

Unlike styles in word processing and page layout programs, CSS are either made part of the HTML code of a web page or kept in a separate CSS file, where they can be referenced by many web pages. This CSS file (with the extension .css) is just a text file that holds specifications for the typographic details of any HTML tagset you want to format. By changing an external CSS style sheet, you can alter the look of an entire website in an instant, since the changes will "cascade" to every page in the site that invokes that style sheet.

While a full discussion of CSS is beyond the scope of this book, here is a brief summary of how it works for typographic design. For example, the HTML tagset for typical body paragraphs (<p> . . . </p>) could be set up in the CSS file as follows:

p {font-size: 10pt; font-family: Tahoma; margin-left: 20}

This style specifies that every paragraph using the <p> . . . </p> tagset should be in 10-point Tahoma and indented 20 pixels from the left margin. Text for any of the other standard HTML tagsets can be specified the same way, including that for headings (</h1> . . . </h1>, <h2> . . . </h2>, <h3> . . . </h3>), table cells (<td> . . . </td>), and list items (<li> . . . </li>).

You can also apply a cascading style to structures smaller or larger than the contents of a single HTML tagset. Doing so requires creating a *class* that contains the characteristics you want to apply. A class might look something like this:

.update {font-size: 9pt; font-family: Tahoma; background-color: #330000; color: #FFFFFF; font-weight:bold}

This class specifies Tahoma Bold 9-point, in white (FFFFFF) text on a dark red background (330000). You can invoke this class style in an HTML page by putting it within paragraph tags (<p> . . . </p>) for individual paragraphs, division tags (<div> . . . </div>) for multiple paragraphs, or span tags (<span> . . . </span>) for a few words—like this:

<span class="update">Content goes here.</span>

Any text content you put between these tags will appear formatted as you specified in the .update class style.

There are many, many resources for more information on CSS, but a great place to start is the World Wide Web Consortium's site. The World

Wide Web Consortium (W3C) "develops standards and guidelines to help everyone build a web based on the principles of accessibility, internationalization, privacy and security" (2024). Search on their site (https://www.w3.org/) for any number of resources and guides, and you won't be disappointed.

DIGITAL TYPE

Typography on a computer can be complex, but fortunately, computers do most of the work for us. Digital type technologies use mathematics to describe the curves and shapes of each letter so it can be resized and still look good. These technologies also must make letters display correctly on low-resolution computer screens (about 100 pixels per inch, or ppi), as well as on high-resolution printouts (above 300 dots per inch, or dpi). Type technologies must also work with a variety of computers, operating systems, and software programs.

Software makers have worked hard over the past two decades to make using digital type a relatively straightforward task. Still, knowing something about digital type technologies can help you choose a typeface that will work well with the particular document you are designing, as well as with your production method. It will also help you understand more about the possibilities and flexibilities of using type in word processors and page layout programs.

This is one area where type terminology can get hazy. Many people in desktop publishing and commercial printing refer to typefaces simply as "fonts," using that term to cover everything from the typeface design to the computer files used to hold the typeface design. We'll use the term *font file* to describe computer files and stick with the traditional definitions of typeface and font described at the beginning of this chapter.

POSTSCRIPT AND TRUETYPE

Type technologies involve complex standards that require investment from a variety of stakeholders, including software companies and operating system companies—most prominently Apple, Microsoft, and Adobe. As a result, the desktop publishing period has seen several competing standards come and go.

PostScript, a standard created by Adobe in the 1980s, was particularly good at describing type for output to printers, but it didn't work as well on screens. So Apple (later in partnership with Microsoft) developed

TrueType, which was designed to work well both on-screen and in print. Today, TrueType fonts are much more common than PostScript, especially for basic business and technical documents. But PostScript is still preferred by some designers and commercial printers.

TrueType was further developed as OpenType (also known as True-Type Open), which can accommodate both PostScript and TrueType glyph descriptions. It also provides good support for high-end features such as ligatures and swashes (decorative strokes added onto certain characters); specialized fonts such as italic, boldface, and small caps; and optics (specially drawn fonts for very small or very large sizes). Because it also incorporates Unicode (see the next section), OpenType also works well with other writing systems, including right-to-left languages such as Hebrew and Arabic.

More recently, Web Open Font Format (WOFF and WOFF 2.0) have emerged as formats specifically designed for use in web design. These typefaces are essentially compressed versions of TrueType and OpenType fonts with added metadata.

UNICODE

TrueType and PostScript work fine for Western languages, which use a relatively limited alphabet. But what about all of the other languages and their writing systems?

Unicode is an international standard addressing this problem. Strictly speaking, Unicode is not a type standard because it describes *characters* (abstract signs with linguistic meanings) as opposed to *glyphs* (the many possible visual representations of characters). But it's still a dramatically broader way of describing writing systems than earlier type technologies. TrueType and PostScript limited a digital font file to 256 characters, or 8-bit encoding. Unicode's 16-bit encoding, however, allows it to hold 64,000 characters—enough space to provide a unique code for every character in most of the world's written languages, including those that use ideographs, pictographs, or lots of diacritical marks.

OpenType takes advantage of Unicode to apply glyphs to the many different characters available in different writing systems. On most systems, you can access Unicode characters with key combinations on the numeric keypad of your keyboard. The key combination for the Cyrillic letter Э, for example, in Windows, is 042D, Alt+X. The key combinations aren't always easy to remember, but fortunately many programs provide a dialog box from which to insert the appropriate character or find its key combination. You may also add language specific keyboards by adding them from system preferences.

ASCII, RTF, AND XML

Page layout programs don't make very good word processors, and not everyone knows how to use them. So the text for most documents is written and edited in word processors first (sometimes by several different authors) and then dropped or *placed* in a page layout program for design.

This process can lead to some complications, primarily because of the hidden coding in many word processing files. So you should know something about the three most common file formats in which text will likely arrive in your inbox: Microsoft Word files, Rich Text Format files, and American Standard Code for Information Interchange (ASCII) files. You can recognize these file types by their filename extensions:

- Microsoft Word: .docx

- Rich Text Format: .rtf

- ASCII: .txt

Microsoft Word files are based on XML (Extensible Markup Language), a behind-the-scenes coding language that describes the structure and appearance of screens and pages. Unfortunately, Microsoft's implementation of XML is idiosyncratic, so it's often best to translate DOCX files into a simpler format like RTF or ASCII before placing them in a page layout program. In most programs, you can do so by choosing Save As from the File menu, specifying the desired format, and saving the file. This strips out all the hidden XML code.

Often called just "text" or "plain text," ASCII (.txt) boils text down to the basic numbers, punctuation, and letters used in English, leaving behind no hidden markup that might affect later formatting applied in a page layout program. That means, however, that any text settings such as bold or italic will be stripped out of the file. You'll have to reinsert those settings manually in the page layout program.

RTF is a compromise between ASCII and DOCX/XML files. It incorporates all of the basic ASCII characters but retains basic formatting elements such as typefaces, fonts, bold, and italic.

TYPE ON THE SCREEN

The easy implementation of color and the flexibility of on-screen type can make design for the screen simpler in some ways than design for print, increasing our options for typographic design. But type on the screen also has some significant limitations. In this section, we'll discuss those limitations and some strategies for addressing them.

RESOLUTION

Screen resolution refers to the size of pixels computers can turn on or off to show images on a computer screen. The higher the resolution, the smaller the dots and the smoother and clearer the type and other images are likely to look to a viewer. The lower the resolution, the larger the dots, making it harder for readers to make out intricate shapes like type.

As we discussed in chapter 4, computer screens have a much lower resolution than most printers do. The pixels on most computer screens are relatively large—about 100 ppi, although this can vary considerably from system to system and screen to screen. This compares poorly to the resolution of even cheap computer printers, which can usually print at least 300 dpi. Higher-quality laser printers can print at 600 or even 1,200 dpi (even higher, in some cases), and the printers and typesetters used in production printing are typically 2,400 dpi or higher.

The low resolution of computer screens is a particular problem with the curved or diagonal portions of letter forms. Because the rectangular pixels lie in a grid, horizontal and vertical lines render clearly, with good contrast and smooth edges. But curved and diagonal lines don't line up with the rectangles of the pixel grid, so they end up looking jagged.

Two primary technical innovations have helped improve this situation: antialiasing and screen typefaces.

ANTIALIASING

As the number of grays and colors that could be used on screens increased, digital type standards began to use *antialiasing* to smooth out these rough edges. Antialiasing creates a halo of gradually lightening gray pixels around the black pixels of the type. TrueType offered the further refinement of *font smoothing*, which antialiases only the curved and angled portions of letter forms, leaving vertical and horizontal strokes untouched because they already align with the pixel grid. However, both of these techniques can leave the edges of letter forms looking somewhat fuzzy and indistinct, decreasing legibility and readability.

More recently, OpenType has incorporated a subpixel-level smoothing technique called ClearType. Previously, pixels could be turned only on or off. But color screen pixels are actually created with three sub-pixels: one red, one green, and one blue. (See chapter 8 for more information on how screens make color.) ClearType can turn on two subpixels—for example, the red subpixel of one pixel and the blue subpixel of a neighboring pixel—to draw letter forms that extend *across* the rectangular grid of pixels. In effect, this creates antialiasing on a scale one-third the natural resolution of the screen. This technique reduces, but cannot entirely eliminate, the haziness of antialiasing.

SCREEN TYPEFACES

One of the biggest challenges of type technology is making what we see on the screen similar to what we see when we print out a document. In the early days of desktop publishing, designers had to use entirely different typefaces for on-screen and on-paper designs, often called *screen fonts* and *printer fonts*. PostScript still includes separate font files for screen and printers. Later, TrueType allowed designers to create typefaces in single files that work well on-screen and in print. Unfortunately, sometimes a single typeface doesn't work ideally in either medium—and often it's the screen version that suffers most. This became a significant problem as we began to design more and more screen documents, such as web pages.

In response to this dilemma, some typefaces, known as *screen typefaces*, are specifically designed to be used on-screen rather than in print. Common screen typefaces include the following:

- Georgia: abcdefghijklmnopqrstuvwxyz1234567890

- Arial: abcdefghijklmnopqrstuvwxyz1234567890

- Tahoma: abcdefghijklmnopqrstuvwxyz1234567890

- Verdana: abcdefghijklmnopqrstuvwxyz1234567890

- Trebuchet MS: abcdefghijklmnopqrstuvwxyz1234567890

These typefaces render well at a variety of screen resolutions with minimal jaggedness. You'll notice that most of them are sans serif typefaces, which have less ornamentation and therefore greater clarity on-screen.

FONT AVAILABILITY

In addition to the disadvantages just discussed, other problems can arise when a font file isn't available on the computer on which a digital document is viewed. Say you design a website using a typeface from a font file on your computer. It looks great on your screen. But if other users don't have the same font file on their computers, they won't see the design you intended. Three common workarounds try to solve this problem: font substitution, font embedding, and font images.

FONT SUBSTITUTION

If a computer doesn't have a font file the design calls for, the computer will have to replace it with either a default typeface (usually Times New Roman or Arial) or the best available matching typeface. This technique, known as *font substitution*, can often mess up a screen document's design.

To avoid this problem, most web designers stick primarily with common typefaces, including Times New Roman, Arial, Tahoma, and Verdana. But not all of these typefaces are available to Apple computers, which typically substitute Times for Times New Roman and Helvetica for Arial, Tahoma, and Verdana.

FONT EMBEDDING

Another solution to the limitation of font file availability is *font embedding*. This technique actually includes the original font file specifications in the electronic document file itself, making the document work as designed regardless of whatever font files the user's computer might have.

However, font embedding isn't available for all electronic document types. Adobe Acrobat can embed most kinds of fonts in PDFs, and Microsoft Word can embed TrueType fonts in DOCX files. In some cases, you may also embed fonts on your website by either hosting the font file on your site and using CSS to direct to it, or by using something like Google Fonts. There can be disadvantages to this approach, however, in that font embedding can significantly increase the size of the digital file, which now has to include the information for the embedded font files as well as the text, images, and layout information it would normally include. Some browsers also may not support particular font types—TrueType and OpenType, along with Web Open Font Format 1.0, are generally supported on more browsers than Web Open Font Format 2.0, or Scalable Vector Graphics-based faces (which are newer versions of OpenType).

A final but very important disadvantage to font embedding is that not all type designers allow their font files to be embedded. Although type *designs* cannot be copyrighted (you can generally create a new typeface that is visually similar to an existing one without infringing on copyright as they are typically not deemed as having achieved a high enough level of creative authorship; see copyright.gov, 2024), a computer *font file* is a piece of software protected by a copyright. Embedding a font in a document (or worse, distributing the font file directly to users) can be a copyright violation, since it publishes a new copy of the font over which the copyright owner has no control. Be sure to read and follow the terms of use of any font before you decide to embed it in a document.

FONT IMAGES

Finally, you can work entirely around the font file availability problem by creating image files on your computer that use exactly the typeface you want and then save them typically as GIF image files (or less commonly as JPGs or PNGs). You can then insert the image files into your on-screen design. The type in the image will look pretty much the same on your screen and on the user's. Designers sometimes use this technique for the navigation buttons on websites.

This approach has a couple of disadvantages though. First, it increases bandwidth requirements; images are entirely separate files that must be loaded into a web page, while type can be part of the HTML coding itself. Second, it can cause accessibility problems for users that rely on screen-reader technologies. HTML-coded text can easily be read by the text-to-speech software that many users rely on to read web pages to them. A font image, however, doesn't really have any text in it; the file includes only visual information, not textual information. One solution to this difficulty is to describe each font image within an HTML alt (alternative text) attribute, like this:

<img src="home_button.gif" alt="Link to Home Page">.

Overall, however, use font images sparingly, when the on-screen visual design really needs to look the same on your computer and on those of your users.

EXERCISES

Now that you've read this chapter, you likely know a lot more about type than when you started! You've worked through typographic terms and general history, and thought about how culture shapes our perceptions of type. You've thought about how type shapes enable use, and then thought about various typographical systems and how we think about type in different design spaces like physical pages and screens. Now it's time to practice what you learned. The following exercises should help.

1. Take an old copy of your favorite magazine and snip out a sample of each different typeface you find. Tape or glue these samples onto a sheet (or several sheets) of paper and bring them to class. Compare your samples with those of your classmates and discuss the following questions:

    o Which typefaces do you think would work particularly well for reading text, headings, or display?

    o What rhetorical "voice-over" or persona do you think these typefaces convey? Are they businesslike, fun, dramatic, playful, stern?

    o Which typefaces do you find most interesting to look at? Using the vocabulary you've developed from reading this chapter, what exactly do you like, in particular, about these typefaces?

2. Explore the typefaces available in one of the many design spaces now available. Google Fonts and Adobe Fonts are good starting places, but there are many, many others. Find at least four different faces you've never seen before, then discuss with your class what makes them stand out to you and where you might use them.

3. Type this phrase in a word processing program in 24-point type, copy it, and paste it three extra times: *The quick brown fox jumps over the lazy dog.* Then change each of the four versions to four different typefaces—two that look very similar and two that look very different. What specific differences do you see in the letter forms? Look in particular at the serifs (or lack of serifs), unusual letters (like *a*, *g*, and *q*), stroke width, x-height:cap-height ratio, stress, counter shape, and other features covered in this chapter.

4. Analyze the typographic hierarchy of a document. How did the designers use typographic characteristics such as size, shape, value, and color to create this hierarchy? Do you see any potential problems in the design of the typographic hierarchy in terms of how well it reinforces the structure of the document?

5. Analyze the typographic design of a print or electronic document for its legibility, readability, and usability. How well

does the document's typography match its purpose and the users' needs and expectations? What changes or improvements would you make if you were to redesign the document?

6. Redesign a page of the document you just analyzed, changing the typographic design to better fit the document's mission and intended use. Be ready to explain and justify your design choices.

7. Explore the typographic identity guidelines of an organization. To find a suitable organization, search for "identity guidelines" (in quotation marks) in a search engine such as Google. What aspects of typography does the organization specify that its employees must use? How do the typographic guidelines differ for screen and paper documents? Why do you think the organization is making such a big effort for typographic consistency in its documents? What explanations does the organization give for its efforts in creating a stable typographic identity?

REFERENCES AND FURTHER READING

Arah, T. (2003). *Web type expert: All that you need to create fantastic web types*. Friedman/Fairfax.

Brumberger, E. R. (2003). The rhetoric of typography: The persona of typeface and text. *Technical Communication, 50*(2), 206–223.

Carson, D. (2003). *Design and discovery*. TED Talks, https://www.ted.com/talks/david_carson_design_and_discovery/transcript?language=en

Chen, C. J., Wang, Y. W., & Fang, W. P. (2014). A study on captcha recognition. *Proceedings: 2014 10th International Conference on Intelligent Information Hiding and Multimedia Signal Processing, IIH-MSP 2014*, 395–398. https://doi.org/10.1109/IIH-MSP.2014.105

Connare, V. (2017, Mar. 28). How we made the typeface Comic Sans (interviews by Ben Beaumont-Thomas). *The Guardian*. https://www.theguardian.com/artanddesign/2017/mar/28/how-we-made-font-comic-sans-typography?CMP=share_btn_url

Copyright.gov. (2024). What visual and graphic artists should know about copyright. https://www.copyright.gov/engage/visual-artists/

Garfield, S. (2012). *Just my type: A book about fonts*. Avery.

Heller, S., & Meggs, P. B. (Eds.). (2001). *Texts on type: Critical writings on typography*. Allworth.

Hinds, A. (2023). Underline your links. *Accessibility Weekly, The Admin Bar*. https://theadminbar.com/accessibility-weekly/underline-your-links/

Keyes, E. (1993). Typography, color, and information structure. *Technical Communication, 40*(4), 638–654.

Larson, K. (2004, July). The science of word recognition. *Microsoft Typography*, 1–14. http://www.microsoft.com/typography/ctfonts/WordRecognition.aspx

Lupton, E. (2004). *Thinking with type: A critical guide for designers, writers, editors, & students*. Princeton Architectural Press.

Lupton, E., & Miller, J. A. (2016). Period styles: A punctuated history. In M. Goldthwaite (Ed.), *The little Norton reader: 50 essays from the first 50 years* (1st edition, pp. 315–322). Norton.

Mackiewicz, J. (2006). Audience perceptions of fonts in projected PowerPoint text slides. *Technical Communication, 54*(3), 295–307.

McLean, R. (1980). *The Thames & Hudson manual of typography*. Thames and Hudson.

Swanson, G. (Ed.). (2000). *Graphic design and reading: Explorations of an uneasy relationship*. Allworth Press.

Velasco, C., & Spence, C. (2018). The role of typeface in packaging design. *Multisensory Packaging: Designing New Product Experiences*, 79–101. https://doi.org/10.1007/978-3-319-94977-2_4

Wang, J., Qin, J., Xiang, X., Tan, Y., & Pan, N. (2019). CAPTCHA recognition based on deep convolutional neural network. *Mathematical Biosciences and Engineering, 16*(5), 5851–5861. https://doi.org/10.3934/mbe.2019292

Warde, B. (1956). The crystal goblet, or printing should be invisible. In B. Warde, *The crystal goblet: Sixteen essays on typography* (pp. 11–17).

Williams, R. (2015). *The non-designers design book* (4th edition). Peachpit Press.

World Wide Web Consortium. (2024). *Making the web work*. https://www.w3.org/

World Wide Web Consortium (W3C). (n.d.). *Cascading style sheets*. https://www.w3.org/Style/CSS/

Wyatt, C. S., & DeVoss, D. N. (Eds.). (2017). *Type matters: The rhetoricity of letterforms*. Parlor Press.

7

GRAPHICS

LEARNING OBJECTIVES

Upon completing this chapter, you will be able to:

- Discuss graphics in terms of perception, culture, and rhetoric

- Critique graphics for good human factors

- Read graphics critically and recognize when they are misleading

- Understand different image file types

- Describe the differences between vector images and bitmap images

Graphics are a powerful way to convey information and ideas, and thus are an essential part of any document designer's toolkit. Graphics are also easier to create today than ever before, leading to a somewhat different approach to incorporating them in documents. Years ago, designers composed the text of a document and *then* decided what graphics they needed. Today, most designers first consider what must be conveyed visually and then write the text to support those images—if text is necessary at all. In other words, designers now work in what Robert E. Horn (2002) called "visual language": using graphics either without prose or as a full partner with prose. And while in this chapter we focus primarily on "graphics" as they relate to non-typographic elements, don't forget that type itself is graphic, as we discussed in chapter 6. The very shapes of words and letters convey messages beyond the sounds we typically associate with them.

Graphics have two main functions in document designs: conveying information and influencing users through visual rhetoric. Some graphics lean more toward one function than the other, but most graphics play both roles in some way. Consider the graphic in figure 7.1, which appeared on the packaging for a microbrewed beer. This graphic's primary purpose is to convey useful information. If this beer were a mass-market variety, consumers might know what to expect from it. But consumers will likely want to know how this unusual, small-batch beer tastes before they buy it. The graphic, therefore, characterizes the taste of the beer on two scales: light vs. dark and malty vs. hoppy (sweet vs. bitter). Although the graphic uses some text, it delivers this essential information primarily through visual communication, providing *X*'s to show where this beer falls on the two scales. The light/dark scale even reinforces this information by using a bar that gradually shifts from light to dark along its length. As flavors, "malty" and "hoppy" are pretty difficult to describe, but the kind of people who'd buy such an esoteric beer might already know what these terms mean and can imagine the taste that this graphic shows visually.

Figure 7.1. Conceptual graphics on microbrewed beer packaging. Notice how the graphic conveys information about the taste of the beer on two scales: from light to dark and from malty to hoppy. Also notice the visual style of the graphic. Like an informal sketch, the graphic fits both the product image and the user's expectations. *Source*: Flying Dog Brewery, "Flying Dog IPA."

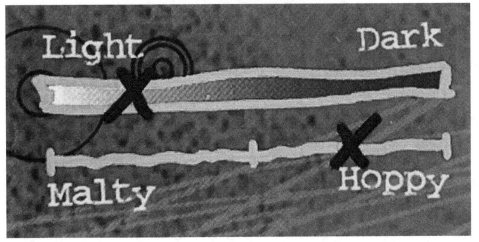

Graphics also set a visual tone—the casual attitude and playfulness of this image set by the slightly off-kilter appearance, hand-drawn wobbliness of the lines, and hand-marked *X*'s implies that this beer is not some stodgy, snobby beer. There's another aspect of this image that's important as well—it complements the rest of Flying Dog's brand imagery, which uses art and design by renowned illustrator Ralph Steadman. Steadman is perhaps most famous for the work he did with Hunter S. Thompson in *Fear and Loathing in Las Vegas* and is a prolific artist and caricaturist. Those recognizing his work and the style of his work then associate Flying Dog beers with the gonzo attitude imbued in Steadman's work, further building a brand identity through cultural rhetoric (see chapter 3).

This visual style and conscious choice of designers makes it clear that the information provided through the graphic should help you enjoy the beer, rather than read statistics about it. The graphic doesn't tell us precisely *how* hoppy or *how* light the beer is—just that it generally tends toward lightness and hoppiness. Brewers might need precise measurements for color and bitterness, but most beer drinkers don't care about that level of precision. Overall, the visual rhetoric of the graphic reinforces a casual, yet stylish approach to discriminating beer drinking.

In this chapter, we discuss a variety of techniques for creating and using graphics in document designs, in terms of both conveying information and presenting effective visual rhetoric. We'll use the term *graphics* generally to describe design objects intended to *show* ideas or information rather than to *tell*, which is the primary function of prose. However, keep in mind that graphics can also include text to help users understand what a graphic means.

Graphics are often intimidating to new document designers, who might be more familiar with working with prose. But most of us are already experienced consumers of graphics, even if we haven't created very many yet. This chapter will build on your own experiences to give you a basic introduction to the most common forms of graphics, while discussing the rationale for using graphics in your document designs and the ethics and practicalities of doing so.

THREE PERSPECTIVES ON GRAPHICS

As users and consumers of graphics, we tend to take them for granted. A picture *is* sometimes worth a thousand words, but sometimes it takes more than a thousand words to explain why a picture works as it does. Scholars have approached this challenge through the perspectives of visual perception, visual culture, and visual rhetoric.

PERCEPTION

Scholars interested in the perceptual aspect of how we respond to graphics tend to work on the assumption that images are made up of universally recognizable visual elements that require active perception but no particular training to understand. For example, scholars suggest that we interact with visual images cognitively by assembling their basic features into complex shapes, recognizing graphics first by their component parts—Irving Biederman's concept of the *geon* (1987). Geon theory suggests that we perceive all objects as made up of abstract, three-dimensional (3D) objects. Geons are mostly geometrical, such as cylinders, cones, blocks, and wedges. We assemble these basic shapes to recognize more complex shapes—a tree being a collection of branching cylinders or long, narrow cones, for example. Once we recognize the basic building blocks and their relationship to each other in a complex object, we can recognize the object from many viewpoints—below, above, beside, near, or distant.

According to geon theory and other theories based on cognitive psychology, the ability to turn basic visual perceptions into the recognition of an image is more or less inherent in the human visual system, supported by both our physiology and psychology. In the case of geons, we don't need to learn anything to recognize these primitive objects or see how they assemble into complex objects; we can do it from birth. Our mind simply assembles complex objects from the bottom up, starting with the basic building blocks of perception.

CULTURE

Approaches from this bottom-up perspective, however, tend to underestimate the power of training and culture on what we see. This is true particularly in the consideration of graphic images because graphics are by definition *created* by people for other people to see. Although these created images might use our inherent perceptual abilities, they are tied just as strongly to social and cultural factors—particularly in terms of how we ascribe meaning to images. We learn how to read images through experience and association, not just through raw perception.

Consider the cover of a leaflet guiding visitors to the Helen and Peter Bing Children's Garden at the Huntington Library, Art Collections, and Botanical Gardens in San Marino, California (figure 7.2). The title of this leaflet uses two images: a fern and a hummingbird. If we consider these images only perceptually, they show us nothing other than a plant and a bird. But our training, education, and experience help us recognize

Figure 7.2. Culture and experience influence how we see graphics. Most users will recognize the graphics of a fern leaf and a hummingbird in the title of this leaflet not only as representative images but also as a stylized letter *C* and apostrophe. *Source*: The Huntington Library, Art Collections, and Botanical Gardens, "The Helen & Peter Bing Children's Garden at The Huntington."

them in this context as a stylized *C* and apostrophe. Undeniably, cultural values and expectations play a role in our interpretations of what we see. Without visual culture, users would have a difficult time creating meaning from images.

RHETORIC

This relationship between perception and culture suggests that we should think about both factors in our designs, taking into account our inherent ability to see images and the cultural values users might apply to those images.

Consider another example from the Huntington, a general "Information Guide" leaflet distributed to visitors (see figure 7.3).

Figure 7.3. Graphics help form a visual rhetoric for documents. This leaflet uses a broad variety of graphics to present information to visitors, including photos, maps, visual and textual callouts, pictograms, and a legend. These graphics also give users a positive impression of the site, allowing them to build an attractive mental picture as they begin their visit. *Source*: The Huntington Library, Art Collections, and Botanical Gardens, "Information Guide."

Side A

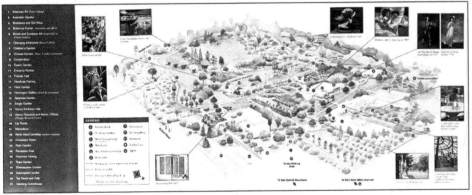

Side B

Detail: Legend with pictograms

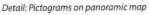

Detail: Pictograms on panoramic map

Detail: Visual callouts with captions

The leaflet presents a lot of information about the Huntington using a consistent visual rhetoric that encourages users to look forward to their perambulations through the grounds and galleries. The leaflet uses several graphic forms both to convey information and to present a visually coherent view of the multifaceted institution:

- *Decorative photographs* of many attractive aspects of the collections and gardens, including photographs of the founder, paintings, sculptures, galleries, buildings, flowers, cacti, people walking in gardens, and children playing in fountains.

- A *panoramic map* of the grounds, showing the various facilities, paths, roads, and parking within the landscape.

- *Visual and textual callouts* from the panoramic map, showing visitors what they can expect to find and see in different locations.

- *Pictograms* marking useful facilities on the panoramic map, such as information kiosks, telephones, water fountains, restrooms, and access for the disabled.

- A *legend* identifying what the pictograms mean.

These graphics provide essential information to users about the Huntington. The leaflet's visual images also present a consistent visual rhetoric that encourages people to think that this is a nice place for individuals or families to visit, with many things to see, all arranged for easy access and enjoyment.

Because the Huntington is a private institution, it relies on visitors' fees and donations to remain open and to support the work of many researchers who use its extensive library. That means this leaflet carries a persuasive burden as well. It must convince visitors that they're welcome at the Huntington and that it's a friendly, comfortable, and beautiful place to visit—well worth their financial support. So the graphics, and in turn the whole document, must be presented in a very high-level design: beautiful, clear, and useful. Without graphics, this leaflet could scarcely do as good a job of both informing and persuading.

GRAPHICS AND PRINCIPLES OF DESIGN

One of the most challenging aspects of communicating through graphics is to present a clear and accurate idea of the information or ideas. But what exactly is clarity, and how do we design graphics with this quality?

As always, the first consideration is the user and their situation, needs, and limitations. The better you know your user, the more likely you will create graphics that meet their needs successfully. Ideally, the decision to use graphics at all should arise from your analysis of users and their needs (see chapter 9 for more information on researching users).

But once you've decided a graphic is necessary, your best guide is the principles of design we've outlined as SEACOP:

- Similarity

- Enclosure

- Alignment

- Contrast

- Order

- Proximity

As you'll recall, document design is object-oriented; a good designer pays attention to the qualities of various visual objects on the page and then uses principles of design to show clear relationships between those objects.

The simple bar chart in figure 7.4 uses all of these principles to show increasing enrollment at Auburn University between 2012 and 2022. The

Figure 7.4. How a bar chart uses the principles of design. This bar chart uses the principles of design to imply relationships between various design objects. *Source*: Auburn University.

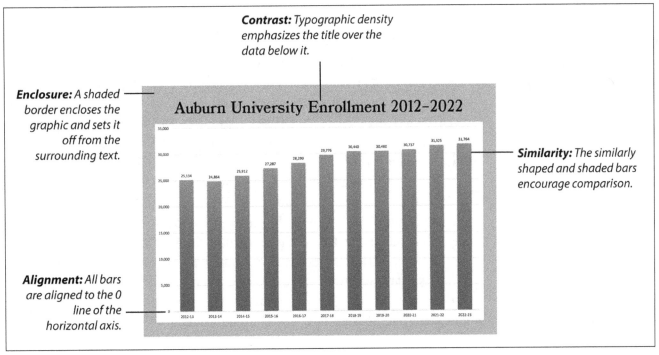

graphic is *enclosed* in a border to set it off from any surrounding text. It uses *alignment* within the framework of an *x-y* coordinate grid: the *x*-axis shows different years, and the *y*-axis shows enrollment numbers. The bars are arranged in *order* of years. *Proximity* helps users associate labels with the appropriate design objects within the graphic. How the graphic is presented is as important to the narrative as the information—think back to the first example in this chapter, a design drawn using Ralph Steadman's style. Would this chart convey the same tone if it were presented in that style?

In all your work designing graphics, try to keep in mind how principles of design can help you tell a story, reveal a truth, or show a good way to do something. Graphics are a powerful tool for communicating complex information and ideas.

GRAPHICS AND ETHICS

It's often said that people believe what they can *see.* But as a result, they're also sometimes less skeptical about what they see than about what they hear. That tendency makes it particularly easy to mislead users with graphics, whether intentionally or unintentionally. As a designer, you'll need to be cautious about and aware of the ethical implications of what you show users through graphics. Three important ethical aspects of using graphics involve distortion, viewpoint, and copyright.

DISTORTION

Mark Twain was said once to comment that there are three kinds of lies: "lies, damned lies, and statistics." Statistics themselves are easy to manipulate to fit many arguments and ideas. When expressed graphically, statistics are even more prone to distortion. Consider what is perhaps the most commonly distorted statistical graphic: the line graph. The graph in figure 7.5 accompanied a story about a survey of British opinions of the then-new European Union currency, the euro. In 2002, Great Britain was involved in considerable debate about whether to join the European Union in its new unified currency. The line graph on the left shows the exchange rate between the euro (€) and the British pound (£). It suggests that the value of the euro had dropped dramatically compared to the pound. The combined line graphs on the right show the results of the survey questions "Have you used the new euro notes and coins yet?" and "Do you think it would be a good idea to join the euro in

Figure 7.5. A distorted bar graph. This newspaper graphic visualized British opinions during the controversy over whether the United Kingdom should take part in the European Union's new currency, the euro. What distortions do you see in this graphic? *Source:* Created by author from *The Times*, "Anti-Euro mood hardens among Britons," July 31, 2002.

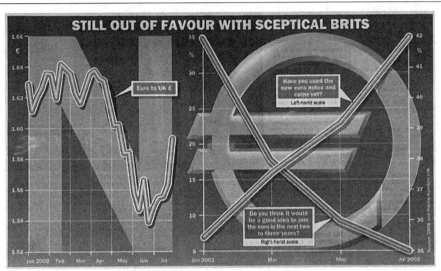

Original graphic, showing euro-GBP exchange rate on the left and responses to survey questions on the right ("Have you used the new euro notes and coins yet?" and "Do you think it would be a good idea to join the euro…?").

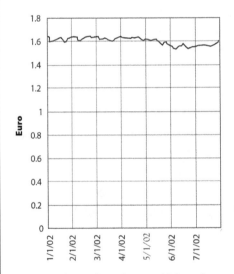

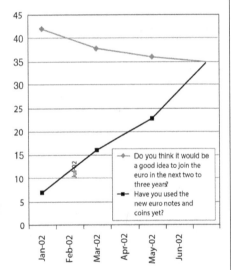

Revised graphic of euro-GBP exchange rate using a normal zero point. Notice that the dips in the exchange rate appear much less drastic now.

Revised graphic of survey responses using the same zero point for both graphs. Notice that the graph no longer forms the "X" of the original graph, which helped support the idea that the euro was being received negatively.

the next two to three years?" Taking advantage of the shape of the line on the left-hand graph, the designers placed an *N* behind the left-hand graph and a euro symbol (€) behind the right-hand graph. "N€" seems to form the word "NO" behind the combined graphic; the open ends of the euro symbol are even joined to make it look even more like an *O*. The lines of the right-hand graph also form an × over the euro symbol. Together, these visual cues emphasized the idea that Britons were strongly rejecting the currency.

But notice that on the left-hand graph, the scale for the vertical axis starts at 1.52 rather than at 0. Compare this to what a version of the graphic would look like if the vertical axis started at 0, as in the first redrawn graph in figure 7.5. What's going on here? In the original graph, the line seems to show a significant drop in the value of the euro, but in the redrawn version, the drop seems much less drastic, the value of the euro holding more or less steady at around £1.60. This distortion is called an *elevated y-axis*. By starting the vertical axis above zero, the graph exaggerates the changes in currency exchange rates.

The right-hand graph in figure 7.5 shows what we mean. The two lines of the graph are marked against different vertical scales—one on the left, one on the right—neither of which starts at zero. The second redrawn graph in figure 7.5 shows what the graph would look like if the lines shared a common scale starting at zero. This redrawn version seems to suggest that although confidence in the euro had dropped, the number of people using the new currency increased. This story differs significantly from the story told by the original graphic, which suggests that people strongly rejected the euro.

These distortions are remarkably easy to make, particularly with the statistical graphics tools commonly available in office software. Often, distortions are unintentional; the designer made a misleading graphic out of ignorance or inattention. Just as often, however, distortions arise from a desire to make data say something it really doesn't. Figure 7.5 tells a dramatic story of Britons rejecting the euro, but it does so only by manipulating the visual display until the line graphs form "No" and × over this issue.

Another common distortion derives from an innocent desire to make data look fancy. Microsoft Office products, among others, allow you to apply a 3D look to information graphics, even if the data doesn't really need that many dimensions. In figure 7.6, a pie chart has been made 3D, even though the third dimension doesn't add any information to the graphic. This use of perspective distorts users' perceptions of the data because the pie wedges nearest the viewer can seem unduly large, while those farthest away can seem smaller than they really are.

Figure 7.6. A pie chart in 3D and 2D. Meaningless 3D effects in graphics can distort user perceptions and make it difficult for the user to compare segments. Those sections closest to the viewer can appear even larger than intended, those farther away smaller. *Source*: Created by the author.

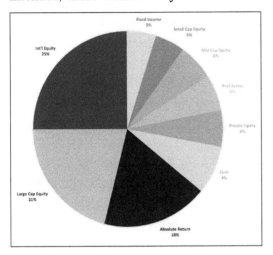 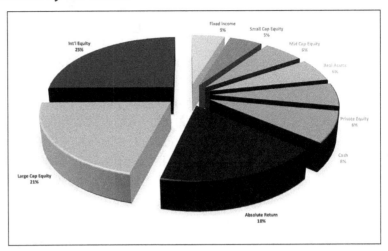

Nor do we find distortion only in abstract images; distortion can appear just as easily in representational images. Through the wonders of Photoshop and similar image editors, it's very easy to alter photographs to make something appear to have happened that never did—and artificial intelligence (AI) image creators have only made such alterations even easier. Image alteration is nothing new, however. For example, in 2000, the University of Wisconsin photoshopped the face of Diallo Shabazz, a Black man, into an all-white crowd of students at a football game on the cover of their application packet in order to make the school look more diverse than it really was. Diallo was surprised—he had never been to one of their games (see Pritchep, 2013). This case has become a standard example of how organizations use "fauxtography" (fraudulent photography) to market an image that differs from reality.

> Design graphics that portray the truth as accurately as possible, without bias.

Scholars such as Edward R. Tufte (1983), Stephen M. Kosslyn (1994), and Nigel Holmes (1984) have cataloged many of the most common distortions, particularly of information graphics. But the most significant point is to design graphics that portray the truth as accurately as possible, without bias.

VIEWPOINT

Graphics inevitably show the world from a particular viewpoint or perspective. Unfortunately, this means that they can conceal as much as they reveal, through either abstraction or outright exclusion. For example, Sam Dragga and Dan Voss (2003) have criticized the tendency of graphics in federal accident reports to exclude figures of human beings. Diagrams and statistical graphics in these reports use abstraction to exclude any reference to humans, using geometric lines and shapes to describe the accidents or to measure human pain and loss. Similarly, photographs tend to focus on vehicles, facilities, roads, and equipment, leaving human beings outside of the frame. This practice might be intended to avoid alarming images, but it has the effect of making the accidents seem less significant than they are.

It's particularly easy to use the viewpoint of photography to conceal important information. A photograph asks users to look at whatever is pictured from a particular perspective. That perspective might easily hide aspects of the scene that are unattractive or counter to the client's agendas. One photographic technique particularly sensitive to concealment is *cropping*. Cropping a photograph to exclude important information and mislead users can be unethical (figure 7.7).

Of course, viewpoint can also be a great guide to users, showing them how to look at information from a useful perspective. But make sure that the viewpoint you use in your graphics is factual and honest, not concealing or excluding important information.

Figure 7.7. Cropping an image to conceal information. Cropping can leave out the whole truth. The cropped version of this image might make a very appealing picture of an old-fashioned home on a real estate website. But the uncropped version shows large construction cranes and semitrailers behind the house. Would you buy the house if you saw the untouched photograph? *Source*: Created by the author.

COPYRIGHT

A common first impulse when using graphics in documents is to borrow someone else's graphics, rather than to create your own. The existing graphics built into many word processing and design programs are often called *clip art*, and today most of these programs come with at least a small collection of clip art, such as cartoons, drawings, and photographs—others are linked to entire searchable, metadata-tagged collections. Designers are also sometimes tempted to use graphics copied from websites or scanned from previously printed documents. And, of course, as we've briefly mentioned already, the growth of AI image generators adds yet another layer of complexity to the equation.

Using borrowed graphics isn't necessarily wrong. Modifying or combining existing graphics can be a good way to start using graphics in your document designs—and it can save a lot of time. But like all created works, graphics are protected under *copyright*. Copyright is a legal status in which the creator or owner of the graphic has the right to say whether and how the graphic can be reproduced by others. This system makes sense from the creator's viewpoint: if you created a graphic, you wouldn't want someone else to use it without your permission.

Copyright is granted by US law as soon as a graphic is created. It's not necessary to register the graphic to gain copyright. Under current US law, simply creating the graphic (or text, for that matter) gives you the copyright for your lifetime (and your heirs for decades after that). Copyright does expire, typically a century or more after the creation date, but it can be renewed. With AI-generated images it's not really all that more complicated . . . yet. If a human didn't generate the image, it's not owned by a human. At the time of this writing, as described by the Copyright Alliance (2024), "the U.S. Copyright Office will refuse to register any work that is solely generated by AI. A federal district court, in upholding the Office's decision to refuse registration for such a work, affirmed these principles. In other words, a work generated by the AI is in the public domain." If you add to AI-generated material, the bits *you* create are copyrighted, but not the bits created by AI—anyone can use those, anytime. *Public domain* means just that—we're all free to use it, as materials in the public domain are unprotected by copyright, trademark, or patent. Great for a quick fix, but maybe not a great idea if you're trying to create the next major international brand. Issues regarding copyright and AI are evolving at a truly remarkable rate, however. So whatever you do, whenever you do it, check the current state of copyright laws!

So, questions of generative elements aside, how do we use borrowed material within the boundaries of copyright? It's possible to use a portion of a created work under a principle called fair use. However, this concept works better for text rather than graphics. The idea is that if you quote only a small part of a text—for example, to comment on what an author said—you are not preventing the author from profiting from their text as a whole. Though various percentages of a work have been considered as fair use in the past (e.g., less than 10 percent of a complete work), a percent-based approach is no longer legally reliable. Instead, courts consider four factors in fair use:

1. The purpose and character of the use

2. The nature of the copyrighted work

3. The amount and substantiality of the portion used

4. And the effect of use on the public market

These factors ask those who would use another's images, words, music, and so on, to think about *why* they want to use another's work; whether that work is common knowledge or factual vs. creative; whether the amount used of a work constitutes a portion substantial enough as to represent the complete work as a whole; and how replication of that work affects factors like economic impact for the rights holder.

Unlike text, *every single graphic* is counted as a whole work, even if the document or website that it comes from contains many graphics. And graphics are difficult to use any way except as a whole—even 10 percent of a graphic isn't often very useful, and if you are able to determine the original work from the piece copied, you've probably used enough to count as a "substantial" portion, no matter how small the piece. Accordingly, you can assume that any graphic you scan from someone else's document or copy from the internet probably requires the copyright holder's permission before you use it in your own designs. Using copyrighted graphics without permission can hold serious legal consequences; don't do it. If you have any questions at all, you may it useful to contact your local or university library. University of Maryland's Global Campus, for example, offers an extensive guide to understanding fair use, as do many others, and the US government maintains a site offering the general outline of fair use, along with links to examples, at copyright.gov.

Using copyrighted graphics without permission can hold serious legal consequences; don't do it.

Fortunately, getting the copyright holder's *permission* generally involves contacting the copyright holder, asking if you can use the graphic in your document, and explaining how you intend to use the graphic. If at all possible, get the permission in writing. Some copyright holders will give you permission for free, but others will ask for a fee. The more famous the graphic or artist, the higher the price, usually. You can also purchase more generic pieces or collections of clip art from businesses called *design bureaus*, or you can look for works that are in the public domain, but again, anything that anyone can use anytime may not be suitable for your design purposes, so think carefully about your audience, context, and purpose, as always.

Permission usually comes with a few strings attached. For example, copyright holders might restrict the number of times you can use the graphic, or they might restrict its use to nonprofit purposes. These restrictions are often expressed in a legal document called the *terms of use* or *conditions of use*; follow the requirements of these documents explicitly. Creative Commons, for example, the central site for searching for and distributing Creative Commons licensed productions, offers exceptionally clear terms of use. Lawrence Lessig and Eric Eldred created Creative Commons licensing in 2001 as a way to offer more explicit licensing options for creators and developers. Their multitiered approach lets designers choose how they want their work shared and what users of any given work are allowed to do with that work (see figure 7.8).

Figure 7.8. Creative Commons licensing types. Creative Commons licensing offers very explicit licensing tags that help designers and users understand how work may be shared and/or altered. *Source*: Creative Commons.

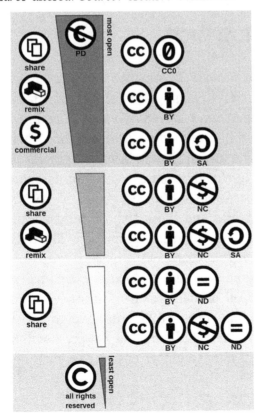

Even if it's not explicitly required in the conditions of use, always give credit to the copyright holders for their work. There are many ways to do so, depending on the document's context. In corporate documents, it's common to provide source information in a source line—a caption directly above or below the image. For published documents, you might also need to create a credits page that lists the copyright owners of all the images you used, along with the page on which each image appeared in your document. For academic documents, including student projects, provide a full citation and bibliographic entry.

Copyright law can be ambiguous, so err on the side of caution. If you have any doubts about the legality of using a borrowed graphic, look for a firmer alternative.

Design Tip: Copyright Checklist

- Find out who owns the copyright to the graphic you want to borrow.

- Obtain the copyright owner's *written* permission to use the graphic.

- Read and follow the terms or conditions of use.

- Give appropriate credit through citations (footnotes, endnotes, image credits, source lines).

WHY USE GRAPHICS?

Graphics can both provide information and encourage people to act or think in particular ways. Most graphics, however, fall toward one or the other end of the spectrum between *information* and *persuasion*. Robert L. Harris suggests that *information graphics* are those "whose primary function is to consolidate and display information graphically in an organized way so a viewer can readily retrieve the information and make specific and/or overall observations from it" (2000, p. 198). He distinguishes between these graphics and those "whose primary functions are artistic or for purposes of entertainment, promotion, identification, etc."—what we'll call *promotion graphics* (198).

Information graphics, such as charts, diagrams, maps, and illustrations, were developed primarily to help users see meaning in a mass of data. William Playfair (1759–1823), the inventor of the pie chart, once commented that "figures and letters may express with accuracy, but they can never represent either number or space. A map of the river Thames, or of a large town, expressed in figures, would give but a very imperfect notion of either, though they might be perfectly exact in every dimension; most men would prefer representations, though very indifferent ones, to such a mode of painting" (quoted in Costigan-Eaves & Macdonald-Ross, 1990, p. 324). As Playfair suggests, information graphics can help users recognize both quantitative information ("figures") and nonquantitative information (geography, spatial relationships, and concepts) at a single glance—often much more quickly than we could with text or numbers.

Promotion graphics, such as logos and decorative graphics, are used primarily to convince users of the ethos of a document. Logos, for example, are carefully designed to give users a particular idea of the company or organization they indicate. They are intended as a graphic shorthand for everything the company or organization represents—an approach often called *branding*. Corporations and organizations take branding very seriously, spending millions of dollars creating and protecting their logos and their corporate "look." Other promotion graphics can create a visual tone for a document. For example, opening each chapter of a manual with a repeated decorative element can make the document look sophisticated or casual, serious or playful, professional or trendy, as the rhetorical situation demands. Although promotion graphics might not convey as much information as information graphics, they help create a visual tone for the document, which can influence how users respond to it.

This difference between information and promotion graphics isn't exclusive. The *primary* purpose of information graphics is to convey information, but information graphics can also play a role in visual rhetoric. For example, information graphics can convince audiences by simplifying complex information—something that is harder to do with prose. Likewise, promotion graphics can convey useful, although generally limited information; a logo can help users quickly recognize what organization a document speaks for and where it comes from. For example, consider the logo of the US Department of Agriculture (USDA), used on the labels of certified organic foods (figure 7.9). This logo both represents the USDA and conveys important information to consumers.

Table 7.1 lists some of the most common purposes and genres (types) of information and promotion graphics. Naturally, we can't cover all the myriad kinds of graphics commonly used in document designs. (Robert L. Harris's *Information Graphics: A Comprehensive Illustrated Reference* [2000] is a particularly useful encyclopedia of the many different options for information graphic design.) But the next sections discuss each of the categories in the table, with examples and explanations of the dynamics of the most common genres.

Figure 7.9. A logo with promotional and informational purposes. The logo of the US Department of Agriculture (USDA) appears on the labels of certified organic foods. This logo has both promotional and informational purposes. It promotes the USDA and organic foods, while conveying important information about organic standards to consumers. *Source*: USDA, https://www.usda.gov.

Table 7.1. Purposes and Genres of Information and Promotion Graphics

a. Information Graphics

Purpose	Common Genres
To show how something looks	Representational illustrations, such as photographs and line art
To show how things are related in space	Maps
To show how actions or occurrences are related in time	Process diagrams, such as Gantt charts, flowcharts, and PERT charts
To show ideas and relationships visually	Concept diagrams, such as Venn diagrams, network diagrams, and organizational charts
To show quantitative relationships	Statistical graphs, such as bar/column graphs, line graphs, pie graphs, and scatter graphs
To provide a simple mark for an idea	Pictograms

b. Promotion Graphics

Purpose	Common Genres
To mark identity	Logos
To establish a visual tone	Decorative graphics

Source: Created by the author

INFORMATION GRAPHICS

ILLUSTRATIONS: SHOWING HOW SOMETHING LOOKS

Often, users simply need to know what something *looks* like. Illustrations help us to show users how objects such as buildings, tools, interfaces, or product features *appear*, as if the user were looking at the objects themselves. This feature of illustrations is known as *representation*, because the graphic temporarily stands in for the real object. Though many think of "illustration" as meaning "something done by hand," this is not technically correct. Instead, illustrations draw a viewer's attention to particular details. Four kinds of illustration are particularly common in technical documents: photographs, line art, scientific illustrations, and screenshots.

More about . . . Graphic Terminology

The terms applied to even common types of graphics vary from field to field. For example, some designers call pie charts *pie graphs* or even *circular area graphs*. More broadly, even the terms *chart* and *graph* are difficult to pin down. They're often used interchangeably, and *chart* is also a common term for sea navigation maps.

We'll try to be consistent with our terminology in this chapter, but don't be surprised if you encounter some of these other terms in reading other books.

PHOTOGRAPHS

One of the most common ways to illustrate how something looks is to photograph it—either chemically on film or digitally with a digital camera or scanner. For quick illustrations, photographs are very convenient. With little skill or forethought, you can take a digital photo and insert it into a document in just a few minutes. Photographs are often one of the first graphics that beginning designers consider adding to their document designs.

However, photographs are more complex than they might seem. To print well, photographs require careful color management and a full awareness of how production methods will affect how a photo will appear in a document. Photographs used on-screen also present technical challenges. Due to the relatively low resolution of most screens, it's difficult to get good detail in a photograph in electronic documents such as web pages or PowerPoint presentations. If the resolution is high enough, digital photographs can take up too much computer memory and network bandwidth, or they may simply be too large to show on the fixed resolution of a computer monitor.

In addition, sometimes photographs convey *too much* data to be understandable. The level of detail in photographs can distract users from their immediate task. For example, think about navigating on a road trip: Would you rather have a road map or a satellite photograph to help you get to your destination? Most people would prefer the road map because it strips out the extraneous information that doesn't help them understand how to get where they're going. Ironically, *less data* can mean *more usable information*.

Finally, photographs can represent only what the camera captures. It's difficult to use a photograph to show how something looks from different perspectives—inside, outside, top, or bottom. Getting a camera into those positions to take a picture can be difficult or impossible, but sometimes these perspectives are exactly what users need to see (for more information on using photographs, see "Digital Photography" in this chapter).

LINE ART: DIAGRAMS AND DRAWINGS

Diagrams and *drawings*, together called *line art*, can sometimes do what photography cannot. As the term suggests, line art usually employs lines or other abstract shapes to show something. It presents an abstracted version of what someone might see, stripping away all the extraneous details and concentrating the user's attention on the important details.

Line art also allows us to show users things that a camera can't access, such as the inside of an object. *Cross-sectional drawings*, for example, take an imaginary slice off of an object to show us the interior. Similarly, *cutaway drawings* peel a section of the "skin" from an object to let us see the inside. And *exploded drawings* show how the individual parts of an object fit together by separating them a short distance from each other. None of these things are easy to show with a photograph (figure 7.10).

Line art can also have disadvantages—the most significant being that drawings require more skill and time to create than photographs do. For this reason, designers often save line art for the most important points in a document and reuse or modify existing drawings as much as possible. Another common technique is to use photographs as the basis for line art. Using a drawing program, it's easy to place a photograph on one layer and then trace the important parts of the photograph onto another layer. When the photograph layer is deleted, only the line art remains.

Figure 7.10. Examples of common line art. Line art (drawings and diagrams) can come in many forms. Three common genres are cross sections, cutaways, and exploded diagrams. *Source:* Created by author from Lewis, P., Tsurumaki, M., & Lewis, D. J. Manual of section. Princeton Architectural Press, p. 83; NASA; and Cristian Enache, https://www.jcstudio.ro/.

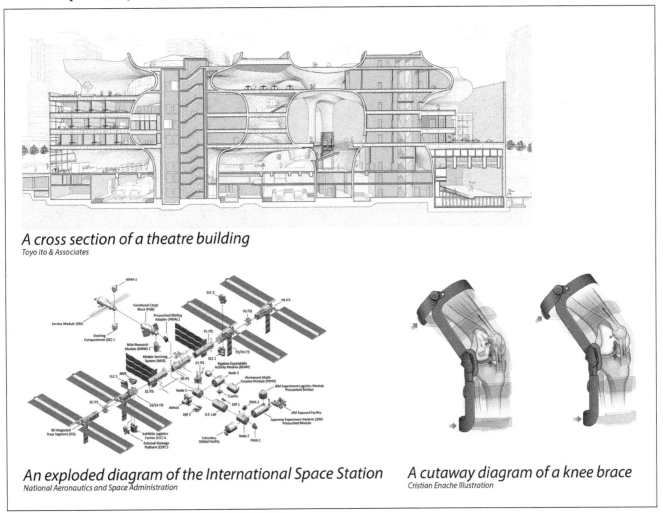

A cross section of a theatre building
Toyo Ito & Associates

An exploded diagram of the International Space Station
National Aeronautics and Space Administration

A cutaway diagram of a knee brace
Cristian Enache Illustration

SCIENTIFIC ILLUSTRATIONS

Similar to line art, *scientific illustrations* are often employed to do what photography cannot. They may include line art, drawings, and sketches, but also much more. Scientific illustrations are art in the service of science—a representative painting of a dinosaur at a museum might be much more highly detailed and realistic than line art, serving to illustrate a scientist's perspective of what an extinct species looked like, or a complex visual depiction of a virus may help scientists understand how a new vaccine might work.

Figure 7.11. Illustration of pigmentation differences between the speckled mad-tom catfish and the freckled madtom catfish. These fish are small, just under 6 inches at their largest, and difficult to photograph because of their coloration and the way light reflects off their skin. An illustration lets the users of this image more readily appreciate the pattern differences between the two. *Source*: In Ross, S. T. (2001). *The Inland Fishes of Mississippi* (D. G. Ross, Illustrator). University Press of Mississippi.

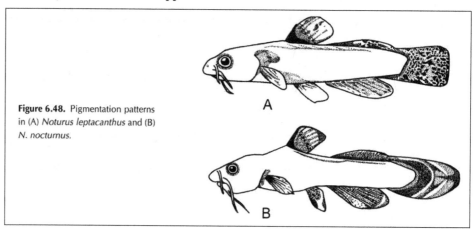

Figure 6.48. Pigmentation patterns in (A) *Noturus leptacanthus* and (B) *N. nocturnus.*

Scientific illustrations may serve to depict archetypical representations of nature that a camera cannot capture—a scientific illustrator might study several fish of a particular type, for example, then draw a single, representative specimen that would allow scientists working in the field to identify other fish of the same species (figure 7.11). The archetypical representation can avoid individual ambiguity—a broken fin, unique coloration due to environmental factors or disease, and so on—without adding "noise" (unnecessary visual clutter) to the image.

SCREENSHOTS

Because so many technical documents show users how to use software, screenshots have become very popular. Screenshots are a snapshot of the image a computer screen is showing when the screenshot was made. They are particularly useful for showing software interfaces and dialog boxes in manuals.

Screenshots are so easy to produce that they're sometimes overused. Remember, every graphic must have a purpose. Do you really need a screenshot for every single step of a procedure or only for the most important or potentially confusing steps? (For more information on creating and working with screenshots, see "Screenshots" in this chapter.)

MAPS: SHOWING HOW THINGS ARE RELATED IN SPACE

Mapmaking is one of the earliest developments of graphic communication, and an amazing one at that. Remember that for most of human history, people had never seen Earth from an aerial view. Whoever discovered that a sheet of paper, parchment, or bark could represent an abstract version of Earth's surface must have been a genius indeed.

Maps are now an essential tool of modern life, as well as a common part of documents. Maps are usually 2D representations of 3D spaces. In documents, maps are commonly used for three significant roles: to convey geographic information, to show us how to get from one place to another, and to show the geographic distribution of statistical information.

Geographic maps are perhaps the most common, showing the 2D spatial relationship between objects and places from an overhead view. Literally meaning "world writing," geographic maps can represent any portion of the world, from the layout of a single building to the entire globe. In this regard, geographic maps are particularly useful for showing *where* something or someplace is in relation to other things or places with which the user is already familiar.

Closely related to geographic maps are *wayfinding maps*, which are specifically designed to help users navigate through geography.

Wayfinding maps include road maps and facilities maps, which help us find important information such as the path out of a building in an emergency or the path to the nearest restroom. Facility wayfinding maps can be incorporated into other kinds of documents, or they can be posted in a useful position in a facility—for example, in a kiosk near the entrance or next to the elevator on each floor of the building. (Figure 7.3 shows a wayfinding map.)

Thematic maps show the geographic distribution of statistical information. There are hundreds of kinds of thematic maps, but one common kind is the *choropleth map*, which uses shading to indicate the statistical density of phenomena in a geographic region (figure 7.12). Thematic maps can be put to just about any purpose, but we see them often in depictions of risk, for example, in models of disease outbreaks, as Candice Welhausen (2024) describes in relation to yellow fever, or environmental change such as rising seas and flood projection, as described by Sonia Stephens and Dan Richards (2020) and Dan Richards and Erin Jacobson (2022).

Figure 7.12. A choropleth thematic map. This choropleth thematic map shows the total rainfall in Alabama in November 2011 by county during Hurricane Rita. The different shades show differing rainfall amounts. *Source*: National Weather Service.

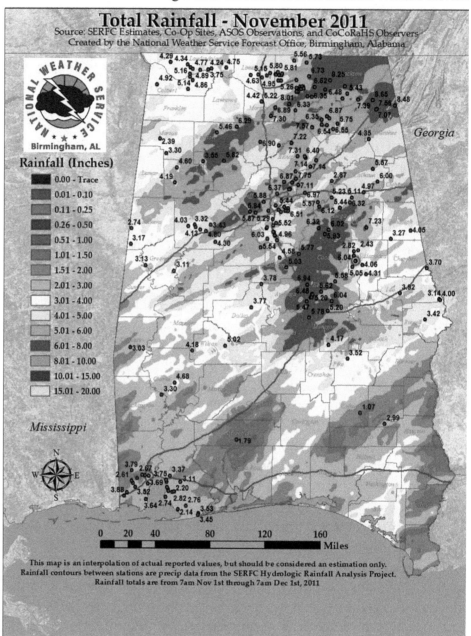

PROCESS DIAGRAMS: SHOWING HOW ACTIONS ARE RELATED IN TIME

Process diagrams focus on showing how actions or steps are related in time. They're most often useful in documents that explain how something works, such as process descriptions and procedures, or in those that show how something is *planned* to work, such as proposals and project planning documents.

The most common kind of process diagram is probably the *flowchart*. Flowcharts use a small, relatively standardized geometric vocabulary to indicate the starting and ending points of a process, as well as the steps and decision points along the way (see figure 7.13). Steps are connected

Figure 7.13. A flowchart. This flowchart uses shapes and arrows to show the steps in implementing the National Environmental Policy Act and Executive Order 12114. *Source*: NASA.

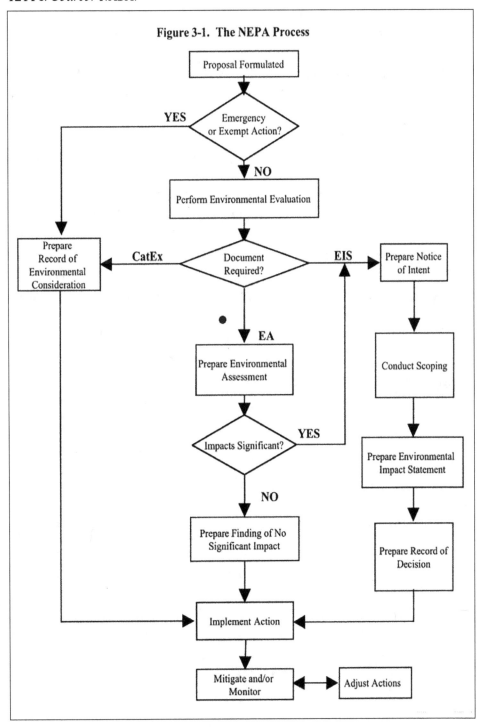

by lines, often with arrows to show the direction of time. Flowcharts are useful for showing complex procedures graphically, and they're easy to create with the drawing tools in most office software.

Gantt charts and *PERT charts* were developed in the twentieth century as graphic means for planning and tracking progress on large projects. Gantt charts were first developed by Henry Laurence Gantt (1861–1919). They present each activity in a project as a separate horizontal line, measured on a horizontal scale of time and marked by beginning and ending points. These charts make it easy to see relationships between simultaneous activities. For an example of a Gantt chart, see chapter 10, figure 4.4.

PERT (program evaluation and review technique) charts (figure 7.14) are also organized chronologically, but they do not use a consistent time scale. PERT charts are formed by simple lines joining labeled shapes. The shapes represent milestones—important points in the project timeline. Although Gantt charts are great for planning simultaneous actions, PERT charts work best for identifying the relationships between actions. They're most often used to analyze *critical paths*—essential activities that must be completed before others can commence. Critical paths are sometimes difficult to see in Gantt charts, which separate activities into separate lines rather than joining them, as do PERT charts.

Both Gantt and PERT charts are often posted in project offices, but they're also typically included in project documentation and proposals.

Figure 7.14. A PERT chart. Teams use PERT charts to organize their work on complex projects. This PERT chart shows the tasks involved in building a house. The critical path of essential tasks is represented in red in the original, lighter gray in this representation. *Source*: Archtoolbox.

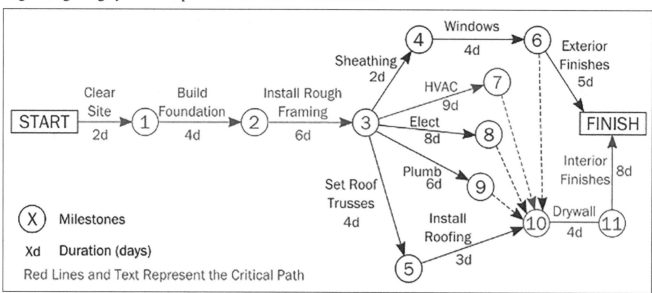

Because project plans sometimes change dynamically, a variety of software programs have been developed to help implement and automate Gantt and PERT charts.

CONCEPT DIAGRAMS: SHOWING IDEAS AND RELATIONSHIPS VISUALLY

Concept diagrams are perhaps the most abstract of diagrams because they focus on showing otherwise intangible ideas in graphic form. One common type of concept diagram is the Venn diagram, named after the Cambridge University logician John Venn (1834–1915), who first proposed them in 1880. Venn diagrams are particularly effective for conveying how overlapping groups of concepts or objects are related to each other (figure 7.15).

Network diagrams are frequently used in planning computer networks to show the logical relationships between the different computers, databases, routers, and nodes used in network design. Network diagrams use simplified icons to represent these different parts of the network and lines to show how they connect to each other. A similar conceptual diagram, the *site map*, is used to plan websites, showing how the various web pages link to one another. As websites grow increasingly dynamic, using information from databases to make web pages on the fly, site maps and network diagrams have begun to look more alike.

Figure 7.15. A Venn diagram. Venn diagrams show the relationships between overlapping categories. In this example, a path toward sustainable development is shown via overlapping environmental, social, and economic elements. *Source*: ConceptDraw.

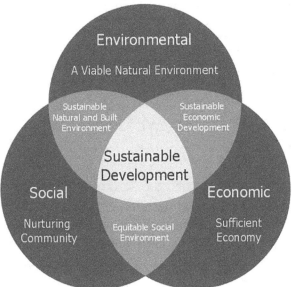

Figure 7.16. An organizational chart. This organizational chart shows the management structure of the Virginia Department of Labor and Industry. Organizational charts and others like them are known as tree-root diagrams because they show entities in a branching hierarchy. *Source*: Virginia Department of Labor and Industry.

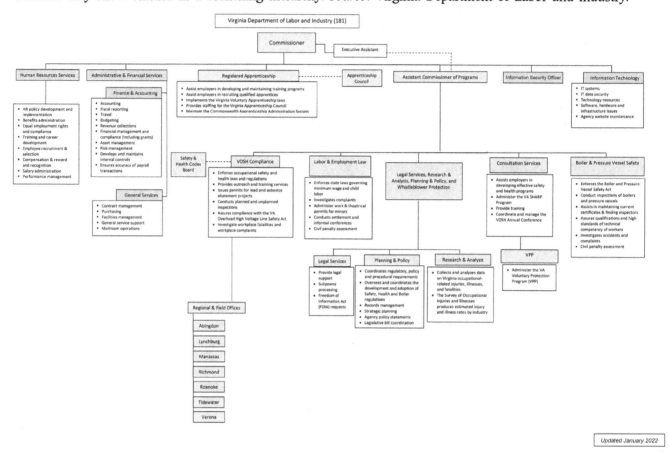

Organizational charts and other *tree-root diagrams* (figure 7.16) show hierarchical relationships. An organizational chart usually starts with the highest and broadest position of responsibility at the top and divides into lower and narrower areas of responsibility. Organizational charts are useful for showing how different positions are related to one another in terms of supervision and responsibilities.

STATISTICAL CHARTS: SHOWING QUANTITATIVE RELATIONSHIPS

In this book, we can only scratch the surface of statistical charts, perhaps the most varied and complex information graphics used in communication. Statistical charts are the product of a field of statistics once known as *graphic statistics*, which uses geometric forms to show the relationships between different kinds of statistical information. Common statistical charts include forms such as bar/column graphs, line graphs, and pie charts, most of which you probably already know well. If you don't, consult a basic technical communication or statistics textbook such as Markel and Selber's *Technical Communication* (2021), or Freedman, Pisani, and Purves's *Statistics* (2007).

Many information graphics use a 2D coordinate grid arranged on a *horizontal axis* (also known as the x-*axis*, or *abscissa*) and a *vertical axis* (also known as the y-*axis*, or *ordinate*). Often, the horizontal axis shows consistent divisions of time, and the vertical axis shows divisions of quantity, although this relationship can be reversed. Data points are plotted on this coordinate grid and marked or connected with lines or rectangles. Other chart forms show quantity by area, angle, or distance from a central point. Pie charts, for example, use the angle of pie slices to show the comparative relationship of parts to a whole.

Designers are always innovating new ways to visually display statistical information, often combining or modifying the basic forms of statistical charts. To examine some of these innovations, see in particular the many books of Edward Tufte and Alberto Cairo.

When designing anything other than basic statistical charts, you will probably need to work closely with content experts, statisticians, and even software programmers (for dynamic charts). But you will be the one to decide how to present the statistical data. Content experts may disagree on which kind of chart would work best or whether charts are even the best way to show statistics. But because *you* are the ambassador between the content providers and the users, you must explore ways to help users *see* the complex relationships between different kinds of data.

PICTOGRAMS: SHOWING A COMMON IDEA WITH A SIMPLE MARK

Pictograms (sometimes called *pictographs*) are simplified, abstract marks used to express common ideas such as *restroom, fire hose, stairway, home page*, and so on. Pictograms are especially useful for wayfinding and signposting, as they don't rely on language to convey an idea. But they can also be used effectively to mark repeated elements in technical documents, such as helpful tips, cautions, and warning statements. In computer interfaces, they're often used to mark buttons on software toolbars or navigation buttons on websites (figure 7.17). In computer interfaces, such pictograms are often called *icons*, but don't confuse this with Charles Saunders Peirce's concept of icons, as discussed in chapter 3. Speaking in Peircean terms, pictograms are closer to indexical marks than to icons, since they indicate something but don't always represent

Figure 7.17. Pictograms as navigational tools. Pictograms are used in signposting, technical manuals, and software interfaces. Here, pictograms show the main links in an internet radio website. *Source*: Pandora, https://www.pandora.com/.

(look like) what they point to. And some are even closer to symbols: the international pictogram for recycling, for example, has an entirely abstract relationship to this complex concept.

The biggest limitation of pictograms is that they are easy to misunderstand or misrecognize. Pictograms are meant to be self-explanatory, but frequently users must be trained (or train themselves) to recognize what a pictogram means. For example, the pictograms in figure 7.17 include an image of a thumbs-up and a thumbs-down. Without some prior knowledge, it's difficult to recognize that one means "don't play songs like this," and the other means "play more songs like this."

PROMOTION GRAPHICS

Although graphics used for promotional purposes might seem less important than information graphics, they still play an important role in marketing and conveying information about ideas, products, and organizations.

LOGOS: MARKING IDENTITY

Logos are similar to pictograms in that they serve as indexes to more complex ideas, such as corporate identity and product quality. Unlike pictograms, however, logos can and often do include text or initials. As we discussed earlier, corporations spend many millions of dollars to establish a consistent and meaningful graphic identity, in which logos play a significant part. Corporations pay large fees just to own the copyright to a well-known, easily recognized logo. For example, the Italian firm Bugatti, renowned before World War II as a maker of luxury automobiles, was entirely defunct for decades; its only remaining property was its distinctive logo, a backward *E* against a *B* for Ettorio Bugatti, the company's founder. Volkswagen purchased the rights to the logo in 1998 before they had even developed a car to put it on. VW paid a high price to own not only the logo itself but also the cultural ideas the logo represents: luxury, exclusivity, and high performance. Logos are so important that legal disputes about them can drag on for years. For example, Apple Computer (now Apple Inc.) and Apple Corps (the music production company started by the Beatles) contested the use of their similar-looking apple logos for more than a decade.

The challenge of designing logos lies in creating a relatively simple graphic that will fit the organization, its identity, and its products while making a distinctive statement that users can recognize instantly. Logos often take the form of the stylized name or initials of the organization, as we can see in the logos for corporations such as GTE, Verizon, Coca-Cola, and Ford. They can also be a graphical play on words, such as Apple Inc.'s stylized apple with a bite taken out of the side. Or they can be entirely abstract, such as Chevrolet's "bowtie" or Nike's "swoosh" trademark, which have nothing to do with cars or athletic shoes except for the associations created by successful marketing.

And in fact, a good part of logo recognition *must* come from marketing—training users to associate the logo with a desired organizational identity or ethos. To build brand identity, designers use logos wherever they can, including on products, product labeling, documentation, and advertisements. The more the public sees the logo associated with the identity the organization wants to promote, the more likely the logo will grow in meaning and value.

You may not be professionally involved in designing logos, but if you design documents in a corporate environment, you should at least expect to display your corporation's logo prominently and repeatedly. For example, title pages, headers, and footers of printed documents are common display spaces for logos, since those positions indicate to users what organization the document speaks for. Navigation bars or headers of websites commonly include logos as well.

Consistent use of logos is one key to successful branding, so be sure to check if your organization has stated requirements for using its logo. Organizations often include information in an organization-wide style guide about how logos should be used in documents from business cards to websites.

DECORATIVE GRAPHICS: ESTABLISHING A VISUAL TONE

The least information-intensive promotion graphics serve a primarily decorative function. *Decorative graphics* can include graphics in any form, including photographs, drawings, and cartoons. They are typically distinguished by their purpose: to give a visual style or tone to a document and to capture the user's attention. Designers typically use decorative graphics to create an appealing thematic connection with the subject or

Figure 7.18. Decorative graphics. Decorative graphics often carry a thematic connection to the purpose or subject of the document, but they don't usually contain much direct information. *Source*: Luce Foundation Center for American Art, "Explore"; American Academy of Allergy, Asthma, and Immunology, "Childhood Asthma."

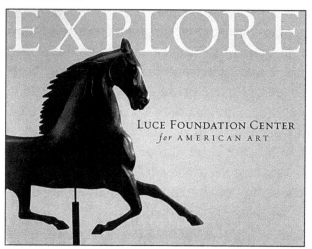

This guide to an art museum uses an image of a horse statue to emphasize the theme of the document: The horse seems to invite us to "Explore." Because the statue is also in the museum's collection, this image could also be considered an illustration that shows visitors what kind of art to expect at the Center. But its primary function here is to attract attention and interest.

This leaflet uses toy blocks to create a visual link to the document's subject by spelling out "asthma." The blocks create a sympathetic tone, but they don't tell us much about childhood asthma.

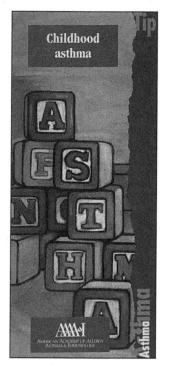

purpose of the document (figure 7.18). The style or tone of a decorative graphic can influence how users react to a document and create a sense of ethos for the organization that created it. Decorative photographs can also model a desired response to the document.

Given this definition, decorative graphics are more likely to be used in documents with a primarily promotional focus, such as flyers and advertisements. But they can also appear in more information-driven documents such as reports, proposals, and manuals, especially if those documents are intended to persuade the user or evoke a particular tone in keeping with corporate branding. And any document that needs to capture a user's attention is a good candidate for decorative graphics.

Decorative graphics have legitimate functions, but using them successfully requires good taste and an artistic eye. It can also be difficult to anticipate consistently how users will react to decorations, and decorations can overwhelm or contradict the message or essential information of the document. Images are inherently ambiguous, and decorative graphics are especially so because they hold little informational weight. So for the most part, use decorative graphics sparingly and strategically.

Use decorative graphics sparingly and strategically.

CREATING AND MODIFYING GRAPHICS

Even if used within the bounds of copyright, clip art and other borrowed or reused graphics have certain disadvantages. Borrowed graphics are so easy to use that people don't always think carefully about incorporating them into a document design. All borrowed graphics were designed for a particular context (or in the case of clip art, for no particular context), so they can often look strange, conspicuous, or awkward in a different design. Users also can usually recognize clip art when they see it (figure 7.19), which can give the impression of a cheap shortcut rather than a thoughtful, integral component of the design. Fortunately, we have alternatives: you can learn to create your own graphics or at least to modify existing graphics in creative and rhetorically effective ways. All it takes is time and a willingness to experiment with software that might be new to you.

Figure 7.19. Clip art. Clip art might be well executed, but it's often an obvious mark of amateur design. If you use clip art, think about how you might modify it to match the purpose of your design (if the terms of use allow doing so, of course). *Source:* Pixabay.

BITMAP GRAPHICS VS. VECTOR GRAPHICS

Before we go on, you'll need to understand one central distinction: the difference between *bitmap graphics* and *vector graphics*. These represent the two main approaches for creating digital graphics. Each approach has its own advantages and disadvantages, abilities and limitations. In short, bitmaps describe graphics pixel by pixel, while vectors describe graphics with mathematics.

BITMAPS

Most digital photographs and scans are originally created as bitmaps (also known as *rasters*, the German term for a grid). A bitmap graphic file describes each individual pixel of information in the graphic.

Bitmaps allow minute control of each pixel so you can specify its color and brightness precisely. Bitmaps work particularly well for photographs and scans because these graphics use a wide variety of colors with smooth blends between them. Most photo retouching software packages focus on manipulating bitmap images (although some of these programs can also make limited vector graphics). These programs also allow us to "paint" smooth, flowing forms by altering a swath of pixels with simulated tools such as a paintbrush or pencil with a mouse or tablet.

> Bitmaps describe graphics pixel by pixel, while vectors describe graphics with mathematics.

Because all graphics (vector and bitmap) are eventually displayed on-screen or printed as bitmaps, nearly all software capable of displaying graphics can display bitmaps. Bitmap file formats include Windows Bitmap (BMP), Tagged Image File Format (TIFF or TIF), Joint Photographic Experts Group (JPEG or JPG), and Graphics Interchange Format (GIF).

Bitmap graphics saved in these standard file formats can be read in many kinds of software on most computer platforms, making bitmaps a very common format for exchanging graphics.

Common design programs often work with both vector and bitmap graphics, but some focus more fully on one than the other. Adobe Photoshop, for example, works more effectively with bitmap graphics, while Adobe Illustrator is aimed at designing with vectors.

The biggest limitations of bitmaps are their size, shape, and links to resolution. Size is the most obvious problem: saving an image as a bitmap typically makes for a large computer file. It just takes a lot of data to describe every pixel, even in a small image. The higher the resolution, the more data the file must contain to describe in an image of a given size. And the greater the color depth—the number of colors possible for each pixel to display—the more data is necessary to describe each pixel. In response, computer scientists have developed a number of compression algorithms, such as JPG, that allow computers to describe some pixels and "guess" at the remainder. However, such compression typically degrades image quality.

A second limitation is that all bitmaps are essentially rectangular. This shape results from the rectangular grid of rectangular pixels that makes up a computer display. Even if the object pictured by a bitmap is some other shape—a circle, for example—the file itself remains rectangular, and the remaining pixels in the corners must be filled with some color. For years the exception was GIF graphic files, which can incorporate transparent pixels, effectively allowing us to "see through" the corners of a nonrectangular graphic. GIFs, however, are limited to 256 colors, making them more suitable for line art, logos, and pictograms than photographic images. This capability of transparency has spread into formats like PNG and JPG in what is now called the *alpha channel*. The alpha channel allows these file formats to specify pixel transparency.

Because bitmaps describe the individual pixels of a digital image, they're also closely tied to *resolution*, a measurement of the density of pixels in the image. Resolution is usually measured as pixels per inch (ppi). Graphics that are intended for computer monitors have a relatively low resolution, usually less than 100 ppi. Graphics intended for print typically have a higher resolution, as determined by the production method. Most laser printers print at 600 dots per inch (dpi), while most commercial imagesetters work around 2,400 lines per inch (lpi). The higher the resolution, the smoother and more lifelike the graphic looks, but the larger and more computer-intensive the graphic file must be (for more information about resolution, see chapter 4).

Because bitmaps are inherently tied to resolution, they limit modification. Resizing (also known as *scaling*) is a particular problem. If you make a bitmap larger, each original pixel is actually spread to many pixels, leading to a distorted, jagged look known as *pixelation* (see figure 7.20). If you make a bitmap smaller, you force the software to *downsample* the image, or pack the average data from many original pixels into a smaller number of pixels—with inevitably poor results. Similar problems arise from other manipulations, such as skewing or rotating the image.

Finally, it's difficult to modify individual parts of a bitmap graphic, such as a particular object in a photograph. You can select and modify an area of pixels, but for a bitmap, "object" doesn't really mean anything because bitmaps record information only in terms of individual pixels. Even if those pixels *appear* to form an object when we look at the photograph, it's difficult to change the size or shape of the object by manipulating individual pixels.

More about . . . Common Bitmap File Formats

This table lists the bitmap file formats you'll encounter and use most often. You can always recognize graphics file formats by their filename extensions.

Extension	Name	Description	Common Uses
BMP	Windows Bitmap	The basic file format for bitmaps. Easily read by most Windows and Apple programs, but very large file sizes.	Any bitmap graphic editing; transferring uncompressed images
TIF (TIFF)	Tagged Image File Format	A basic uncompressed file format that can be read by most graphics programs. Less common than BMP overall, but more common in publishing and document design.	Incorporating images into page layout programs
JPG (JPEG)	Joint Photographic Experts Group	A compressed format for bitmap images. JPG uses lossy compression, meaning that some information is discarded in the process. Most bitmap graphics programs allow you to specify the amount of compression (and thus the resulting image quality) on a sliding scale from 12 (little compression = large file, great quality) to 0 (maximum compression = small file, poor quality). Some programs may have slightly different scales (e.g., 0–10).	Creating photographic images for websites; archiving photographic images
GIF	Graphics Interchange Format	Another compressed format. GIF files can include a maximum of 256 colors, so they're best for images with consistent areas of color (as opposed to color gradients). GIFs compress images by recording every color used in the image, and then recording in a table which pixel gets which color. Because there are so few colors possible in a GIF, this method can describe every pixel in a very small file. GIFs can also include transparent pixels, allowing nonrectangular graphics.	Creating line art such as logos, pictograms, and diagrams for websites
PNG	Portable Network Graphic	A compressed format designed as an alternative to GIF. PNG files can incorporate up to 48-bit color, far beyond the 256 colors allowed by GIF files. PNGs can be used successfully for both line art and photographs.	Line art and photographs for websites
RAW	Raw, unprocessed data	Various proprietary image file types designed for use on digital cameras. Many digital cameras save images as RAW and then translate them into BMP, TIFF, or JPG for transfer to a computer. Increasingly, bitmap graphics programs allow direct editing of RAW files. The lack of standards makes using RAW files more complicated; not all programs can read all types of RAW files.	Capturing images on digital cameras; image editing in some programs

Figure 7.20. Pixelation in a bitmap graphic. Here is a bitmap graphic in original size, then scaled up 450 percent. In the scaled-up version, you can clearly see fuzziness and pixelation in the petals of the flower. *Source:* Created by the author.

Original

Scaled 450%

VECTORS

Fortunately, changing size and shape is the strength of vector graphics. While bitmaps describe every pixel in the graphic, vectors use mathematics to describe the lines, shapes, patterns, and colors of drawing objects. Vectors work best for drawing images with sharp edges and consistent areas of color, such as line art, diagrams, maps, and statistical graphics.

Look at the rectangle filled with a color in figure 7.21. A bitmap would record every pixel covered by the rectangle; in a sense, it doesn't record "rectangle" at all, but just the characteristics of each pixel in the graphic. A vector version of this object, however, would describe it simply as having a certain width, height, position, line width, and fill color. This vector information is much more efficient than describing every pixel, making for significantly smaller file sizes. The same goes for any object, whether it's a triangle, circle, oval, octagon, curve, line, or complex polygon.

Because vector objects are described by their basic characteristics, you can change those characteristics easily. This control allows you to manipulate the objects in vector graphics with great precision and flexibility.

You can change the size, position, and color of an object as much as you like; you can rotate or skew it at will; and you can change the width of the line and apply patterns or textures to its surface. In short, vectors let you treat each design object *as* an object.

The vector shape is described by its essential characteristics (width, height, position, fill color, border width) rather than by recording every pixel. Because the vector shape is described this way, it can easily be resized and reshaped.

Vectors are also resolution-independent, meaning they scale up or down very smoothly without jaggedness. What's more, while bitmap graphics must be manipulated as one big piece, vector graphics can include many individual design objects, each of which can be manipulated separately. We can combine (*group*) vector design objects to build up more complex graphics, and we can *align* objects precisely and *distribute* objects evenly across a given area.

Vectors do have some disadvantages. They don't manage smooth gradations of color very well—one of the strengths of bitmaps. The files for vector graphics are also often proprietary to the software used to make the graphics, such as Adobe Illustrator (AI). Proprietary file formats mean that for the most part those graphics will be editable as vectors only within the program used to create them. There are a few standardized file types for vector images, such as Windows MetaFile (WMF), Enhanced MetaFile (EMF), and Scalable Vector Graphics (SVG). In addition, one very common proprietary format is used primarily for exchanging vector graphics from one program or platform to another: Adobe's PostScript (PS), which is also the page description language used by Adobe Acrobat and many computer printers. However, although PostScript saves graphics as vector data, it doesn't let you modify the graphic as easily as if you'd stayed

Figure 7.21. A rectangle drawn with a vector graphics program. *Source*: Created by the author.

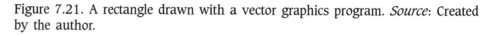

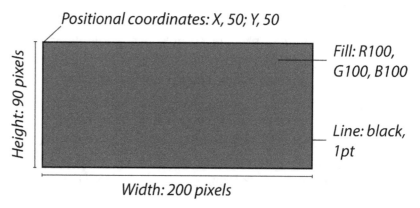

with the native file format of the program with which you created the graphic. PS and PDF (also created by Adobe) are inexorably linked—a PDF is just PS code that does not require a full compiler or editor for access (see Adobe, 2024).

Keep in mind that although vectors are great for *creating* graphics, they will eventually need to be *output* as bitmaps. Printers and computer monitors both require bitmap images for image display. Once you've exported a vector graphic to a bitmap format, you will no longer be able to modify the new bitmap file as a vector. Instead, you'll need to go back to the original vector file, make your changes, and export the file to bitmap format again.

More about . . . Bitmaps and Vectors

Bitmaps	Vectors
Describe graphics pixel by pixel	Describe object characteristics with mathematic
Best for photography and scans	Best for diagrams, maps, and statistical graphics
Always rectangular	Any shape
Minute control over individual pixels	Flexible control over individual and grouped objects
Mostly standardized file types (BMP, TIFF, JPG, GIF, PNG) with a few proprietary ones (PSD, PSP)	Mostly proprietary file types (AI, CDR, FH, PS) with a few standardized ones (EMF, WMF)

WORKING WITH PHOTOGRAPHS AND OTHER BITMAP GRAPHICS

ACQUIRING BITMAP GRAPHICS

Before you can modify bitmap graphics or incorporate them in documents, you'll first need to acquire raw versions of the graphics. The methods you'll find most useful are digital photography, scanning, and making screenshots. All of these approaches use hardware to digitize visual data one pixel at a time.

Digital photography. Digital photography is so common and inexpensive now that it needs little explanation, but one issue is particularly important for document designers: the relationship between camera resolution, measured in megapixels, and image resolution, measured in ppi. Digital camera resolution is usually measured in megapixels, a raw counting of the number of pixels in the digital image created by the camera. A megapixel is one million pixels, so a one-megapixel camera would make a theoretical image resolution of 1,000 by 1,000 pixels (1,000 × 1,000 = 1,000,000). (In practice, of course, most digital cameras take rectangular rather than square photographs.)

More about . . . Bitmaps and Resolution

Keep in mind that the most important issue in creating bitmaps is the relationship between *native resolution*, the resolution at which the image was captured, and *output resolution*, the resolution at which it will be printed on-screen or on paper. (Resolution refers to the size of pixels and their density in the display area, usually measured in ppi.) So as you create bitmaps, remember the following two general principles.

Resolution affects not only the detail level of the image but its output size as well. The actual size of the output device's pixels (for monitors) or dots (for printers) determines how big a bitmap image will be produced in physical dimensions. For example, an image with a native resolution of 300 ppi displayed on a 100 ppi screen will appear three inches wide, while the same image printed on a 600 dpi laser printer will appear one-half inch wide. That's a big difference!

You can decrease resolution, but you can't increase resolution. If you create an image at a particular resolution, you can always *downsample* to a lower resolution. But you can't go the other way and "upsample" to a higher resolution. Doing so would require using more data than the image contains. The only way to obtain that data would be to reacquire the image at a higher native resolution.

Downsampling has some advantages, especially because very-high-resolution images take up a lot of computer memory—and certainly, there's not much point in using higher-resolution images than you intend to output. You'll discard some of the bitmap image information through downsampling, but you can manage exactly how far to go without losing important details.

Design Tip: Bitmaps and Compression

Once you count all the linked and embedded graphics a document can contain, it's very easy to create a document with a combined file size of a gigabyte or more. These files can take considerable time to render or save, and they may take up more memory space than you'd like. In these situations, compression can be a great bonus—but it also reduces the quality of your images, making it seem more like a necessary evil. Here are some tips for handling compression:

- In most cases, use JPG for compressing photographic images and GIF or PNG for compressing line art. The exceptions are small, low-resolution thumbnails of photographs, which are often more efficient to save as GIFs or PNGs.

- Make all editing changes to bitmaps at native resolution in a non-compressed format, such as BMP, TIFF, or RAW. Compress your graphics to JPG or some other compression format only at the last moment—just before incorporating them into your design.

- Learn how to manipulate the level of JPG compression. Most bitmap graphics programs allow at least ten levels of JPG compression, and some divide the scale into one hundred levels. The higher the number you choose, the less compression applied and the higher the resulting quality.

- For GIFs, try reducing the number of colors recorded in the image file. GIFs can include up to 256 colors, but they don't have to include that many. If the graphic you're working on is black and white, a GIF holding it would need only two colors. Most bitmap and vector drawing programs allow you to specify how many colors are included, from two (black and white) to 256. You can also specify whether you want a restricted palette of colors or a palette indexed as closely as possible to the RGB values in the original image.

- Choose the highest compression settings that look acceptable for your particular project. Sometimes, saving digital file space is more important than including beautiful but space-intensive graphics. If possible, check how your graphics will look using your final output method. If you intend to laser print and photocopy the output, for example, you can easily and cheaply experiment with finding a compression level that works best for that production method. Of course, it's a little harder to do so with offset lithography, but get as close as you can to the final output.

The good news is that the resolution of digital cameras has increased remarkably over the years, from the early, one-megapixel cameras of the 1990s to cameras capable of hundreds of megapixels. Some smartphone cameras are even capable of hundreds of megapixels. At the same time, the cost for high-resolution images has plummeted, as has the cost of digital media storage for those images. If your means and equipment permit, always take the highest possible resolution digital photographs. You can always downsample them later, but you can't create a higher-resolution image from a lower-resolution one.

Scanning. Scanners are the most common hardware for digitizing existing paper images into bitmaps. Most scanners are easy to use and increasingly inexpensive to buy. Since the quality of scanned images varies, if you're going to be working with scanned images frequently, invest in a high-quality scanner. When choosing a scanner, also pay attention to the scanning software. Some scanning software usually comes free with the scanner, but there are special third-party programs that give greater control over the image capture. Even if you use your bitmap graphics program to import scanned images, you will still be relying on the scanner software to run the scanner itself.

But even with the free scanning software that comes with a scanner, you get a considerable amount of control. You can set the resolution easily, usually with a slider or dropdown menu to choose whatever ppi level is appropriate to the job. You can scale the image to a particular output size, or you can adjust the orientation of the image. You can also set the color depth—the number of colors each pixel is capable of showing—anywhere from black and white (1-bit) to gray scale (usually 8-bit) to millions (or more) of colors (24-bit, 32-bit, or 64-bit). The greater the color depth, the more data that will be captured, so the higher the image file size and the longer the scan will take to complete.

As a result, it's usually best to scan images at a resolution around 1.5 times the desired output resolution. Any higher resolution will waste your time, since the extra data will have to be discarded in downsampling the image to the output resolution. Getting a good scan involves some significant challenges. If you're going to be scanning halftone images, you'll probably run into problems with *moiré* (see figure 7.22). Halftones are usually used

Figure 7.22. An example of moiré. Moiré can occur when you scan a halftone image on a digital scanner. The raster (grid of pixels) created by the scanner interferes with the halftone screen of the original, creating odd patterns in the resulting digital image. *Source*: *Independence Examiner*, March 27, 2006, p. 3B.

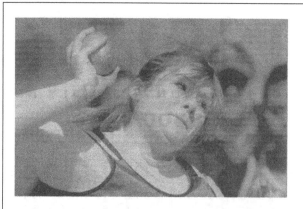

This image was scanned from a newspaper halftone at 100 dpi.

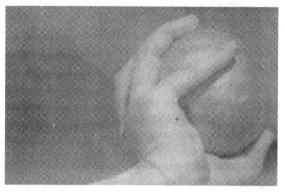

A close-up of the upper left-hand corner of the original shows moiré interference.

to create photographic images in offset printing. If you're scanning an image from a newspaper, magazine, or other offset document, you'll probably be scanning a halftone. Moiré refers to the interference pattern that can build up between the grid of pixels the scanner creates in making a bitmap and the pattern of dots that make up the halftone image. These two patterns can interfere with each other and create distracting patterns and artifacts in the final image scan. The problem is more common with the lower-lpi halftones used in newspaper graphics, but it can also occur when scanning high-resolution, glossy documents such as magazines.

The most common cure for moiré is applying a digital filter (often called a *descreen* filter) to blur the input somewhat. Most scanning software includes such a filter and can apply it as the image is scanned (which, of course, takes more time). You can also apply descreen or blur filters in bitmap graphics programs such as Photoshop after having created the scan. If you don't have access to such filters, you can use a low-tech solution: turn the original on the scanner glass so it sits at a slight angle. You can straighten the image again in a bitmap graphics program. With a little experimentation, this can decrease moiré considerably, but again, with the loss of time trying to find the right angle.

Scanners are also prone to hardware problems such as *barring*, *speckling*, and *artifacting*. Barring happens because the scanner head of some scanners must take multiple passes over the original to build up the digitized bitmap. It physically moves back and forth across the image on a carriage mechanism. If the mechanism is worn or sloppy, these passes can lead to visible bars across the bitmap, particularly in dark areas of the image. Speckling and artifacting usually occur because of dust, smudges, or scratches on the scanner glass or on the original itself. These imperfections refract and reflect the light the scanner shines on the original, leading to some odd effects. Most of these problems can be overcome with judicious use of blur and unsharp mask filters in a bitmap graphics program (see "Using Common Filters" in this chapter).

Screenshots. Screenshots are a common part of software documentation, but creating them brings up some of the same challenges as making digital photographs with low-resolution cameras. Screenshots are very simple to make. The operating system software for both Microsoft Windows and Apple computers can take basic screenshots of windows or dialog boxes:

- *In Windows*, press Ctrl+Print Scrn to take a screenshot of the whole screen and Alt+Ctrl+Print Scrn to take a screenshot of the active window or dialog box. After you take the screenshot, it is copied to your computer's clipboard; you can then paste it directly into bitmap editing software or a word processing file.

• *In MacOS*, press Command+Shift+3 to take a screenshot of the entire screen as a PNG file; press Command+Shift+4 to take a screenshot of a portion of the screen.

Specialized software can help you create more complex screenshots, such as allowing a time delay before the shot is taken, capturing cursors on the screen, and taking an image of a web page that's larger than your actual screen.

Screenshots are limited to taking an image of the screen at its current viewing resolution. With some computer screens, resolution can be set as low as 72 ppi—far too low for print production and pretty low even for screen production, since most screens are now at higher native resolutions. Screenshots often look blurry and indistinct as a result, defeating the purpose of showing a screenshot in the first place. Another problem arises in the images that screenshots capture: computer screens, windows, dialog boxes, and text. These images are not only resolution-dependent, but they also use lots of straight lines and large areas of single colors. Type is a particular challenge because of antialiasing, which blurs the edges of screen type slightly to make it look better against its background (see chapter 6 for more information about antialiasing). This blurring might look good on a computer screen, but it can make screenshots difficult to read.

These issues make resizing screenshots particularly challenging. Resized screenshots often end up fuzzy and difficult to read. Both of these problems are exaggerated when a screenshot is printed as a halftone on paper, since the halftone grid pattern can interfere with the bitmap grid and create moiré (see "Scanning" in this chapter).

The best strategy for dealing with these problems is to assess your output methods before taking the screenshots. If you intend to output to screen, match your screenshot resolution with the expected or most common output screen resolution. If you intend to output to paper, take screenshots on a system using the highest possible native screen resolution, and then resize carefully in a bitmap graphics program.

In moving from screenshot to output, also keep in mind that the raw size of pixels is not a constant from monitor to monitor. System software measures monitor resolution in terms of the number of pixels in width and height, but we're more interested in the pixel *density*: the number of pixels per actual square inch of screen. Although monitors can typically be set at a variety of resolutions, each monitor has a native resolution; at that resolution, some monitors have bigger or smaller pixels than others. So a very large monitor, even set at its highest monitor resolution settings, can still have relatively large pixels, resulting in a fuzzy screenshot image.

Design Tip: Screenshots

- For screen output, match screenshot resolution to expected output resolution.

- For print output, capture the highest possible native screen resolution.

- Experiment with turning off or modifying screen font smoothing options before taking the screenshot, especially if you expect users to read the text in dialog boxes.

- If you take a screenshot using the Print Scrn button in Windows, it may look like nothing happened. But in fact, the image was copied to your clipboard; just paste it into the program you'll be using.

- For complex screenshots, consider purchasing specialized screenshot software.

EDITING BITMAP GRAPHICS

We can only begin to discuss the many techniques available for editing bitmap graphics, but the following are some of the most common ones you'll need to work with photographs, scans, and other bitmaps. Because bitmap graphics programs work differently and change constantly, we'll stick with general concepts. You'll also need to check out the help files for the program you have access to for specific instructions. But don't be afraid to experiment with some of the many tools, settings, and filters these programs offer. Sometimes the best effects are the results of serendipity, and experimentation is a great way to learn a new program.

Remember: CMYK for print output; RGB for screen output.

Saving the original file. Before you do anything else to a newly acquired bitmap, use Save As to create a new version of the file. Archive the original in a safe place. If you make changes to the original and save it, you're stuck with the changes you've made, even if they were mistakes. But if you make mistakes on a copy, you can always just go back to the original.

Setting the output mode: CMYK or RGB. Your next step should always be to set the output mode as either *CMYK for print output* or *RGB for screen output.* (You can read more about these terms in chapter 8.) Most programs will even prompt you for this information in a dialog box when you open a new document. Do nothing else to the file until you've completed this step. If you make changes to the file in the wrong mode, you'll find the image doesn't come out as you expected when translated to the correct output mode. In most programs, you can also choose to change a graphic to gray scale if you're going to be printing in black and white.

Adjusting image quality. Almost every bitmap graphics program will allow you to adjust at least the brightness and contrast of your bitmap images. But some provide more complex and flexible tools:

- *Levels.* This tool allows you to adjust the highlights, midtones, and shadows of the image to appropriate levels. It also allows you to set the black point (the darkest pixel) and the white point (the brightest pixel) of the image, which helps you use the entire available range of tones.

- *Curves.* This tool provides even more flexibility, allowing you to adjust many levels of brightness. It also allows you to adjust the red, green, and blue channels separately.

Many programs also include automated, one-step commands for adjusting levels, contrast, and color to the best generic settings. In Photoshop, these are Auto Levels, Auto Contrast, and Auto Color. They might not give you the best results, but they're often a good place to start.

Using common filters. Once you've adjusted the image quality, you can tackle other problems, such as moiré, speckling, striping, and blurriness. Most programs include *filters* that will fix these problems for you, such as *descreen, despeckle, unstripe,* and *sharpen.* One particularly useful general-purpose filter for modifying photographs is the *unsharp mask filter.* Unsharp mask sharpens the edges of color areas, creating greater distinctness and density in the image.

For the wide variety of other filters available, you'll find the best results simply by experimenting. But keep in mind a cautionary note: bitmap graphics programs can come with some pretty wild and crazy filters. They're fun to play with, but make sure that whatever graphic you create is rhetorically appropriate for the document you're designing, its users, and your client.

Cropping. One of the most powerful tools available to you is the ability to crop images. Cropping, as it might sound, involves cutting off unwanted parts of an image. But this definition makes it sound less important than it really is. With cropping, you have the opportunity to guide the user's eye to what's most important in an image.

Consider the first photograph in figure 7.23, a yard scene. Cropping the image slightly brings the focus to the foliage and the statue. Now, rather than thinking "yard," viewers might think "tree" or "statuary." Crop the image even further, and the focus changes to the statue's face and the flower behind its ear. In this way, cropping can allow you to adjust the rhetorical focus of an image.

Figure 7.23. Cropping a photograph. Cropping a photograph can help you adjust an image to focus on whatever you think is most rhetorically important. The first image seems be a yard scene. The second image highlights the flowers and full statue. The third image focuses on the statue's countenance. *Source*: Created by the author.

As we said before, you should always be ethically sensitive about what you crop out of a picture. You have an obligation to show users the truth (see "Viewpoint" in this chapter).

CREATING DRAWINGS, DIAGRAMS, AND MAPS WITH VECTOR GRAPHICS

The tools for creating vector graphics differ remarkably from those for bitmaps, but there's always some overlap. For example, vector programs can typically import bitmap graphics to be used as part of a vector design, and they typically use similar approaches for color, layers, and filters. You may already have experimented with the limited vector drawing tools available in the Draw toolbar of Microsoft Word. Dedicated vector graphics programs are not so very different, but they are more flexible and precise.

Again, this section isn't intended to replace software documentation but to give you a sense of some useful techniques common to most programs. Most of the functions of vector graphics programs start with a collection of tools. The techniques outlined in the following sections use one or more of these tools, so it's a good idea to experiment with them.

DRAWING COMMON SHAPES

Vector graphics programs excel at creating regular and irregular geometric shapes. Most will include tools that draw rectangles, triangles, ellipses, and polygons of any shape or size. Just click on a tool and start drawing. You can change the shape of what you've drawn by using a selection tool and grabbing one of the "handles" that will appear on the shape. Odd shapes or curves are a bit more complex. Most programs allow you to use a pencil tool to draw freely, simply by clicking and dragging the cursor wherever you want the line to go. For more precise shapes, another common tool is the pen tool, which works by clicking where you want the line to begin, and then clicking again where you want the next connection point to be. If you click and drag on the second connection point, the line between the two will become a curve that you can adjust to your liking (also known as a *bézier curve*). Finally, you can apply colors, gradients, patterns, or even bitmap images to the different surfaces of the objects you've created.

GROUPING OBJECTS

All of these shapes can become hard to keep track of, so you can also combine objects by *grouping* them together into larger assemblies. To do so, in many programs you typically select all the objects to be grouped by pressing down the shift key on the keyboard and clicking on each object you want to select. After all the objects are selected, choose the group command to group the objects into a single composite object.

Grouping allows you to solidify one part of the drawing while you work on the rest. It also allows you to move and manipulate several related objects as one piece. If you need to make further adjustments to individual objects in a group, you can always ungroup the objects and regroup them when you're finished making changes.

ARRANGING OBJECTS

One of the most powerful advantages of vector graphics programs is the ability to arrange and rearrange objects to your liking. Three important tools are ordering, distributing, and aligning objects. These tools give you tremendous control over the design relationships between your design objects—especially in terms of proximity and alignment.

- *Ordering* refers to how the objects are "stacked," much as you can stack playing cards on top of one another. Most programs allow you to change this order to bring one object in front of or behind another.

- *Distributing* objects allows you to spread out several objects equally. It usually works by selecting all the objects to be distributed (remember, Shift and click to select multiple ob-jects) and then selecting a command to distribute the objects in equal spacing vertically or horizontally. The objects will be equally spread out between the far left and right objects (for distributing horizontally) or the far top and bottom objects (for distributing vertically).

- *Aligning*, as the term might suggest, lines up a series of objects. You can choose whether to align objects horizontally or vertically. For horizontal alignment, you can also choose to line them up along their tops, bottoms, or middles; for vertical alignment, you can choose to line them up on their left sides, right sides, or centers.

With these techniques, you can manage the relationships between individual design objects easily and consistently.

USING PATHS AND TYPE

Most vector graphics programs create type within text boxes. You can change the width or height of the text box just as if it were any other drawing object. You can also modify the type in many of the ways you would in a page layout or word processing program, controlling typeface, font, leading, kerning, tracking, and so on. Most programs allow you to draw a path (using the pen or pencil tool), or convert an object into a path, and then apply text to it (see figure 7.24).

Figure 7.24. Typing on a path. Most drawing programs let you draw a path with the pen or pencil tool and then apply type to the path. If you need to adjust the path afterward, the type will conform to the new path. This technique gives you a lot of flexibility, especially for titles and other kinds of display text. *Source*: Created by the author.

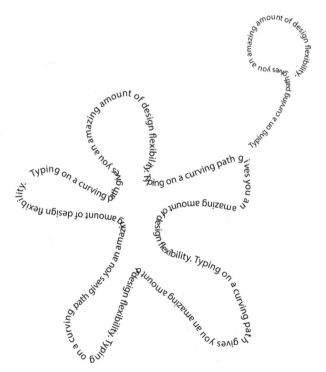

CREATING STATISTICAL GRAPHICS

Statistical graphics are particularly challenging to prepare, primarily because they must both work well visually and represent statistical data faithfully. Four approaches are most common: using office software, drawing graphics manually with vector drawing software, using diagramming programs, and using specialized data visualization software.

USING OFFICE SOFTWARE

The most common approach for very common genres such as pie charts, bar/column charts, line charts, and scatter plots is simply to rely on the statistical graphics tools in office software such as Microsoft Word or Excel. The biggest advantage of preparing statistical graphics with office software such as Microsoft Word is that nearly everyone has access to this software; sometimes in team-based document development, that access is the key to a successful project. Even if you're working alone, creating graphics for business documents that don't require unusual graphic forms is easy and convenient with office software.

In most office software the chart design tools allow you to follow a wizard to help import or input the chart data, choose a chart type from a limited series of common genres, and modify design features such as labeling, legends, titles, borders, and shading. The software then places the resulting graphic as a design object in the office program you're using. The graphic will be resizable and editable, and you can quickly change or correct the base data to automatically redraw the chart.

However, office software typically constrains you to the most common graphic forms, and it encourages some graphic practices that should generally be avoided, such as meaningless 3D effects (see figure 7.6). In addition, the default settings on colors and shading are often ineffective.

USING VECTOR GRAPHICS SOFTWARE

Vector graphics software avoids some of the limitations of word processing software, but it has its own limitations. Naturally, vector graphics software is very flexible, allowing you to draw pretty much whatever you need. You can create graphics with a much higher level of polish with vectors than with office software. You also have considerably more output options to import files into other programs, such as a page layout program.

However, drawing statistical graphics with vectors takes a lot of time and introduces some statistical inaccuracy, particularly if the graphic is complex. Rather than letting the software determine data plots automatically from a spreadsheet, you will probably be approximating data points manually. Vector software allows you to size and position objects with great accuracy, so this isn't an impossible task—just one you must often manage yourself.

For example, if you're drawing a bar chart, you'll have to figure out exactly how long the bar from the 0 point to the data point would be. To do so, you'd need to divide the total height of the vertical scale by the total increments in the scale and then multiply the result by the data point. If your scale will run from 0 to 50 and it's going to be four inches long, that means each step on the scale will be 0.08" apart (4 ÷ 50 = 0.08). A bar at a data point of 38 would then be 0.08" × 38 = 3.04".

Fortunately, some vector graphics programs, such as Adobe Illustrator, provide a graph tool with functionality similar to that of the automatic graphic generation of office software. But these tools can be just as inflexible as those in office software, besides being rather fiddly to work with.

USING DIAGRAMMING PROGRAMS

If your project calls for creating a lot of conceptual diagrams, such as flowcharts or network diagrams, consider using a specialized diagramming program. Programs such as Microsoft Visio, Gliffy, LucidChart, and many, many more use a standard library of ideograms and shapes that you can assemble into useful diagrams quickly. Most of these programs allow diagrams to be exported to a variety of graphics formats and to a variety of other kinds of software.

For more complex data visualization, you can also use a specialized software package. Some of these packages come tied to statistical programs, such as Mathcad and SAS/GRAPH. Others are dedicated plotting programs, such as Gnuplot, R (The R Project for Statistical Computing), Plotly, and more. These programs offer very flexible and powerful graphing tools, but they often take considerable expertise to use.

INCORPORATING GRAPHICS INTO DOCUMENTS

Once you've created your graphic, how do you incorporate it into your document in a way that makes clear the relationship between the graphic and the rest of the document? The key is to use both explicit labeling and the implicit design principle of proximity to forge a clear relationship between graphics and text—or other graphics, for that matter. This combined use of genre, proximity, and textual information deliberately constructed to motivate an audience is known as *scripto-visual rhetoric* (Dawkins, 2009; Ross, 2017).

USING PROXIMITY

The simplest way to show a relationship between a graphic and its surroundings is with proximity—just putting it close to what it's related to. Try to design your page layouts so the graphics are as close as possible to the text that discusses them.

Make every effort to place graphics on the same page or at least in the same spread as the text that refers to the graphic. There's nothing more annoying to users than having to flip back and forth through a printed document or scroll around through a web page to see the graphic being discussed.

Consider using *text wrapping* options to embed the graphic in the text. Most page layout programs give you considerable control over exactly how text will wrap around the object. You can set the object's *wrap points* to give a lot of space or just a little. Use the principle of proximity to good effect by editing your wrap points carefully.

Design Tip: Tracing Layers

You can start to create stylistic graphics with office software or specialized data visualization software and then use the layering capabilities of vector graphics software to create the finished graphic.

In general, here's how to do it:

- Create your graphic with office software or specialized data visualization software.

- Make a screenshot.

- Paste the screenshot into a new layer in your vector graphics program.

- Create a second layer above the layer with the screenshot.

- Use the vector drawing tools to trace a new graphic over the top of the screenshot version.

- Delete the screenshot layer and save and export the final graphic.

This approach works equally well for other kinds of graphics. For maps, you can import a screenshot from online mapping software or a scanned paper map. For line art illustrations, you can import a photo and trace line art over it. In all cases, make every effort to be as accurate as possible in your tracing so that unwanted distortions don't creep into your graphics.

USING ALIGNMENT AND ENCLOSURE

Just as you did with proximity, you can use alignment and enclosure to show the relationship between text and graphic. Many product manuals align diagrams or screenshots beside each step, giving users direct visual instruction or feedback. Or you can actually use a box or a shaded area to show that a portion of text is related to an accompanying graphic.

USING EXPLICIT REFERENCES

You can also incorporate explicit references between the text and the graphic. First, you have a couple of traditional options for labeling the graphic so you can refer to it from the text:

- *Titles:* Strong and descriptive titles on your graphics give you a good marker to refer to and help readers recognize what your graphic is about.

- *Numbers:* In formal documents with lots of graphics, it's common to use figure and table numbers as a reference system. Traditionally, figures (anything that's *not* a table) are numbered separately from tables. You can also create a separate *list of illustrations* or *list of figures* to help users access your graphics more immediately. Such lists are usually included after the table of contents or at the end of a document.

Of course, after giving your graphics a handle, you'll need to grab onto it by providing clear references from the text to the appropriate graphic. You can do so parenthetically "(see figure A)" or work the reference into the text itself, for example, "as you can see in figure B."

MAKING GRAPHICS SELF-SUPPORTING

Although you might refer from the text to the graphics, graphics should also be self-supporting. Users will often skim through a document and just look at the pictures. Your graphics should be able to stand on their own, while also being well integrated into the text.

To make your graphics work independently of text, consider these techniques:

- *Callouts:* Incorporate plenty of callouts in your graphics to explain important points or point out significant information.

• *Captions:* Include a caption with each graphic to explain what it is and what it means.

These techniques won't undercut the connection between the graphics and the text—they'll simply give users more points of entry into the document, letting them decide how they want to read the document.

As with any aspect of technical and professional communication, remember to make your graphics *accessible.* That means that no matter how complex the image, from a picture of puppy to a complex flowchart, you need to provide full alt text (alternative text) if your work will be shared electronically or placed in an online environment. There's a lot more to accessibility than just alt text, of course. Whole bodies of scholarship are devoted to this line of thinking, and you may want to look at work on accessibility and disability studies as a way to get started—you might begin with Lennard Davis's *The Disability Studies Reader* (2016). Here, though, to start small, your alt text allows screen readers to describe your images to users with visual impairment and provides users a description of your design even if images fail to load, while also allowing your images to be read by search engines, if search engine optimization (SEO) is part of your design plan. The combined alt text for all of the images in this book, for example, is as long as one of the chapters!

EXERCISES

In this chapter you've gone from thinking of graphics from the purely theoretical sense to getting into the nuts and bolts of image manipulation and file types. You've worked through graphics in terms of perception, culture, and rhetoric; thought about human factors; thought a bit about the ethics of graphic design; and worked through various file types and even the programs and program types (e.g., vector vs. bitmap) used to create or manipulate images. Now it's time to practice what you've read. The following exercises will help you work through a lot of those complex elements.

1. Find a document with lots of graphics. What kind of graphics do you see, according to the categories discussed in this chapter?

    o Information graphics (illustrations, maps, process diagrams, concept diagrams, statistical charts, pictograms)

    o Promotion graphics (logos, decorative graphics)

2. Imagine that you're designing a document to be printed, and you want to use a graphic from one of the following websites:

   o https://www.cnn.com/

   o https://xkcd.com/

   o A personal blog at https://medium.com/

3. What ethical, legal, and technical issues would you have to deal with before using the graphic in your own document? How would you resolve these issues?

4. Create an information graphic of the required coursework in your degree program for the use of prospective students. The graphic can use text as well as graphics, but it should focus on showing how the curriculum works visually.

5. Import a digital photo into a bitmap editing program such as Adobe Photoshop and perform the following actions:

   o Adjust the color, levels, and contrast—first by using the automatic commands (in Photoshop, Auto Color, Auto Levels, and Auto Contrast) and then by hand.

   o Crop the photo. You can do this either by using the built in crop tool in many programs or by selecting the part of the image you want to remove and deleting it.

   o Apply one or more filters from the Filter menu to change the look of your photo. If you don't like what a filter does, you can always choose File -> Undo (for one step back), or in Photoshop, choose File -> Step Backward for multiple steps back.

   o Change the image mode to CMYK. In Photoshop, choose Image -> Mode -> CMYK Color.

   o Save the file as a TIFF. To do so, choose File -> Save As, and change the file type to TIFF.

6. Draw a map of your hometown or campus to use for directions to your home for visitors:

   o Go to a mapping website, such as Google Maps or MapQuest and search for a map to the appropriate area of town.

   o Create a screenshot (Windows: Alt+Print Scrn; MacOS: Command+Shift+3).

o Open a new drawing in a vector drawing program such as Adobe Illustrator, and paste the screenshot into the drawing.

o Create a new layer above the map layer.

o Use the drawing tools to trace a new map, focusing on only the important details. Be sure to label streets and landmarks!

o Delete the bottom layer (the one with the map screenshot on it) and save the file as a GIF or PNG.

REFERENCES AND FURTHER READING

Adobe. (2024). Best-in-class RIP technologies providing office printing and commercial printing solutions. https://www.adobe.com/products/postscript.html

Biederman, I. (1987). Recognition-by-components: a theory of human image understanding. *Psychological review, 94*(2), 115–147.

Butts, J., & Walwema, J. (2021). Rhetorical hedonism and gray genres. *Communication Design Quarterly Review, 9*(2), 15–26.

Copyright Alliance. (2024). *Who owns the copyright to AI-Generated works?* https://copyrightalliance.org/faqs/artificial-intelligence-copyright-ownership/#:~:text=Applying%20these%20principles%2C%20the%20U.S.,is%20in%20the%20public%20domain

Costigan-Eaves, P., & Macdonald-Ross, M. (1990). William Playfair (1759–1823). *Statistical Science, 5*(3), 318–326. https://www.jstor.org/stable/2245819

Davis, L. (Ed.). (2016). *The disability studies reader* (5th edition). Routledge.

Dawkins, H. (2009). Ecology, images, and scripto-visual rhetoric. In S. I. Dobrin & S. Morey (Eds.), *Ecosee: Image, rhetoric, nature* (pp. 79–94). State University of New York Press.

Dragga, S., & Voss, D. (2003). Hiding humanity: Verbal and visual ethics in accident reports. *Technical Communication, 50*(1), 61–82.

Freedman, D., Pisani, R., & Purves, R. (2007). *Statistics* (4th edition). W. W. Norton.

Gigante, M. E. (2018). *Introducing science through images: Cases of visual popularization.* University of South Carolina Press.

Harris, R. L. (2000). *Information graphics: A comprehensive illustrated reference.* Oxford University Press.

Holmes, N. (1984). *Designer's guide to creating charts and graphs.* Watson-Guptill.

Horn, R. E. (2002). Visual language and converging technologies in the next 10–15 years (and beyond). In M. C. Roco & W. S. Bainbridge (Eds.), *Converging technologies for improving human performance: Nanotechnology, biotechnology, information technology and cognitive science* (pp. 141–149). NSF.

Markel, M., & Selber, S. (2021). *Technical communication* (13th edition). Macmillan.

McCloud, S. (1994). *Understanding comics: The invisible art.* Harper.

Melançon, L. (Ed.). (2014). *Rhetorical accessibility: At the intersection of technical communication and disability studies*. Routledge.

Oswal, S. K. (2019). Breaking the exclusionary boundary between user experience and access: Steps toward making UX inclusive of users with disabilities. *Proceedings of the 37th ACM International Conference on the Design of Communication* (pp. 1–8).

Pritchep, D. (2013). *A campus more colorful than reality: Beware that college brochure*. NPR: Weekend Edition. https://www.npr.org/2013/12/29/257765543/a-campus-more-colorful-than-reality-beware-that-college-brochure

Kosslyn, S. M. (1994). *Elements of graph design*. W. H. Freeman.

Richards, D. P., & Jacobson, E. E. (2022). How real is too real? User-testing the effects of realism as a risk communication strategy in sea level rise visualizations. *Technical Communication Quarterly*, *31*(2), 190–206.

Ross, D. G. (2017). The role of ethics, culture, and artistry in scientific illustration. *Technical Communication Quarterly*, *26*(2), 145–172. https://doi.org/10.1080/10572252.2017.1287376

Stephens, S. H., & Richards, D. P. (2020). Story mapping and sea level rise: Listening to global risks at street level. *Communication Design Quarterly Review*, *8*(1), 5–18.

Tufte, E. (1983). *The visual display of quantitative information*. Graphics Press.

Venn, J. (1880). I. On the diagrammatic and mechanical representation of propositions and reasonings. *The London, Edinburgh, and Dublin Philosophical Magazine and Journal of Science*, *10*(59), 1–18.

Wainer, H. (2013). *Visual revelations: Graphical tales of fate and deception from Napoleon Bonaparte to Ross Perot*. Psychology Press.

Ward, M. (2016). *Deadly documents: Technical communication, organizational discourse, and the Holocaust: Lessons from the rhetorical work of everyday texts*. Routledge.

Welhausen, C. A. (2023). Wicked problems in risk assessment: Mapping yellow fever and constructing risk as an embodied experience. *Journal of Business and Technical Communication*, *37*(3), 281–306. https://doi.org/10.1177/10506519231161617

Zdenek, S. (2015). *Reading sounds: Closed-captioned media and popular culture*. University of Chicago Press.

8

COLOR

LEARNING OBJECTIVES

Upon completing this chapter, you will be able to:

- Discuss color as it relates to perception, culture, and rhetoric

- Describe the differences between creating and using color on screens vs. printed materials

- Discuss how to create emphasis and focus using color

- Create effective color schemes

Designers have worked on how to use color effectively for many centuries. Paleolithic hunters recorded their hunts on cave walls in pigments of ochre, sienna, and carbon. Mayan cartographers painted their linear maps in color, laying out the paths between cities. European document designers, beginning with the medieval period, used red ink, or *rubrication*, to mark the beginnings of sections of text, or illuminated capitals to draw attention to the beginnings of text. After the development of printing, printers learned how to use color in their documents more consistently, and they developed new technologies such as *process printing* to expand the range and uses of colors in document design. Color in a document became a mark of the importance and value of its contents.

With the digital age, color has become steadily more accessible and cheaper to print. In screen documents, using color involves no added expense, opening up new opportunities for design. But despite these changes in the cost of color, users still find color attractive and impressive—making color a good way to promote a strong ethos.

Figure 8.1 shows the US Department of Energy (DOE) site. The designers chose a color scheme using bright green and several different blues above the fold, and high value black text on white background for most of the written content, which has several advantages. It creates a coherent visual impression for the entire site, implying that the DOE is environmentally conscious, but also that they are dedicated to clearly presenting factual news. The blue and green palette has cultural significance, with blue implying clean air and water and green representing the land. The colors also call directly back to the DOE's logo, which uses green, blue, and yellow. In the footer, the contrast is reversed, with white text on a black background, which clearly delineates footer content from news content while maintaining high contrast and legibility. This careful attention to color helps the site express the organization's values and agenda with a coherent rhetorical voice.

Using color successfully sometimes seems mysterious, a matter more of taste than of principle. But the principles and concepts discussed in this chapter will help you make good choices about color in your design, helping you to:

- Understand color perceptually, culturally, and rhetorically

- Use common color systems for print and screen

- Use color contrast for emphasis and focus

- Create effective color schemes

Although this chapter covers basic technical issues about color, the focus of the chapter is on designing *with* color. (For more information on managing color through the printing process, see chapter 11.)

THREE PERSPECTIVES ON COLOR

PERCEPTION

To use color effectively in design, you must understand how people perceive color both physiologically and cognitively. This will help you lay

Figures 8.1 and 8.2. A color design shown in color (8.1) and gray scale (8.2). Take a moment to compare the two versions. What rhetorical effects would be lost if a user chose to print this page in black and white? In figure 8.1, notice the consistent, high-contrast, vibrant color scheme. The US Department of Energy's site uses greens and blues to suggest an environmental consciousness while maintaining high contrasts for ease of use. *Source*: US DOE, https://www.energy.gov/.

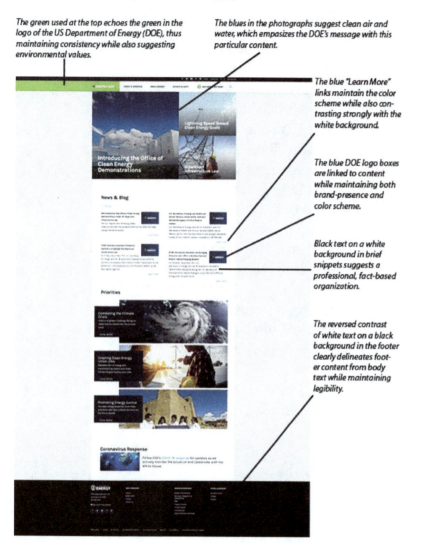

The green used at the top echoes the green in the logo of the US Department of Energy (DOE), thus maintaining consistency while also suggesting environmental values.

The blues in the photographs suggest clean air and water, which empasizes the DOE's message with this particular content.

The blue "Learn More" links maintain the color scheme while also contrasting strongly with the white background.

The blue DOE logo boxes are linked to content while maintaining both brand-presence and color scheme.

Black text on a white background in brief snippets suggests a professional, fact-based organization.

The reversed contrast of white text on a black background in the footer clearly delineates footer content from body text while maintaining legibility.

the groundwork for much of the management of color in printing and computer applications.

As we experience it, color is not a quality of objects or light but a complex of physical, neural, and mental processes that respond to light. Typically, humans can perceive light in a small range of wavelengths between 380 and 700 nanometers (nm)—what is called the *visible spectrum* (figure 8.3).

What we perceive as color, however, relies on more than just a mechanical response to different wavelengths of light, which make up

Figure 8.3. The visible spectrum and tristimulus values. Notice that the wavelengths of light we perceive as red, green, and blue (RGB) overlap considerably. As a result, some colors we perceive, such as brown and orange, exist not as a particular wavelength but only as a combination of wavelengths. *Source*: Based on Williamson and Cummins. (1983). *Light and color in nature and art*. John Wiley and Sons (p. 64).

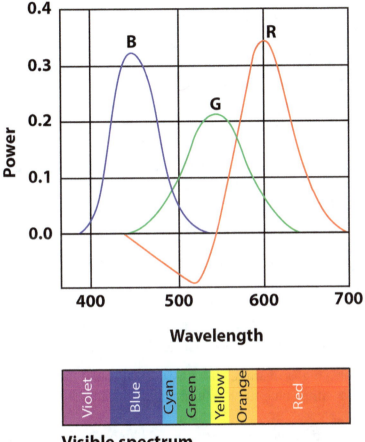

Visible spectrum

only part of our perceptions. Specifically, we can perceive three interacting qualities of light in the visible spectrum: *hue*, *saturation*, and *brightness* (HSB).

HUE

Hue refers to a human perception of wavelengths of light. What we experience as hue is actually a response to reflected or transmitted light as it hits light-sensitive cells at the back of our eyes, called *cones*. The typical human eye has about six million cone cells. To make the wavelengths of light into what we experience as colors, the eye uses three kinds of cones:

- *Red-sensitive cones*, which react to wavelengths from about 700 to 450 nm, peaking at about 575 nm. About two-thirds of all the cones are red-sensitive.

- *Green-sensitive cones*, which react to wavelengths from about 680 to 400 nm, peaking at about 535 nm. About one-third of all the cones are green-sensitive.

- *Blue-sensitive cones*, which react to wavelengths from about 550 to 400 nm, peaking at about 445 nm. Only a small number of cones are blue-sensitive.

The wavelengths that can be seen by the three types of cones overlap considerably—for example, green cones (680–450 nm) are sensitive to light through most of the wavelengths we perceive as blue (550–400 nm). So what we perceive as a color is often some *mixture* of perceptions of the three different kinds of cones and cannot be directly mapped to the neat layout of the visible spectrum.

These three kinds of cones feed their inputs into a series of nerves called *ganglion cells* (or just *ganglia*), which make a further discrimination in color:

- Some ganglia distinguish between red and green. They can sense either red or green but not both at the same time.

- Other ganglia distinguish between blue and yellow. They can sense either blue or yellow but not both at the same time.

This discrimination between red/green and blue/yellow is why most people would never describe a color as "bluish yellow" or "reddish green." That would be asking a ganglion to do two things at once: sensing blue and yellow at the same time or red and green at the same time. However, red-green ganglia sensations can mix with blue-yellow ganglia sensations, creating colors we might describe as greenish yellow (lime), bluish green (turquoise), or yellowish red (orange). And red-green sensations can overlap in the cone stage of the process; combined red and green cone responses lead to yellow ganglion sensations.

Putting these elements together—the cones that receive input and the nerves that process the information—results in different theories of how we actually end up seeing color. One theory is that color perception is the result of an additive process described as the *trichromatic theory of color vision*, where our perceptions of red, green, and blue at the receptor-level combine to allow us to perceive the colors of the spectrum. Another theory, the *opponent process theory*, suggests that we perceive color at the neural level, when our brain registers the presence of paired oppositional colors, such as red/green and blue/yellow, ultimately "seeing" only one. The trichromatic and opponent process theories are complementary, not oppositional, as taken together they offer a complex explanation of how we both perceive and process color. In fact, a third theory, *zone theory*, suggests that both occur, just in different visual pathways (see, for example, Parthasarathy & Lakshminarayanan, 2019).

We describe color built from perceptions of red, green, and blue as a *tristimulus* system, and we can measure and assess *tristimulus values* on an *x*-, *y*-, and *z*-axis. What all this means for you as a designer is that while we physically process light through our eyes and brain based on red, green, and blue sensitivity, we can also replicate and document those perceptions technologically to reproduce similar perceptions in documents.

Graphically, the hues are often represented on a circle or wheel (figure 8.4). The traditional color circle represents the three basic hues or *additive primary colors* (red, green, and blue) 120 degrees apart from each other. The combination of two primary colors is known as a *secondary color*—for example, red and green mix to create yellow, and green and blue mix to create cyan. Two colors directly opposite from each other (180 degrees apart) are known as *complementary* colors. Colors close together on the same side of the wheel are known as *analogous* colors.

Figure 8.4. A hue-saturation color circle. This hue-saturation color circle shows the additive primary (RGB) and additive secondary (CMY) hues spaced equally around the perimeter. The most saturated colors are on the edge of the circle, while mixtures (desaturated colors) increase toward the center. *Source*: Created by the author.

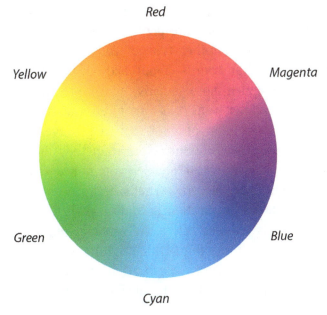

SATURATION

The basic responses of cones to red, green, and blue also lead to our perceptions of color *saturation* (also known as *chroma*), which is our perception of the relative *purity* of color. The more fully saturated a color, the less we perceive mixtures of other colors. We see fully saturated colors as strong and pure perceptions of a single hue, but the addition of other colors reduces the saturation. For example, if we start with red and add in equal portions of blue and green, we see pink—a relatively unsaturated color. Unsaturated colors often look muted, whereas saturated colors look vibrant.

On a hue-saturation color circle (figure 8.4), colors placed closer to the edges are more saturated than colors closer to the middle. At the very center, all colors combine equally to make grays.

BRIGHTNESS

The third component of color perception is *brightness*, sometimes referred to as *luminance* or *value*. Brightness describes the viewer's perception of the intensity of light that is transmitted or reflected from a surface. For projected light, such as light beaming from a computer monitor, we perceive brightness directly—the higher the intensity of the light, the greater the brightness we perceive. But the perceived luminance of an object such as a document printed on paper depends on the intensity of the light shining on the object, the object's own reflectivity, and the reflection angles from the light to the surface and on to the viewer. Highly reflective objects reflect more light than less reflective objects and thus appear more luminous. But even a highly reflective object also looks different in stronger light than it does in less intense light.

Whereas hue and saturation are perceived by the cones in the eye, brightness is perceived by another type of cell called a *rod*. We have considerably more rods than cones—about 120 million per eye, on average—making our eyes much more sensitive to the presence of light than to color. But the sensations from rods can affect the sensations of our cones, and thus our perception of color. You may have noticed this at dusk, when objects take on different colors than they did in the bright light of noon. If you're like most people, you notice that as the sun sets, everything gradually turns gray. That's because as the light fades, our perceptions of hue and saturation become less distinct, whereas our perception of brightness remains active. As a result, *brightness* can best be described as a range from white (full brightness and equal mixture of colors) through a range of grays (less brightness but still equal mixture of colors), and finally to black (the absence of light). Brightness is also linked to hue because some hues are simply brighter than others, regardless of saturation. Red and yellow, for example, have a naturally higher brightness than blue and green at the same level of saturation. For this reason, reds and yellows are often called "warm" colors, and blues and greens are called "cool" colors. Some people perceive warm colors as advancing toward them and cool colors as receding away from them, but this perception isn't universal.

OTHER GRAPHICAL COLOR MODELS

The traditional hue-saturation color circle does a good job representing hues as degrees around the circle (0 degrees–360 degrees) and saturation as the distance from the center, but it doesn't do as good a job of describing brightness. More complex graphical models have tried to supply this information by describing color as a 3D space, in keeping with the three sensations of color (HSB).

One common way to represent this 3D relationship graphically is based on the work of Albert H. Munsell (1858–1918), an influential color theorist. In essence, Munsell-based color models represent brightness by adding a third axis that projects at right angles from the disc of the color circle—like a spindle piercing the center of a spinning top (figure 8.5). This third dimension arranges the amount of brightness from no brightness

Figure 8.5. A 3D color circle. This 3D color space shows the interactions of hue, saturation, and brightness (HSB). The hue and saturation dimensions are arranged the same as in fig. 8.4 but tilted back. The brightness dimension extends in a range from black (no brightness) through grays (some brightness, equal mixture of hues) to white (full brightness, equal mixture of hues). *Source*: Created by the author.

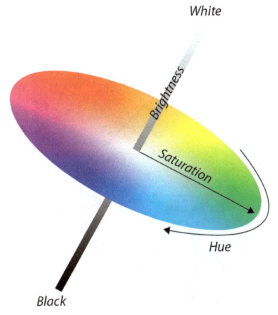

(black) at one end of this axis to full brightness (white) at the other end of the axis.

What many accept as the most accurate graphical model of color perceptions is the CIE L*a*b model (figure 8.6), sponsored by the Commission Internationale de l'Éclairage (the International Commission on Illumination, or CIE for short). CIE L*a*b (often simply called "Lab color") is a tristimulus model that measures color on three dimensions but takes into account the distinction color theorists have made between red/green and blue/yellow. The three dimensions are as follows:

- *L*, or *luminance*, which corresponds to perceptions of brightness

- *a*, the range of sensations from red to green

- *b*, the range of sensations from blue to yellow

Graphically, this 3D space has the same disadvantage as the Munsell space: it buries most of the colors inside the space, which is referred to as the "Munsell color solid" (Munsell Color, 2024). So it's often represented in two dimensions as a curved line with blue and red at the ends and green in the middle. The area inside the curve holds all the colors that viewers report as being perceptible, ranging from full saturation at the edge to white in the middle.

Figure 8.6. CIE L*a*b color space. CIE L*a*b describes the full range of colors perceivable by a typical human being. This color space shows the gamut of possible perceptions of color with additive (RGB) and subtractive (CMYK) gamuts. RGB and CMYK technologies can reproduce only a subset of the perceivable hues. *Source*: Based on Williamson and Cummins. (1983). *Light and color in nature and art*. John Wiley and Sons (p. 60).

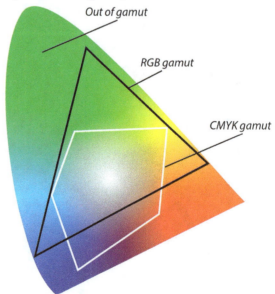

SIMULTANEOUS CONTRAST

People rarely see a color isolated from all other colors, so another factor comes into play in color perception: simultaneous contrast. Simultaneous contrast refers to the fact that multiple colors in the viewer's field of vision can interfere with each other and alter our perceptions of the individual colors (figure 8.7). For example, a yellow square on a blue background takes on a subtly green cast when compared to a yellow square on a white background. A circle of blue on a background of red will look slightly more purple than it would by itself. Accordingly, it's important to consider colors in their context.

COLOR VISION DEFICIENCY

For some people, differences in color perception rise to a level described as *color vision deficiency* (commonly known as *color blindness*). About 5 to 10 percent of men and about 0.5 percent of women experience color vision deficiency because of a genetic difference from the majority of human beings.

Color vision deficiency is rarely a complete inability to see colors; almost always, it's a limitation in the ability to perceive one or two hues. By far the two most common conditions are *protanomaly*, a limitation of the ability to perceive red, and *deuteranomaly*, a limitation of the ability to perceive green. In both of these conditions, the perception of color is shifted away from the hue that's hard to detect and toward hues that are easier to detect. So for someone with protanomaly (red weakness), purple (red + blue) will just look like a kind of blue, and most pure reds will look green.

Obviously, color vision deficiency can present significant problems for designers using color in communication. In general, desaturated colors

Figure 8.7. Simultaneous contrast. Our perceptions can change when we set colors against other colors. A color surrounded by another color is particularly susceptible to this effect. Here, the small squares are identical in each set, but the surrounding colors make the sizes of the small squares look different from set to set. *Source*: Created by the author.

 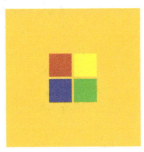 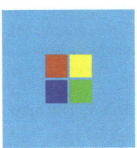

are more difficult for people with color vision deficiency to perceive than fully saturated colors. Colors that are close to each other are more difficult to distinguish than highly contrastive colors. And due to the common occurrence of protanomaly and deuteranomaly, colors in the red or green parts of the spectrum are particularly problematic. As you work, you may find it valuable to check your design in one of the many color simulation tools available online (a simple search for "color blindness simulator" should yield many options). Many of these allow you to upload an image or link a site, then view your design through simulated color perception types, including protanomaly, deuteranomaly, and more. While online simulations are not likely to be 100 percent accurate, they should give you a general idea of what your design looks like through another's eyes. You should also consider checking your design against something like the Web Accessibility in Mind (WEBAIM) contrast checker, which allows you input foreground and background colors and then lets you know if your work will be generally accessible in various formats.

CULTURE

Despite the general accuracy of the models of color perception discussed above, the models remain attempts to universalize color perception. CIE L*a*b, for example, is based on the idea of a "standard observer"—a generalized, average human being. However, color perception varies considerably from person to person. What looks blue to you may look purple to someone else; their orange may be your red. Different cultures ascribe different values and significances to colors. So determining what a particular color means to a particular person in a particular culture is a difficult issue.

However, people *do* commonly assume that colors have meaning, or at least general associations. You can probably name some of these assumed general color connections off the top of your head: red associates with blood, black with a dark negativity, yellow with a sunny positivity, greens and blues with nature (for reference, take a look back at figure 8.1), gray with ambiguity. But these associations are cultural, and even within the same culture, colors can have multiple connotations, depending on context. The association of red with blood can be good or bad, depending on whether we're in love or bleeding, and that association can then be positive (e.g., a Valentine's Day card) or negative (e.g., a Danger Hazardous Area sign). Black can mean sleek instead of evil if it's on a cocktail dress or tuxedo. Yellow can imply cowardice, green can imply illness or envy, blues can be associated with depression, and gray can imply sophistication. What other color associations can you list?

Our cultural background deeply influences how we apply meaning to color. For example, in Western cultures white is often associated with purity or innocence: this association is reinforced every time we see a bride's white wedding dress. But in other cultures, this association can be remarkably different: in Japan, white is commonly worn at funerals.

Colors can be linked to changing and competing ideologies, but these connections are often tenuous, and context is everything. Red can mean communism, when considered in terms of mid-twentieth-century propaganda. But red can mean democratic patriotism when on a piece of red, white, and blue bunting in a Fourth of July parade.

Moreover, cultures change. Avocado green and harvest gold were popular colors for kitchen appliances in the 1970s, but if you see those colors at all today, they're probably considered retro or nostalgic. Because the meanings or values associated with colors differ so much between and within cultures, it makes sense to pay attention to user preferences before making color specifications for a document. In particular, focus groups can help us to assess users' attitudes toward different colors and color schemes. Industrial designers commonly use focus groups to gauge customer responses to color schemes for products from automobiles to household appliances.

It's also a good idea to find out more about the client context of your document. For example, corporations, universities, and agencies frequently make a universal decision about what color scheme their products and documents will use. Sometimes, these color schemes stretch across all aspects of the corporate identity, from packaging to advertising to internal communication. Coca-Cola's use of a consistent scheme of red and white is one example. Corporations make these specifications to encourage users to associate these colors with a corporate identity—a technique known as *branding*. So users may actually expect to see a document published by a company to share its corporate color scheme, and they might be confused if it doesn't.

RHETORIC

Despite the ambiguities of color associations, there are at least three tendencies we can generally rely on for rhetorical purposes.

COLOR CAN CONVEY MEANING

Users will often accept colors as meaning something in a document if we *explicitly indicate* what they mean. For example, if a road map legend indicates that red marks interstate highways, users usually allow that meaning to override any broader cultural meaning of red, such as its common association with caution or danger.

But we can't always expect a specified color association (e.g., a map legend color) to work without being explicitly connected (indexed) to its meaning, and we can't assume the color association to extend beyond that document. In some cases, a broader cultural meaning can overpower a local indexed meaning, no matter how much we indicate otherwise. In Western cultures, for example, documents rarely index male with pink and female with blue, so powerful are the associations of blue with masculinity and pink with femininity. On the other hand, if you are designing a brochure for a conference on gender equity, switching these color associations might be an appropriate reminder of cultural values of gender. And indexing pink or blue to mean something *not* connected to gender wouldn't necessarily activate the cultural association.

COLOR CAN IMPLY VALUE

The very presence of color suggests expensiveness and therefore a certain cultural status. Purple, for example, was for centuries the most expensive pigment in the world (the color, *purple*, is named after it), and anything colored with purple was considered rare and precious. The value came from rarity—the pigment was made from the spiny dye murex snail (also known as the purple dye murex), and the process to make the dye, Tyrian purple, was difficult and expensive. We get the color name from the dye: in Greek, it was *purphura*; in Latin, *purpura*; in Old English, purpul; and now, of course, purple. Royal robes were dyed with purple, and even the paint used for royal portraits used purple as the pigment. Purple might be easier to reproduce now, but even today, with color inkjet printers as common as they are, color printing costs significantly more than black and white. Users often accordingly assume that a document printed in color is more important than one in black and white and give it more initial attention, if only because its producers thought it was important enough to spend money printing it.

Of course, color is free on screens, so a rhetoric that links color to value might not work as well there. But even on the screen, users recognize color as an extra step in the design and value it accordingly. A lack of color can imply that the information being conveyed is boring, dull, ephemeral, or highly technical.

Which colors say "money" to people depends on broad cultural values and conditions. In the eighteenth century, rich people in Europe and America painted rooms in brilliant and expensive blues, yellows, and greens because these colors advertised wealth; less affluent families painted their rooms with milder, earthier colors like white and buff (if they could afford paint at all). In the mid-nineteenth century, however, the development of new dye technologies made bright colors more widely available and affordable, so by the end of the century, the poor (trying to keep up with earlier standards of taste) painted rooms in bright, garish colors. In the meantime, the rich had moved on to the muted, earthy shades of the arts and crafts movement, and later to the whites and pastels of modernism.

Given these evolutions in color and tastes, many companies today actually pay color consultants to predict what colors will be popular or even cool, hoping to use the information to confer an added perception of value to the user experience.

COLOR CAN ATTRACT ATTENTION

While the specific meaning of a color is nearly impossible to pin down, designers can use contrasts of color to great rhetorical effect. Within a uniformly gray, black, or white field, a small amount of nearly *any* color can make users at least look. For example, imagine a typical black-and-white document that begins with a large dropped cap in bright red. Wouldn't you be drawn to that spot and start reading there?

Of course, this use of color depends on our physical perceptions of figure-ground separation. If the color you use to bring attention to a design object is too close to the background color in hue, saturation, or brightness, it might fade into the background. A dropped cap in bright yellow won't contrast well with a white background because both yellow and white have a high brightness. So when determining color contrast, be sure to take into account all dimensions of color: hue, saturation, and brightness.

More about . . . Color Rhetoric

- Use color to convey meaning.
- Use color to imply value.
- Use color to attract attention.

CREATING COLOR ON SCREENS AND PAPER

Given our perceptual and cultural responses to color, it's clear that color can be a useful technique in visual rhetoric. But reproducing color in documents involves some technological hurdles, both in print and on the screen. The technological approaches to reproducing color in these two media are complementary but significantly different enough to require separate explanations.

In this section, we'll examine two ways to reproduce color:

- *Additive color (RGB)*, which reproduces color on screens

- *Subtractive color (CMYK)*, which reproduces color on reflective surfaces, such as paper

We'll start with additive color because it's the more basic of the two systems—in fact, subtractive color is based on additive color.

COLOR ON SCREENS: ADDITIVE COLOR (RGB)

As we said earlier, the three basic hues that our eyes respond to are red, green, and blue. These colors are known as the *additive primary colors*, commonly abbreviated *RGB*. Our perceptions of light in these three colors combine to create all of the other colors we perceive—hence the term *additive*.

MAKING RGB COLORS

Computer monitors reproduce color by transmitting light in the three additive primary colors. They do so by regulating the light that glows from many small, digitally controlled glowing *pixels*. Each pixel in turn is formed of three *subpixels*: one red, one green, and one blue. Computers can turn these subpixels off or on individually. They can also digitally modulate the brightness of subpixels, mixing RGB to create a perception of hundreds, thousands, or even millions of different colors.

To understand how this works, imagine a single pixel as a tiny cluster of three lightbulbs in red, blue, and green. If we turn off all three of the lightbulbs, we get black, the absence of light. Conversely, if we turn on all three bulbs at full strength, the colors mix to produce white (figure 8.8).

Figure 8.8. Additive color (RGB). Light in wavelengths that we perceive as red, green, or blue mixes to create all the other colors. On a computer monitor, RGB subpixels can be turned on in different strengths (0–100 percent on a 0–255 scale in this illustration) to create all the colors in the RGB gamut. *Source*: Created by the author.

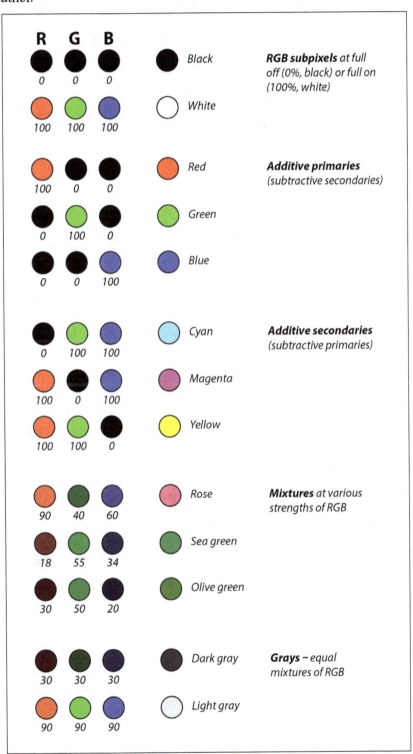

Table 8.1. Lights All Off or All On

% Red +	% Green +	% Blue =	Final Color
0	0	0	Black (no light)
100	100	100	White (all light)

If we turn on one of the bulbs at full strength and leave the other two off, we get the additive primary colors, RGB:

Table 8.2. One of Three Lights On at Full Strength

% Red +	% Green +	% Blue =	Final Color
100	0	0	Red
0	100	0	Green
0	0	100	Blue

If we turn on two of the lightbulbs at full strength and leave the third off, we get the *additive secondary colors*, cyan, magenta, and yellow:

Table 8.3. Two of Three Lights On at Full Strength

% Red +	% Green +	% Blue =	Final Color
100	100	0	Yellow
100	0	100	Magenta
0	100	100	Cyan

If we turn on the lightbulbs at different strengths, we can create all the combinations of colors that the monitor can show, such as these:

Table 8.4. All Lights On at Varying Strengths

% Red +	% Green +	% Blue =	Final Color
90	40	60	Rose
0	100	60	Sea green
30	50	20	Olive green

Likewise, if we turn on all of the bulbs at equal strengths below 100 percent, we get different shades of gray:

Table 8.5. All Lights On at Equal Strengths below 100 Percent

% Red +	% Green +	% Blue =	Final Color
30	30	30	Dark gray
90	90	90	Light gray

Notice that with low strengths, we get darker grays because less light has been emitted. Conversely, the greater the brightness of the subpixels, the lighter the gray (i.e., the closer to white).

Computer programs that allow you to design in color for screen documents will allow you to specify RGB values for the colors you want. Most do so through a *color picker* (see "Using Color Pickers" in this chapter).

SPECIFYING RBG FOR WEBSITES

RGB is also the primary way to describe colors for websites, so lots of web design programs also have color pickers. But Hypertext Markup Language (HTML) uses a numbering system known as *hexadecimal*, or just *hex*, to specify colors on web pages. Understanding hex will allow you to specify web colors more precisely and consistently.

You're probably more familiar with decimal (base-10) numbering, using the digits 0, 1, 2, 3, 4, 5, 6, 7, 8, and 9. As decimal numbers rise above 9, we add a digit to represent the next decimal unit (see figure 8.9):

0	1	2	3	4	5	6	7	8	9
10	11	12	13	14	15	16	17	18	19
20	21	22	23	24	25	26	27	28	29
				. . .					
90	91	92	93	94	95	96	97	98	99

Figure 8.9. Decimal numbering. In decimal numbering (base-10), we add digits after 9 to represent additional decimal units. *Source*: Created by the author.

Hexadecimal numbering works much the same way, but with sixteen digits—0,1, 2, 3, 4, 5, 6, 7, 8, 9, A, B, C, D, E, and F—with 0 as the lowest digit and F as the highest.

Above the first sixteen digits, we can continue numbering with two-digit hex numbers, like this (see fig. 8.10):

00	01	02	03	04	05	06	07	08	09	0A	0B	0C	0D	0E	0F
10	11	12	13	14	15	16	17	18	19	1A	1B	1C	1D	1E	1F
20	21	22	23	24	25	26	27	28	29	2A	2B	2C	2D	2E	2F
30	31	32	33	34	35	36	37	38	39	3A	3B	3C	3D	3E	3F
40	41	42	43	44	45	46	47	48	49	4A	4B	4C	4D	4E	4F
50	51	52	53	54	55	56	57	58	59	5A	5B	5C	5D	5E	5F
60	61	62	63	64	65	66	67	68	69	6A	6B	6C	6D	6E	6F
70	71	72	73	74	75	76	77	78	79	7A	7B	7C	7D	7E	7F
80	81	82	83	84	85	86	87	88	89	8A	8B	8C	8D	8E	8F
90	91	92	93	94	95	96	97	98	99	9A	9B	9C	9D	9E	9F
A0	A1	A2	A3	A4	A5	A6	A7	A8	A9	AA	AB	AC	AD	AE	AF
B0	B1	B2	B3	B4	B5	B6	B7	B8	B9	BA	BB	BC	BD	BE	BF
C0	C1	C2	C3	C4	C5	C6	C7	C8	C9	CA	CB	CC	CD	CE	CF
D0	D1	D2	D3	D4	D5	D6	D7	D8	D9	DA	DB	DC	DD	DE	DF
E0	E1	E2	E3	E4	E5	E6	E7	E8	E9	EA	EB	EC	ED	EE	EF
F0	F1	F2	F3	F4	F5	F6	F7	F8	F9	FA	FB	FC	FD	FE	FF

Figure 8.10. Hexadecimal numbering. Hexadecimal numbering uses sixteen digits as its base instead of ten, with the letters A, B, C, D, E, and F being added after any element ending in _9. In this system, for a sixteen-digit run starting at zero, we start at 00 and end with 0F. The next set begins with 10 (it helps if you read that as one, zero) and ends with 1F. And so on. *Source*: Created by the author.

Computers use hexadecimal numbering because it corresponds more closely than decimal numbering to the 8-bit, 16-bit, 32-bit, and 64-bit color systems used on most computer monitors, mostly because 16 is evenly divisible by 8 and is half of 32. This also matches nicely to the basic mathematics of computer memory, in which each *bit* of memory consists of 8 *bytes*. Using hexadecimals allows computers to specify values in a very consistent set of regulated steps. The preceding hexadecimal grid entirely from 00 to FF includes 256 numbers—16 columns × 16 rows. This allows us to turn on each RGB subpixel of a computer in 256 regulated steps of brightness in 32-bit systems, and 65,536 levels in a 64-bit system.

As a result, HTML specifies RGB values with three two-digit hexadecimal numbers. The first number controls red, the middle number controls green, and the last number controls blue:

RR, GG, BB

These three numbers are run together into one long string: RRGGBB. The higher the hexadecimal number we put in each of these slots, the brighter the light shining out through the corresponding subpixel. So to specify black, we turn *off* all three subpixels, and to specify white, we turn *on* all three subpixels at full strength:

Table 8.6. All Subpixels Off or All On

Color	Code	Red	Green	Blue
Black (no light)	000000	Off	Off	Off
White (full light)	FFFFFF	Full	Full	Full

Making fully saturated red, green, or blue requires turning on the appropriate pixel at full strength (FF) and turning off the other two (00):

Table 8.7. One of Three Pixels Fully On with the Others Off

Color	Code	Red	Green	Blue
Red	FF0000	Full	Off	Off
Green	00FF00	Off	Full	Off
Blue	0000FF	Off	Off	Full

And if we turn on the pixels at different strengths, we get different combinations of colors:

Table 8.8. Pixels On at Different Strengths

Color	Code	Red	Green	Blue
Cyan	00FFFF	Off	Full	Full
Orange	FF6600	Full	Some	Off
Dark purple	660066	Some	Off	Some
Gray (RGB in equal proportions)	999999	Some	Some	Some

Because we can specify 256 strengths for each color with a 24- and 32-bit system, we have access to more than sixteen million colors for websites (256 × 256 × 256 = 16,777,216) at the 24-bit level. This is known as *true color*. The 32-bit level adds transparency (an alpha channel) to the available palette, resulting in upward of four *billion* possibilities. This is referred to as *deep color*, and can extend beyond trillions of colors in higher-bit systems.

You can use these hexadecimal numbers in HTML to specify the precise color you want to display. For example, 036A9E gives a nice sky blue (a tiny bit of red, a little more green, about two-thirds powder blue). But remember that every computer monitor is different. That beautiful 036A9E might not look exactly the same on your monitor as it does on the user's monitor.

Because of the differences among monitors and operating systems, many early web designers followed the practice of relying on 256 basic *web-safe colors*—a subset of the sixteen million allowed by hexadecimals. Sticking with 256 colors also ensured that colors would reproduce on basic 256-color monitors, which were still common in the early 1990s. The web-safe colors were determined by equally spaced values in the hexadecimal grid:

00 33 66 99 AA CC FF

Any combination of these values in the RGB notation (RRGGBB) results in a web-safe color that will look relatively consistent on most monitors. For example, 9900FF creates a web-safe medium purple, and 99FF00 creates a web-safe lime green. As monitors with high color capabilities have become ubiquitous, it's much less common now to restrict web designs to the web-safe colors.

> ## More about . . . Website Colors
>
> Test your website design on a broad range of monitors before making a final decision about colors.

COLOR ON PAPER: SUBTRACTIVE COLOR (CMYK)

Considering red, green, and blue as primary colors might sound odd to you—especially if you remember the primaries from your childhood paint set as red, blue, and yellow. But when we mix watercolors, paints, or inks, we aren't really mixing colors—we're mixing pigments or dyes. These materials *reflect* light in some wavelengths and *absorb* light in others. The reflected wavelengths are what we perceive as tristimulus values, or RGB.

Imagine looking at a piece of red-colored paper in a sunlit room. Because sunlight is an even mixture of all visible wavelengths of light, the light isn't making the paper seem red by itself. Instead, the paper contributes by reflecting light in wavelengths that we recognize as red, while absorbing light in the wavelengths that we recognize as blue or green. The paper isn't inherently "colored"—it merely reflects light in some wavelengths and absorbs others. Likewise, blue paper would reflect blue light and absorb red and green light, and green paper would reflect green light while absorbing red and blue light. This process works for combinations of RGB colors as well. For example, paper that absorbs red and reflects blue and green will give us a perception of cyan; paper that absorbs blue and reflects red and yellow will appear yellow; and paper that absorbs green and reflects red and blue will appear magenta.

This process is called *subtractive color* because some wavelengths of light are absorbed or filtered out, leaving only the remaining wavelengths to be reflected (figure 8.11). When all of the wavelengths of light are reflected, as on white paper, they combine to create a sensation of white. But when all of the wavelengths of light are absorbed, we perceive black—the absence of light.

Subtractive color might seem completely different from additive color. But in fact, the two systems are closely related in that the process of absorption and reflection that leads to subtractive color still creates reflected light in the additive system, as a mixture of RGB. In other words, subtractive color is based on additive color.

Figure 8.11. Subtractive color (CMYK). Subtractive color works when a surface absorbs light in some wavelengths and reflects light in others. The reflected wavelengths mix together additively to create the mixture of light we perceive as a color. In the first example, the surface absorbs red and reflects green and blue, creating cyan. The second example shows green absorption, reflecting red and blue to make magenta. The third shows blue absorption, resulting in a perception of yellow. *Source*: Created by the author.

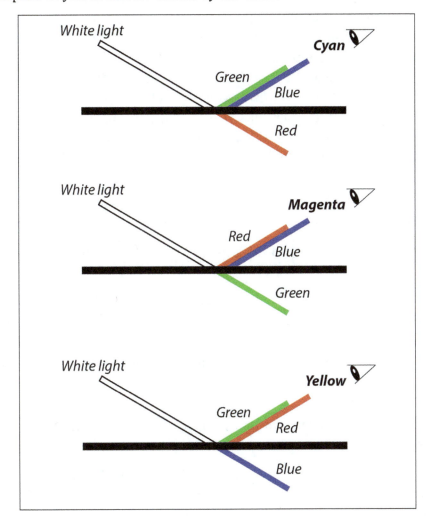

PROCESS COLOR

Subtractive color creates a mechanism for producing color in print. The print medium relies on external light reflecting off the inked surface of the paper to create our perceptions of RGB colors. In other words, we can print in color by using colored inks that absorb light in some wavelengths and reflect light in others. We could simply use lots of different colors of ink to get full-color printing. But except for some special situations (see "Spot Color" in this chapter), using many inks isn't very practical in production. The trick is combining a small number of inks to create all of the colors in a full-color image.

This technique is known as *process color* because it creates a full-color image through a process of overprinting the same sheet four times with four different inks. In this process, the red, blue, and yellow we traditionally think of as the subtractive primaries in painting are actually not quite the right pigments. Instead, process color printing typically uses translucent inks in cyan, magenta, and yellow, or *CMY*.

These subtractive primaries are directly related to the three additive primaries (RGB) that we can perceive: the subtractive (CMY) primary colors are the additive (RGB) secondary colors, and vice versa. This relationship is tied to how inks reflect or absorb wavelengths of light. Each of the process inks absorbs light in the wavelength of one of the additive primaries (RGB) and reflects and combines light of the other two:

Table 8.9. Light Absorption and Reflection with Resulting Appearance for Additive Primaries

Ink Absorbs	Ink Reflects	Ink Appears
Red	Green + Blue	Cyan
Green	Red + Blue	Magenta
Blue	Green + Red	Yellow

Conversely, we can combine the subtractive primaries CMY to produce red, green, and blue:

Table 8.10. Light Absorption and Reflection with Resulting Appearance for Subtractive Primaries

Ink Absorbs	Ink Reflects	Ink Appears
Yellow	Cyan + Magenta	Blue
Cyan	Magenta + Yellow	Red
Magenta	Yellow + Cyan	Green

By mixing CMY in different strengths, we can produce the appearance of many other colors.

In full-color (process) printing, we print dots of the translucent CMY inks at various strengths by using *screens* and *filters* to create *halftone separations*—one to print each color of ink. When combined, these color separations make up a full-color image.

This process happens mostly digitally now, but it's easier to imagine by going back to older photographic methods. Traditionally, printers made separations by photographing an image through a red, green, or blue filter and a sheet of glass that contained a tiny wire grid, like a tight window screen. Taking a picture through the red filter and screen would result in a field of small cyan dots (remember, filtering red allows green and blue light to continue, creating cyan). The dots would be bigger where there was more light in the unfiltered colors and smaller where there was less. Repeating this process with the green and blue filters would result in three separations, one for each of the three subtractive primaries, CMY (figure 8.12). These film separations were then used to make plates for a printing press. Printing every sheet three times—once with the cyan plate, once with the magenta plate, and once with the yellow plate—recreates the entire "full-color" image. (See chapter 10 for more information on color printing.)

Figure 8.12. CMYK color separations in process printing. By filtering out combinations of red, green, and blue, printers create lithographic plates to print cyan (filter out red, leave green + blue), magenta (filter out green, leave red + blue), and yellow (filter out blue, leave red + green). The black or "key" plate (K) usually carries all of the text, plus a black screen to help increase the saturation and deepen the shadows in process images. *Source*: Created by the author.

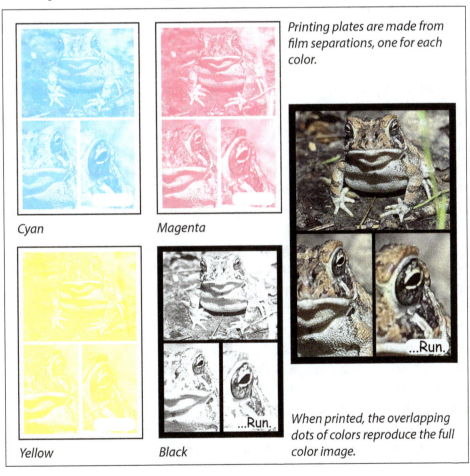

Cyan

Magenta

Printing plates are made from film separations, one for each color.

Yellow

Black

...Run.

...Run.

When printed, the overlapping dots of colors reproduce the full color image.

Or perhaps not the *entire* image—we still need black to make it work. As you'll recall, subtractive color creates black by filtering out or absorbing all of the wavelengths of light, leaving nothing to reflect. You might think that doing so would require mixing together the subtractive primaries (CMY) at full strength; together, they'd absorb all of the RGB light, leaving black. But in fact, doing so creates a dark, muddy, brownish gray—like the color you got when you mixed together all your watercolors when you were a child. So if we print using only the C, M, and Y plates, the shadows tend to be faint and the colors less saturated than we might like.

Adding a black plate to the mix solves these problems. A black screen can be printed in a truly black ink, which absorbs light much better than any mixture of translucent CMY inks. This black plate is where the *K* in CMYK comes in: *K* stands for *key* because black is usually the first color printed and the other three are aligned or "keyed" to it. The key does not have to be black, of course—you could specific any color you wanted, but black is typical. The key plate also typically includes the text, which is usually printed in black ink as well. This plate produces a good, strong black for all of the text, plus a full range of colors and shadows for images.

SPOT COLOR

Process color is a great way to produce many colors, especially for graphics with lots of color variation or gradients, such as photographs. But process color doesn't work nearly as well for creating areas of a single color, such as a solid-colored background, a logo, or colored text. In those situations, process color tends to come out looking weak and inconsistent. This weakness occurs primarily because process color creates color sensations by printing overlapping dots of CMYK inks, all of which, except black, are translucent.

So designers typically specify areas of one color with a *spot color*, which is printed in a particular color of ink, different from CMYK. Spot color gives clear, strong, even color across a whole area. It's not unusual for a high-design document to be designed with six-color printing: CMYK, plus two spot colors. Each additional spot ink, however, increases the complexity and the cost of the print job.

Spot colors come in a wide variety—every color, tint, and shade you can think of, including metallic and high-visibility inks. You will need to coordinate with your print shop personnel when you specify spot colors because they may need to order the particular colors of ink your document calls for. Usually, printers ask that you specify spot colors with a commercial system such as the Pantone Matching System (PMS), which provides standard codes for the different colored inks.

COLOR GAMUT

Why is it so important to understand the difference between these two systems? Because the range of colors we can produce with RGB differs significantly from the colors we can produce with CMYK (see figure 8.6). The range of available colors in a color space is called a *gamut*. A color that cannot be reproduced in the current color space is called *out of gamut*.

Most page layout, drawing, and image editing programs require that you choose whether to work in the RGB or CMYK color space, and they will warn you if you choose a color that's out of gamut according to the color space you have specified. In Adobe color pickers, for example, an out-of-gamut color will be marked by a triangular warning icon in the color picker. Clicking on the warning icon shifts the color you've chosen to the nearest in-gamut color (see "Using Color Pickers" in this chapter).

More about . . . Computer Monitors

All computer monitors use additive color mixing (RGB). If you use CMYK to specify colors on a website, computer monitors may not be able to replicate the color you specified. In other words, the color might be out of gamut. The same applies if you use RGB to specify something that will be printed with process (CMYK) printing: the technology may not be able to replicate the color you specified.

If you choose an out-of-gamut color, what you'll get is the closest color the computer can estimate within the appropriate gamut—and sometimes computers make really poor estimates.

One common mistake when designing for print is to be fooled by a vibrant color on your computer monitor, only to find out that the color prints quite differently in process color. Remember that computer monitors use RGB—even when you're designing in CMYK. So what you see on the computer screen is already the computer's best RGB estimate of the CMYK value you specified. To remedy this problem, see "Matching Colors" and "Calibrating Your Monitor" in this chapter.

DESIGNING WITH COLOR

Now that you have a solid foundation for working with color, let's examine some ways to create color schemes for your designs. At some point early in the design process, it's usually best to specify several colors that work well together rhetorically and aesthetically. Rhetorically, try to choose colors that users will subjectively associate with the tone you wish to set. Aesthetically, try to choose colors that fit with the users' cultural sense of beauty, utility, or professionalism, depending on the purpose of the document.

Because of the many ambiguities of color communication, choices about color schemes can never be entirely dependable or predictable. For subjective or cultural reasons, some readers will inevitably respond differently to the color scheme than you intended. But there are some common tendencies designers usually follow, and by repeatedly encountering documents designed with these tendencies, users become trained to expect certain colors to be associated with certain tones.

COLOR SIMILARITY AND CONTRAST

One of the most common approaches is to use the design principles of similarity and contrast to create color schemes. Design objects with similar colors—colors that lie close to each other in one or more of the dimensions of color (hue, saturation, and brightness)—imply a similarity of purpose, function, or identity. By corollary, contrasting colors can dissociate a design object from the rest of the design, creating emphasis or establishing a contrast in purpose, topic, or function.

For example, we might choose a series of similar colors as the main colors of the design and then add in a single contrasting color to highlight important features—such as a company logo, a dropped cap, a title, or a heading.

Similarity of color is pretty obvious, but creating effective color contrast is a little more complicated because of the different *ways* colors can contrast. Colors can contrast in four primary ways (see figure 8.13):

- *Hue:* The farther apart hues are from each other on a color circle or an HSB color picker, the greater the contrast between them. The highest level of contrastive hues takes one color from one side of the circle and its *complementary color* from 180 degrees opposite.

- *Saturation:* Even within one hue, we can create contrasts of saturation. The farther apart the colors lie in terms of their saturation, the higher the contrast.

- *Brightness:* A relatively dark color will contrast strongly with a relatively bright color.

- *Temperature:* We can make a further distinction between *cool colors*, on the blue-green side of the color circle, and *warm colors*, on the red-yellow side. When cool and warm colors are viewed together, the cool colors often seem to recede away from the viewer, and warm colors seem to advance toward the viewer. Matching a cool color with a warm color creates a contrast of temperature.

You can use any of these kinds of contrasts individually or combine them to create strong color contrast between different elements in your design. Keeping colors relatively close in hue, brightness, saturation, and temperature leads to greater similarity—a terrific tool for showing coherence and unity in your design. But you must also consider the interactions of these different contrasts. For example, saturation affects color temperature: a light pink (desaturated red) might not seem to advance over a highly saturated blue, even though pink is from the warm (supposedly advancing) side of the color circle.

Figure 8.13. Color contrasts. You can form color contrasts by altering hue, saturation, brightness, and temperature. *Source*: Created by the author.

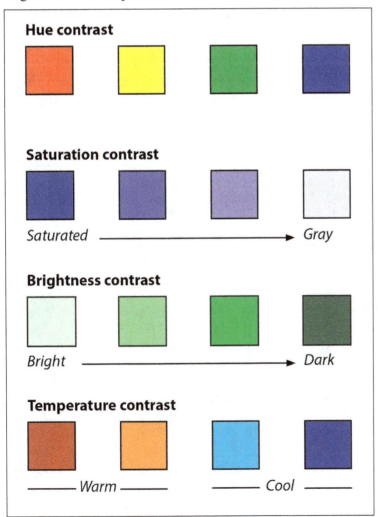

COLOR SCHEMES

Choosing a consistent color scheme can help create a consistent identity within and between documents. It's particularly important to use a consistent color scheme for websites, which can have a myriad of pages; for documentation sets, which include a variety of related documents; and for organizational style sheets, which can regulate the many documents an organization creates.

Designers often rely on several traditional color schemes: monochromatic, analogous, complementary, split-complementary, triadic, and tetradic. All of these schemes are based on choosing hues according to their positions on a color circle. These color schemes are not rules but simply guidelines for choosing hues that work well together. And even after you've chosen a set of hues, you'll still need to make decisions about an appropriate saturation and brightness for each color. Many designers make these decisions by eye, making careful and sometimes subtle adjustments to get just the right set of colors.

MONOCHROMATIC AND ANALOGOUS

Monochromatic schemes use colors restricted to one hue—that is, one point around the color circle. *Analogous* schemes broaden this approach somewhat, using a narrow range of hues contiguous to a central hue (figure 8.14).

Figure 8.14. Monochromatic and analogous color schemes. Monochromatic and analogous color schemes use different brightnesses and saturations of one hue or a small range of hues, respectively. This example is monochromatic, focusing on the hue of green. *Source*: Shaun Inman Design and Development Inc., https://www.haveamint.com.

Monochromatic

Analogous

Monochromatic and analogous schemes often seem very coherent to viewers, and they're particularly well suited to designs restricted by cost or production capability to one color of ink. But because they are concentrated in or around one hue, these schemes offer fewer possibilities for pointing out contrastive elements than other color schemes.

COMPLEMENTARY AND SPLIT-COMPLEMENTARY

Complementary color schemes specify two colors on opposite sides of the color wheel, or close to it. A common design strategy selects one hue from the cool side of the wheel and then a complementary hue from the warm side of the wheel (figure 8.15).

Split-complementary color schemes work similarly but with a little more flexibility. To form a split-complementary scheme, choose one hue, and then choose two or more hues close to the complementary of the first hue (figure 8.15).

Figure 8.15. Complementary and split-complementary color schemes. Complementary color schemes choose two hues from opposite sides of the color circle. Split-complementary color schemes choose one hue from one side of the circle and two or more hues slightly displaced from the complementary hue. This example shows a complementary color scheme. *Source*: Nintendo, *The Legend of Zelda: Majora's Mask*.

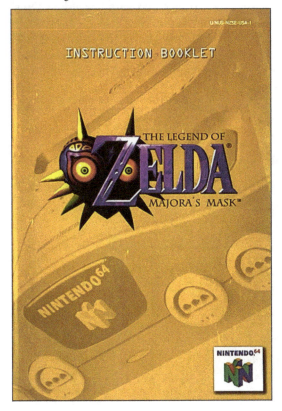

Complementary

Split-Complementary

The orange and purple of this instruction booklet are almost directly opposite each other on the color circle, lending strong contrast to the design.

The downside of complementary and split-complementary color schemes is that they make it easy to create garish designs. If you use one of these schemes, consider using a mostly cool, receding hue for the main color, and then pick out important elements with the complementary or split-complementary hues from the warm side of the circle. (On the other hand, sometimes garishness is just what the rhetorical situation calls for!)

TRIADIC AND TETRADIC

Designers often use hues that are separated at either three (*triadic*) or four (*tetradic*) points around the color circle. A triadic scheme might use red, blue, and green or cyan, magenta, and yellow. A tetradic scheme might use red, blue, green, and orange or cyan, magenta, yellow, and purple. The three or four hues are sometimes spaced equally around the circle, but often the designer adjusts the angles between the hues by eye. As with the other schemes we've discussed, designers often focus most of the design on one hue, using the others to highlight important features (figure 8.16).

Triadic and tetradic color schemes can appear playful and colorful, but they can also look unsophisticated (perhaps because it's common to color children's toys in such widely spaced colors).

Figure 8.16. Triadic and tetradic color schemes. Triadic and tetradic color schemes use hues spaced at three or four points around the color circle. These color schemes often look bright and playful. This example shows the front and back cover of a brochure in a triadic color scheme of orange, blue, and green. *Source*: Library of Congress, "Discovery Guide for Kids and Families."

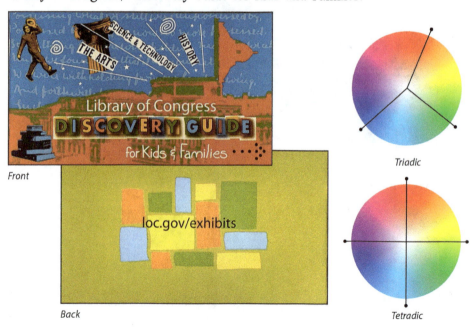

DESIGNING WITH LIMITED COLORS

In print design especially, you might be limited by cost or production factors to designs that can be printed with a limited palette of one or two colors.

DESIGNING WITH ONE COLOR

Designers frequently must work with only one color—and that color is often black. For example, if you know that your design must be printed on a laser printer and photocopied, there's little reason to spend time thinking about grandiose color schemes. But even black can come in different "colors"—the range of saturations we call grays. A design that uses several grays and a strong black can be quite striking and effective. This book, for example, relies mostly on grays and black, due to production costs.

Designing with grays usually involves applying a *screen* to areas of black. In other words, the only ink used is black, but it's applied in small dots; as the dots get smaller and farther apart, they give the appearance of lighter grays. Most page layout and graphic programs allow fine control over screens, from 0 percent (white) to 100 percent (full black).

However, screens can sometimes be visible—enough to distract users from the purpose of the document. This problem arises particularly with photographs and other graphics that use smooth tonal gradients. If your printer resolution is relatively high—at least 600 dpi—most viewers won't be able to see the small dots without careful scrutiny, and most home printers print at resolutions at or above 600 dpi. Professional offset printing will yield an even higher printer resolution, leading to smoother-looking grays. But if your job will be printed on a low-resolution printer, such as a laser or inkjet printer at 300 dpi, the dots that make up the screen will be large enough for most people to see. (Ironically, photocopying a low-resolution printout can smooth out the screens somewhat, making the dots less obvious.)

If your one-color job will be printed professionally on an offset press, you can specify a color of ink other than black, using screens of that color to create your design. This technique can be a good way to create a distinctive-looking document. But remember that black is the darkest, most dense, and most striking contrast with white paper, and users need strong figure-ground contrast to make out the design—especially the text. Any other color will have less contrast with the paper and therefore make text more difficult to read. Some hues, such as yellow, will have an effective luminance too close to white to be usable in one-color designs on white paper.

DESIGNING WITH TWO COLORS

Typically, two-color designs on white paper use black as one of the colors (for text) plus a strikingly different color for important elements of contrast. For example, a company letterhead design might have the text for the address in black and the logo in a fully saturated spot color. Or a technical document might use black text but have headings and other navigational design objects in a different color.

The trick is to choose a color that will contrast adequately with the black text while still providing good figure-ground contrast. A dark blue might have good contrast with the background but too little contrast with the black text. For example, in figure 8.17 the Shakespeare Theatre Company has used a spot ink in purple to establish the layout grid of the page spread, but it contrasts well with black.

It's also possible to use any contrasting two spot inks, such as in the cover of the Freer-Sackler Gallery events calendar in figure 8.18. Here, color "grayscale" images are printed with purple and orange spot inks, a technique known as duotone printing.

Figure 8.17. Designing with two colors. This play program uses purple and black inks not only to build interest but to establish the layout grid. There's enough contrast between the purple spot ink and the black ink that the text stands out clearly from the background. *Source*: Shakespeare Theatre Company, "The Beaux' Stratagem."

Figure 8.18. Duotones. This calendar of events for an art gallery uses a duotone image built up of purple and orange spot inks to give a sophisticated look to the design. *Source*: Freer-Sackler Gallery, Smithsonian Institution, "Discoveries 2006."

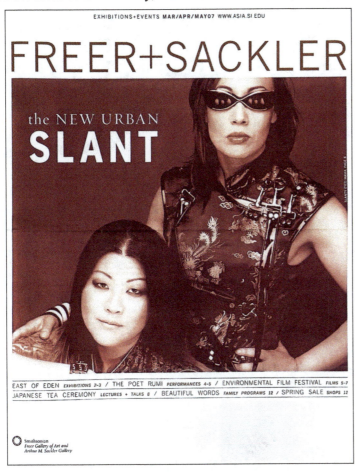

WORKING WITH COLOR ON COMPUTERS

Whether designing for print or for the screen, you will definitely use computers to create your designs and manipulate colors. In this section, we'll look at three important techniques for working with color on computers: using color pickers, matching colors, and calibrating monitors.

USING COLOR PICKERS

Most software color pickers—the dialog boxes in which you specify colors—use both numbers and a graphical model for color. Typically, different programs have their own designs for color pickers, or you can use the system color pickers that come with Windows or Apple OS. As you can see in figure 8.19, what these systems share is an ability to specify color using several different models, including RGB, CMYK, hex, CIE L*a*b, and HSB/L (hue, saturation, and brightness/luminance).

Figures 8.19a and 8.19b. Color pickers in Microsoft Word and Adobe Photoshop. Note that each color picker allows you to specify colors either numerically or graphically using multiple color models. *Source*: Created by the author.

Microsoft's multi-layered color picker offers an HSB color wheel with a slider for saturation (1), sliders and input for grayscale, RGB, CMYK, and HSB (2), choices based on developer, websafe colors, crayons, and more (3), the spectrum (4), and visual representations of colored pencils (5).

Adobe Photoshop's color picker offers a rectangle controlling saturation and brightness (1), a slider that controls hue (2), icons that warn if a color is out of gamut (3), and multiple ways to specify specific colors (4).

More about . . . Color Pickers

Make yourself familiar with the color pickers in the software you use. A little bit of experimentation can give you a lot of confidence in your work with color.

In these color pickers, color specifications are often represented graphically by a 2D space (usually rectangular or circular) and a slider. For example, in basic rectangular HSB/L color pickers, the color picker splits the color circle at the top and flattens it *horizontally*, stretching from left (0 degrees) to right (360 degrees). Because red is traditionally represented on the color circle at 0 degrees/360 degrees, it's represented at the far left and far right sides of the rectangle, with the other colors ranged in between in order of the spectrum. Saturation is represented *vertically* in the rectangular color picker, ranging from full saturation at the top to low saturation (grays) at the bottom. Brightness/luminance is controlled by a vertical slider that ranges from high brightness/luminance at the top to no brightness/luminance at the bottom. You can choose colors in a color picker through three different mechanisms:

- *Clicking in a rectangular or circular color space:* A target mark (usually a small ring or crosshair) will show what spot you've chosen.

- *Adjusting sliders:* Some HSB/L pickers include a vertical slider for brightness/luminance, but many pickers also include sliders that can be dragged to adjust the other dimensions of the color. These sliders are particularly useful for adjusting one dimension while leaving the other settings alone. For example, you can change saturation without altering hue or brightness.

- *Specifying each dimension of the color numerically:* Keeping with the metaphor of the color circle, many color pickers measure hue as 0 to 360 degrees for both additive and subtractive color. RGB values, because they are intended for use in on-screen designs, are often represented as steps from 0 to 255. CMYK pickers typically use percentage values from 0 to 100 percent.

All of these controls are coupled together. For example, if you change the hue numerically, you'll see the target move in the color space and the appropriate sliders shift to whatever color you've chosen.

Different programs use different designs for color pickers. Adobe Photoshop includes a single rectangular color picker that shows numerical values for RGB, CMYK, HSB, CIE L*a*b, and hexadecimals in one view. Microsoft Word offers only RGB and HSL settings. Many software programs use a color picker that is designed to fit the purpose of the program, so a web graphics program will likely put a hexadecimal or RGB picker in the forefront, while a page layout program will default to CMYK.

MATCHING COLORS

When designing documents, you'll want to use colors consistently from page to page. Doing so requires specifying colors precisely as you use the same color in different parts of your design. Accuracy in color specification is particularly important if you're designing with spot colors. Designers use three techniques to make sure their colors stay consistent: keeping track of numerical color settings, applying styles, and using the eyedropper tool.

As we discussed in the previous section, color pickers specify colors not only in a 3D color space but also by specific numbers for colors in HSB/L, RGB, or CMYK. *Keep track of these numbers as you design.* For example, if you want every caption throughout a document to use the same color of type, take note of the color by copying down the corresponding HSB/L, RGB, or CMYK numbers. That way, when you want to make another caption or use the same color for another purpose, you can specify exactly the same color by entering the same numbers. This technique is particularly important if you decide to create multiple documents in the same color scheme, as corporations often do. In fact, you might use or even be involved in writing an organizational style sheet that specifies what colors are to be used in your organization's documents, usually with RGB for screens, CMYK for process color, and Pantone for spot inks.

A second technique, using styles, applies primarily to colored text. You can use a repeatable paragraph or character style to replicate the color of text, just as you can specify its typeface, font, leading, and indentation. Each time you want to use the same color of text, just apply the style to the appropriate words or paragraph. For more information on using styles, see chapter 6.

As a faster and less formal alternative, you can use the eyedropper tool and the palette in many page layout and graphics programs. The eyedropper tool allows you to click on a pixel to pick up its color specification. You can then switch to a different tool to draw objects or enter text using that same color. The palette allows you to save colors in a small window to create a list of your favorite or most-used colors.

CALIBRATING YOUR MONITOR

An additional difficulty arises when we prepare print jobs digitally on a computer: sometimes documents that look great on-screen look significantly different when printed. Why? Because whether you're working in CMYK or RGB, computer monitors display everything in RGB. So the colors you see while designing a brochure or manual in a page layout program are merely RGB approximations of the CMYK colors you've specified. Several steps can help you avoid this problem:

- *Control the computer environment.* Set the background of your computer monitor to a medium gray or a subtle grayscale pattern. Remember, simultaneous contrast can make colors look different; that cheerful photo wallpaper on your desktop can distort your perception of colors in a design.

- *Control the lighting in your workspace.* Light shining on your computer monitor can change perceived color, so indirect workspace lighting is best. Keep the lighting consistent from day to day, and avoid natural light, which can change remarkably throughout the day. Also avoid overhead fluorescent lights, which often lend a blue cast to your perceptions.

- *Color calibrate your monitor regularly.* Most operating systems come with display calibration software, and many sophisticated third-party options for monitor calibration are available. Page layout programs also allow you to specify a particular color set to use, such as Pantone color sets.

These techniques will help you create the greatest possible consistency between what you see on the screen and what you get in print. But keep in mind that there will always be some difference between what you see on your computer screen and what users see after the document is reproduced, whether on paper or on their own computer screen. Ideally, it's best to test out color calibrations by proofing color print jobs (see chapter 11) and by trying out on-screen designs with several different monitors.

EXERCISES

In this chapter you've thought about color as it relates to perception, culture, and rhetoric, thought about color on-screen vs. printed color, learned about using color to create emphasis, and thought about how you might design effective color schemes. You've spent quite a bit of time thinking about color theoretically . . . now it's time to go practice! The following exercises will help you put what you've learned into practice.

1. Software designers create predesigned color schemes for user interfaces and document templates. Their color design choices can be interesting:

   o Explore the color schemes that come with PowerPoint presentation templates. You should be able to find these by looking for the "design" tabs and links. Make a couple of slides to see how the design changes from the title slide to the body slides.

   o Think about and discuss these questions:

     • For what rhetorical purposes would you choose one of these templates? Why?

     • Do the color schemes of the templates follow one of the common schemes described in this chapter?

     • How does the design use color for emphasis?

   o PowerPoint slides are usually projected on a screen or viewed on a computer monitor. What effect do you think the medium had on the designer's color choices?

2. Collect a few documents in a variety of media, formats, and genres, with different color schemes and rhetorical purposes. Include:

   o Paper and electronic media

   o Different formats, sizes, and paper stocks

   o Different rhetorical purposes (persuasion, reporting, instructing)

   o Different genres (reports, brochures, flyers, instructions)

   o Four-color, two-color, and one-color

   o Some with graphics; some with photos; some with both

   Choose two for comparison and analyze their use of color. For each document, think about and discuss the following questions:

   o How would you describe the color scheme? Does it follow one of the common schemes described in this chapter?

o How does the design use color for emphasis?

o How does the design use color to create consistent patterns?

o What rhetorical tone do you think this color scheme was intended to convey? From your viewpoint as a user, is it successful?

3. In PowerPoint, use the Slide Master View to design your own PowerPoint template with a custom color scheme. Choose one of the following tasks to guide your design, or create your own context:

o A presentation for retirees about a retirement community

o A presentation for international students about computer services at your university

o A slide show to accompany an art exhibit.

Be sure to keep in mind the medium, format, and viewing conditions of PowerPoints as you make your design decisions. To start your design, follow these steps:

o Choose Master -> Slide Master from the View menu. You'll see the slide master open in the main editing space. A new toolbar will also appear: the Slide Master View toolbar.

o Start making your design:

  ▪ To design a new title slide, follow the "click to edit."

  ▪ To set a background color, right-click in the margin of a slide and choose Format Background.

  ▪ To draw designs on the background of the slide, use the tools in the Drawing toolbar.

  ▪ To change the color, typeface, or font size of standard text such as titles, subtitles, and bullets, choose the appropriate text in the master view and use the Formatting toolbar to make whatever changes you like.

o When you're finished with your design, click on Close Master View in the Slide Master View toolbar.

o Choose Save As from the File menu, enter a name for your new template, and choose Design Template (*.pot) as the file type. Now you can use this custom-designed template for your own PowerPoint slideshows.

4. Spend some time exploring what designs might look like to people with different abilities to perceive color.

o Go to the Color Scheme Generator at https://paletton.com/ (or any other color scheme explorer that allows you to view color palettes from alternative views) and create a color scheme. (Alternately, you can recreate a color scheme from a sample document.)

o Using the drop-down vision simulation menu at the bottom right-hand corner, change the color scheme to see how it would look to someone with deuteranomaly or protanomaly, the two most common kinds of color blindness.

What differences did you notice between viewing your color scheme with deuteranomaly or protanomaly if this is different from how you personally see color? How would you change your color scheme to make it more easily perceivable for users with these conditions?

REFERENCES AND FURTHER READING

Albers, J. (2013). *Interaction of color.* Yale University Press.
Clair, K. S. (2017). *The secret lives of color.* Penguin.
Gage, J. (1993). *Color and culture: Practice and meaning from antiquity to abstraction.* Little, Brown.
Gage, J. (1999). *Color and meaning: Art, science, and symbolism.* University of California Press.
Itten, J. (1970). *The elements of color* (E. Van Hagen, Trans., and F. Birren, Ed.). John Wiley and Sons.
Macdonald, L. (2023). *In pursuit of color.* Atelier Editions/D.A.P.
Munsell Color. (2024). *Munsell color space and solid.* https://munsell.com/about-munsell-color/how-color-notation-works/munsell-color-space-and-solid/
Parthasarathy, M. K., & Lakshminarayanan, V. (2019). Color vision and color spaces. *Optics and Photonics News, 30*(1), 44–51.
Williamson, S. J., & Cummins, H. Z. (1983). *Light and color in nature and art.* John Wiley and Sons.
Zollinger, H. (1999). *Color: A multidisciplinary approach.* Wiley-VCH.

9

LISTS, TABLES,
AND FORMS

LEARNING OBJECTIVES

Upon completing this chapter, you will be able to:

- Discuss how lists, tables, and forms work to organize information for users

- Articulate the way similarity, enclosure, alignment, contrast, order, and proximity work to shape perception and inform action

- Discuss how design and implementation change based on audience, purpose, and context

Imagine that you work for a property management company and you're reviewing applicants' résumés for a property manager position. Which of the versions of an entry from a résumé in figure 9.1 would you consider most impressive? Both résumés present pretty much the same information, but if you're like most managers, you'd probably prefer to read the version with the bulleted list of job duties. Why? Because it's easier to scan down a list to see what the applicant has to offer than it is to read through a long description. As a paragraph, the job duty items lose visual impact. For example, the item "keeping the books" might be impressive

Figure 9.1. Two versions of a résumé entry. One of the résumés uses a paragraph to describe job duties, and the other uses a bulleted list. If you were an employer, which version would you prefer to read? *Source*: Created by the author.

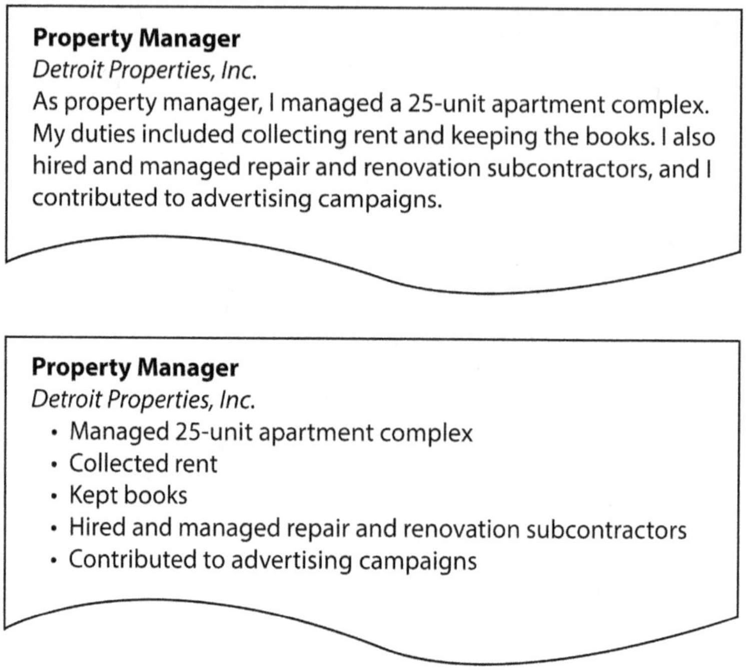

Property Manager
Detroit Properties, Inc.
As property manager, I managed a 25-unit apartment complex. My duties included collecting rent and keeping the books. I also hired and managed repair and renovation subcontractors, and I contributed to advertising campaigns.

Property Manager
Detroit Properties, Inc.
  • Managed 25-unit apartment complex
  • Collected rent
  • Kept books
  • Hired and managed repair and renovation subcontractors
  • Contributed to advertising campaigns

data, but the phrase is lost in the middle of the paragraph. As a bulleted list, however, each item has its own space, helping you concentrate on each item individually.

Like many technical documents, résumés contain many individual pieces of data—names, numbers, addresses, places, and facts. Absorbing all that information can be a challenge—especially if you have piles of résumés to wade through. The bulleted version in figure 9.1 reduces the reader's workload by visually emphasizing the individual pieces of data. And making the reader's job easier gives a better impression of the applicant.

Some of the most difficult tasks a document designer will face are the display and collection of data. By *data*, we mean the myriad small pieces of information—numerical, factual, or visual—that technical documents are often designed to convey or, in some cases, to gather. Users need a clear way to understand the sometimes complex relationships between pieces of data so they can make decisions and take action.

Over the centuries, designers have come up with two highly conventional modes of conveying data to users—the list and the table—and one powerful mode for collecting data from users—the form. These modes are common in technical documents but are not often discussed thoroughly from a document design perspective. So this chapter focuses on these three modes of handling data in technical document designs.

Lists, tables, and forms allow users to scan information quickly and to compare different pieces of information easily, visually representing the relationships between pieces of data. These tasks are important whether the user is scanning to find information, such as in a list or table, or providing information, such as in a form.

We discuss all three modes in one chapter because they all involve the organization of data and because they have many design strategies in common:

- *They all use both the horizontal and vertical dimensions of the page.* Paragraphs take advantage primarily of the horizontal dimension, as word follows word in lines of text. But lists, tables, and forms also make a conscious use of the vertical dimension to convey or gather data.

- *They all use alignment to show relationships between data.* Lists use a vertical alignment, tables use both horizontal and vertical alignments, and forms can use either or both. These alignments give designers an opportunity to show different relationships between pieces of data.

- *They all use enclosure to group data and draw attention to it.* Enclosure (either through positive borders or negative space) groups data into a logical structure so users can more easily compare data.

This chapter provides ideas and guidelines for designing effective lists, tables, and forms. It concentrates on visual design, but it includes suggestions about logical designs for all these modes.

THREE PERSPECTIVES ON LISTS, TABLES, AND FORMS

PERCEPTION

Well-designed lists, tables, and forms depend on our perceptions of visual relationships to express the logical relationship between items of data, each of which stands as a design object in relation to surrounding design objects. More specifically, we can use the principles of design (similarity, enclosure, alignment, contrast, order, and proximity) to show how data points are grouped and connected.

Similarity and contrast can help create a sense of grouping between design objects in lists, tables, and forms. For example, the bullets in a bulleted list imply that the items in the list are logically connected to each other in a relationship of parallelism. Users will typically assume that all of the items in the list have some factor in common. Each entry in the bulleted list in figure 9.1, for instance, is a duty performed at that particular job. On the other hand, we can use contrast to show different levels of data in a hierarchical relationship—for example, by boldfacing the column headings or row headings in a table to separate them visually from the data.

Good uses of alignment and proximity are essential in lists, tables, and forms. Users typically assume that design objects aligned with each other belong with each other, or at least have some sort of parallel relationship with each other. For example, lists are usually aligned to a consistent left-hand margin. Tables rely even further on alignment, since they're aligned both horizontally and vertically. And each column of a table can have its own internal alignment scheme, such as left-aligned, centered, or in some cases (e.g., for currency figures), right-aligned. Forms also rely on alignment: a form with clearly aligned design objects for the instructions, questions, and form fields is easy to fill out and easy to gather accurate data from.

Because lists, tables, and forms can gather many individual pieces of data, we can also use proximity to show how these design objects relate to one another. Proximity pairs with alignment to help users recognize how the individual design objects in a table create a sense of rows and columns. It also helps users associate questions with their answers in a form.

We can also create larger groupings of individual objects through enclosure. Explicit lines or background shading can separate and group data elements, but we can also group data elements implicitly, using negative space. In a table, for example, borders can delineate rows, columns, and cells, though negative space often works just as well.

Finally, lists, tables, and forms emphasize relationships of order between design objects. Lists can be either ordered or unordered, depending on the nature of the data they convey. Tables can be sorted by the contents of particular rows or columns. And forms often start with identification data (e.g., names, social security numbers, addresses) and end with an authorization—a signature. These orderings help users understand the relationships between pieces of data. Without clear visual cues, users may not be able to access information successfully, or in the case of forms, they may not be able to provide all of the requested information accurately and completely.

CULTURE

Lists, tables, and forms have been used in manuscripts for millennia, and in print culture for centuries. Lists are found in very early manuscripts, but they became more common as paper and printing became cheaper. Some of the earliest uses of printed tables were in the astronomical and astrological books of the Renaissance, which recorded the positions and appearances of stars, planets, and comets.

The earliest printed forms were developed for indulgences, documents that were sold by the medieval and Renaissance churches to certify the forgiveness of a sin. We usually think of the first document created with movable type as Johannes Gutenberg's 42-line Bible. But before printing the Bible, Gutenberg tried out his technological innovation by printing indulgence forms, with blanks for a priest to enter the name of the person seeking forgiveness. Governments soon realized the efficiencies of forms for gathering information and for personalizing repetitive documentation. Before printing, repetitive documents such as leases and warrants had to be individually written by hand, but printing could turn out hundreds of copies quickly and cheaply.

These developments in conveying and gathering data were closely linked to changes in cultures and societies. Our culture is drowning in data, making lists and tables essential tools for managing and understanding data. And forms have become an essential device for governments, corporations, organizations, and even individuals to gather data.

RHETORIC

Because lists, tables, and forms typically contain what we might think of as raw data, it might be hard to imagine that they employ a visual rhetoric. But in fact, even these everyday modes for representing data do have a rhetorical power in that they visually imply relationships between pieces of data.

When data is supplied in a list form, it implies that the items are parallel and equal in kind, and potentially related in order. This can have a considerable rhetorical effect, especially because it shows users that the information being represented is organized and clear. Users often complain about the density of information that is expressed in sentences and paragraphs, but they feel much more comfortable about the same information arranged in a list. Lists let users pay attention to one item at a time—for example, completing one step in a procedure before going on to the next.

Tables express two-dimensional relationships in data by arranging data in columns and rows. This layout can have a significant effect on the way people view the data by organizing data in a way that seems logical and reasonable. The same data explained in a paragraph might not be nearly as convincing as in a table. However, tables sometimes over-simplify data by making it look like *only* two-dimensional relationships apply, when in fact most data have relationships in multiple dimensions.

Forms are a special case rhetorically because they often have two sets of users: the people who fill out the form and the people who read and use the data from the forms. Users filling out forms are definitely affected by the visual rhetoric of the form, which represents the ethos of the organization that is requesting the information. People often don't like to provide information on forms, especially if it's personal information. If the visual rhetoric of the form doesn't assure users that the information will be kept carefully and used appropriately, they may refuse to provide any information at all.

To build a positive ethos with these primary users, forms must have a clear organization of data and an open, friendly visual style. Well-designed forms also group questions into logical chunks to help users understand exactly *what* is being asked. The most common relationship in forms is the one between questions and answers.

After these primary users fill out the form, a secondary set of users reads the data from the forms and incorporates it into some larger infor-mation architecture. In the most basic situation, a secondary user might need to enter data from the form into a database. In more automated systems such as web forms, the form must be designed so that the data will enter a database automatically, where it can then be accessed and combined for other purposes. We need to design forms so that these secondary users can gather and use data successfully.

LISTS

Designers often use lists to break up paragraphs of text in a document, such as a report or a set of instructions. But sometimes lists are the main feature of a document. Telephone books, for example, are typically one or several long lists with only a small amount of basal text to intro-duce them. As familiar as lists may seem, however, designing good lists requires some careful thought.

What *is* a list? Strictly speaking, it's simply a series of logically parallel elements in a sequence. Lists *can* be formatted horizontally as text—for example, *apples, pears, mangos, bananas*—but one of the most common visual techniques is to orient lists vertically:

- apples

- pears

- mangos

- bananas

Doing so allows readers to scan the items quickly and efficiently. But this simple list only begins the possibilities that lists offer for designers. In this section, we'll look at some ways we can use the principles of design to make visually useful and appealing lists.

More about . . . Lists and End Punctuation

Don't place any punctuation at the end of list items. The only exception is when each list item is a full sentence, which naturally requires a period to mark a full stop.

LIST ENTRIES

Lists always include at least two entries (logically, one entry by itself can't be a list). The entries in a list can be short (words or phrases) or long (sentences or even brief paragraphs). Lists imply that the items in the list are parallel and have something in common, that they relate to some central principle in a logical manner. Lists often include some sort of heading or introductory text that expresses this central principle succinctly (figure 9.2; see if you can find a small problem in the lists on this example.).

Because lists imply a logically parallel relationship for the items they contain, we typically make the entries both grammatically and visually parallel. In terms of grammar, if one entry in a list is a noun, all of the entries should be nouns; if one entry begins with a verb, they all should begin with a verb. The visual principle of *similarity* emphasizes this logically parallel relationship. Make sure that all entries in the list are formatted in precisely the same way.

GLYPHS AND ORDERING

Lists sometimes have an inherent order, such as the alphabetical order used in telephone directories, but most lists don't have an inherent order.

Figure 9.2. A well-structured list. Lists and list headings need a consistent design to structure data clearly. *Source*: US Postal Service, Publication 614.

List headings *can express a list's central organizing principle. Like any heading, they should contrast with the list items to show hierarchy and dependence.*

CHOOSE ANY TIME-SAVING TOOL

The United States Postal Service® is no longer just behind the counter at your local Post Office™. We offer many convenient shipping tools where you can help yourself and save time. So next time you're looking for a hassle-free way to take care of your shipping and mailing needs, depend on the U.S. Postal Service®.

We've added flat-rate shipping supplies, expanded our services on ***usps.com*** and introduced the Automated Postal Center® (APC®) so you can ship when and where it's most convenient for you and your business. So take a few moments to learn about the convenient ways to access our products and services.

Flat-Rate Shipping Supplies
- Priority Mail® flat-rate boxes (11" x 8.5" x 5.5" and 3.625" x 11.875" x 3.375")
- Priority Mail flat-rate envelopes (document size)
- Express Mail® and Global Priority Mail® flat-rate envelopes (document size)

Online Shipping
- Click-N-Ship® service
- USPS® Insurance online
- Carrier Pickup program

Automated Postal Center (APC)
- Self-service shipping and mailing

In some lists, order simply doesn't matter—none of the items necessarily needs to appear before or after any of the others. Such *unordered lists* usually employ a *glyph* to mark the beginning of each entry (see figure 9.3). These glyphs take the forms of *manicules* (tiny pointing hands),

Figure 9.3. Common glyphs for ordered and unordered lists. You can use many different images for glyphs in unordered lists, but it's usually best to stick with glyphs that users are familiar with. Glyphs should mark the list item and draw users' attention to it but not overpower the text. Notice the alignment of numbered glyphs: although the list entries are left-aligned, the glyphs are right-aligned on the period so multiple-digit numbers (like 10, III, etc.) don't mess up the alignment of the entries. *Source*: Created by the author.

Glyphs for ordered lists		Glyphs for unordered lists	
1.	I.	•	▶
2.	II.	•	▶
3. ...	III.	•	▶
10.	IV.	•	▶
a.	i.	■	◆
b.	ii.	■	◆
c.	iii.	■	◆
d.	iv.	■	◆

flowers, diamonds, arrowheads, and more, but the most common is the printer's bullet (•), which may be so named because it looks like a round bullet hole in the page. Though ubiquitous now, the first documented use of the word bullet to describe this particular glyph didn't occur until 1950, and the systematic use of "bullet points," bullets used to indicate an unordered list, didn't rise to prominence until the 1980s, becoming overwhelmingly prevalent with the release of PowerPoint in 1987 (see Neeley & Alley, 2011). Lists are such an important part of information design that a program making list generation simple changed our lexicon.

In some situations, the order of list items is important, but the entries don't have any inherent order (e.g., alphabetical order). For example, the list of steps in a procedure must be completed in a particular order. In these cases, we commonly create an *ordered list* by using glyphs such as numbers or letters to establish and reinforce the order of items. The glyphs used in ordered lists are most often Arabic numerals, but we can also use capital Roman numerals (I, II, III, IV), small Roman numerals (i, ii, iii, iv), uppercase letters, or lowercase letters. Designers rely on the conventional order of these glyphs to indicate and reinforce the logically ordered nature of the items in the list.

The glyphs for both ordered and unordered lists are typically in the same font size as the entries they introduce, although some designs might use a larger size, a different color, or a contrasting typeface to make the glyphs more prominent. Avoid using glyphs that won't stand out enough to mark the beginning of each entry—but also avoid using glyphs that overwhelm the entries.

LISTS AND ALIGNMENT

By convention and function, lists almost always align on the left. Left alignment gives users an implied vertical line formed by the bullets and the beginnings of the entries; this helps them scan down the list and distinguish each item. Centering a list or aligning it to the right makes the beginnings of the lines uneven and thus makes the list harder to scan, while making it easier for readers to skip lines accidentally (figure 9.4). Align bulleted and numbered lists with a hanging indent so the glyphs (the bullets or numbers) align vertically on one implied line and the entries align on a second implied line parallel to the first (figure 9.5). This makes the glyphs stick out slightly from the items, notifying readers of the beginning of each item. Without a hanging indent, the second or subsequent lines of any item will obscure the negative space around the glyphs, making it harder for readers to see where the next item begins.

Most word processing and page layout programs have a special setting for bulleted and numbered lists that takes care of alignment and indentations automatically. If your software has such a setting, use it rather than making the list manually. This will also give you lots of extra options for modifying the list design.

If you format lists manually, separate the glyphs from the text with a tab rather than by a space or spaces. Tabs stay consistent from line to

Figure 9.4. Using left alignment for lists. Lists need a consistent left alignment to work successfully. Otherwise, we lose much of the advantage of using lists and increase the chances that users will skip items. *Source*: Created by the author.

- Managed 25-unit apartment complex
- Collected rent
- Kept books
- Hired and managed repair and renovation subcontractors
- Contributed to advertising campaigns for the property management company

- Managed 25-unit apartment complex
- Collected rent
- Kept books
- Hired and managed repair and renovation subcontractors
- Contributed to advertising campaigns for the property management company

Figure 9.5. Using hanging indents in lists. Hanging indents clarify the content of the list, allowing users to scan vertically down the list items to find the information they need. *Source*: Created by the author.

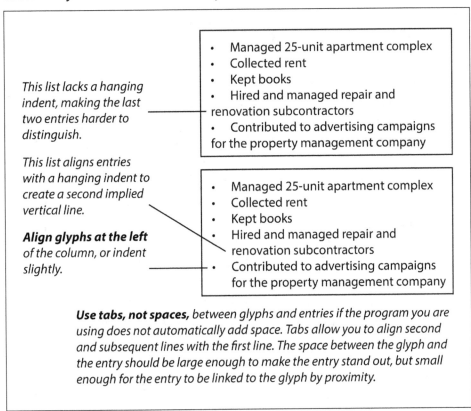

This list lacks a hanging indent, making the last two entries harder to distinguish.

- Managed 25-unit apartment complex
- Collected rent
- Kept books
- Hired and managed repair and renovation subcontractors
- Contributed to advertising campaigns for the property management company

This list aligns entries with a hanging indent to create a second implied vertical line.

Align glyphs at the left *of the column, or indent slightly.*

- Managed 25-unit apartment complex
- Collected rent
- Kept books
- Hired and managed repair and renovation subcontractors
- Contributed to advertising campaigns for the property management company

Use tabs, not spaces, *between glyphs and entries if the program you are using does not automatically add space. Tabs allow you to align second and subsequent lines with the first line. The space between the glyph and the entry should be large enough to make the entry stand out, but small enough for the entry to be linked to the glyph by proximity.*

line, while spaces introduce misalignments. A tab also gives considerable control over the horizontal distance between the glyphs and their entries. This distance should be large enough to make the bullets or numbers stand out but small enough that each glyph still seems visually connected to its entry.

Typically, each glyph in an ordered list is followed by a period or closing parenthesis before the tab. The glyphs are aligned by this mark rather than by the glyph itself. For example, a list ordered with Roman numerals is vertically aligned by the periods so the increasing widths of the numerals don't change the alignment of the entries (see figure 9.3).

MANAGING LONG LISTS

We often use lists to break up long stretches of paragraphs, since these can cause user fatigue and inattention. But long lists can cause as much fatigue as long stretches of paragraphs, so designers try not to make lists longer than around seven to nine items. This is just a general guide-line—lists can be any length you want. A collection of simple items might work well in a long list, particularly if the list is inherently ordered; after all, phone books and dictionaries have thousands of items.

If the data you need to present requires a list longer than seven to nine items, you can subdivide it into smaller sublists—in effect, a list of lists. Designers use the design principles of similarity, alignment, contrast, order, and proximity to group items in a sublist and distinguish that list from the larger list. For example, figure 9.6 shows how a list of items on a menu might be logically and visually divided into appetizers, entrees, sides, and drinks. The sublist items are typographically distinct from the main list items, they're indented further, and they're closer to each other. Designing the list this way might make the list slightly longer, but that increase in length is offset by an increase in usability.

In digital documents such as websites, you can also manage long lists by creating dynamic lists with headings that expand or collapse when clicked. Dynamic lists allow users to see the subheadings as a coherent list in themselves and then click on the subheading they want to expand into a more detailed list. Dynamic lists give users a clear overview of a long list, while still giving them access to individual items.

Figure 9.6. Grouping items increases usability. In paper documents, you can group list items into sublists to create a clearer structure for users. In digital documents, you can even use computer scripts to make dynamic lists that users can expand or collapse to show or hide subitems. *Source*: Created by the author.

Long Undivided List	**Subdivided List**	**Dynamic List**
Chips & salsa	**Appetizers**	▶ **Appetizers**
Fried cheese sticks	Chips & salsa	▼ **Entrees**
Hamburger	Fried cheese sticks	Hamburger
BLT		BLT
Chicken-fried steak	**Entrees**	Chicken-fried steak
Grilled chicken breast	Hamburger	Grilled chicken breast
Green beans	BLT	
French fries	Chicken-fried steak	▶ **Sides**
Side salad	Grilled chicken breast	▶ **Drinks**
Sodas		
Coffee	**Sides**	
Tea	Green beans	
	French fries	
	Side salad	
	Drinks	
	Sodas	
	Coffee	
	Tea	

If your design needs sublists, restrict the design to no more than two or three levels. Make sure the design of each level remains internally consistent, as well as contrasts enough with other levels. For example, to distinguish sublists, you might use a different kind of glyph, a smaller font, or a different color of type.

TABLES

In some ways, tables are just supercharged lists that can show data in two dimensions instead of one. This multiplies the possible relationships we can imply between entries, allowing the user to compare data along a row, down a column, or between multiple rows or columns. Although often used for numerical data, tables can also incorporate text or small graphics.

The table's ability to show multiple relationships, however, means that you must be particularly careful to use the principles of design to establish or even emphasize the relationships you want users to recognize most, or that readers most need to recognize. Like any feature of document design, your primary consideration should be how users will want to use the data in the table.

COMPONENTS OF TABLES

Tables like the one in figure 9.7 include a number of features that you should know about; some are necessary, and some are optional.

NECESSARY PARTS OF TABLES

- *Columns:* The vertical axis of arranging data in a table

- *Rows:* The horizontal axis of arranging data in a table

- *Cells:* The individual area reserved for each data element in a table

OPTIONAL PARTS OF TABLES

- *Column heads:* Labels entered at the top of each column to identify what that column holds. Simple tables with a clear internal ordering of data might not need column heads, but most tables include them.

- *Row heads* (also known as *side heads* or *stub heads*): Labels entered for each row to identify what the row contains. Some

Figure 9.7. Features of tables. This example shows the most common features of tables. *Source*: Created by the author.

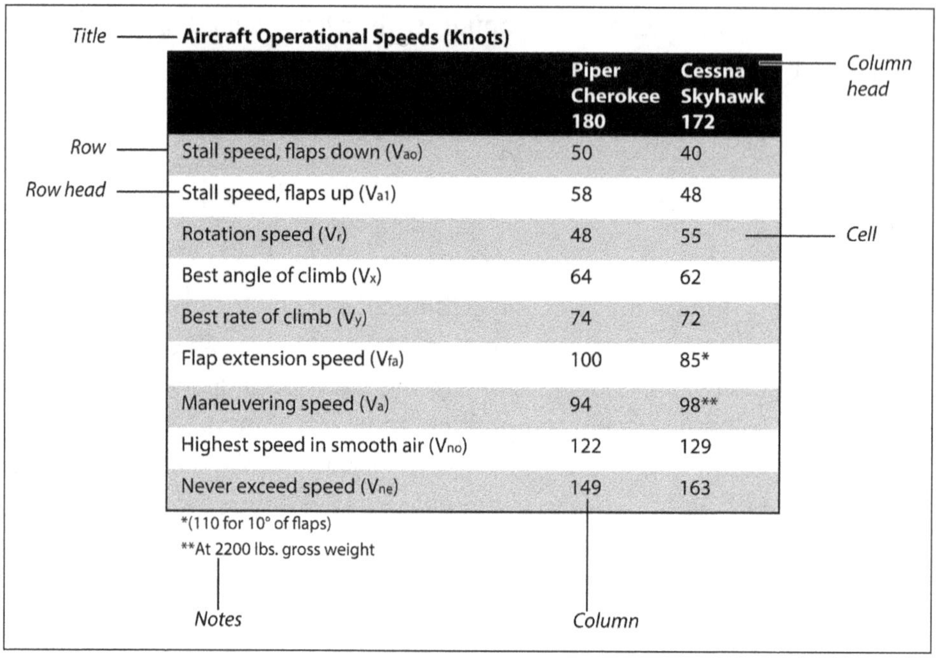

tables use the entries in the first cell of each row to serve both as data and as an indicator of what that row contains, such as in figure 9.7.

- *Titles:* Many tables use a title to identify the purpose of the table.

- *Captions or notes:* Some tables include captions or notes, to identify the source of the table's data.

- *Legends:* Because of the limitations of cell space, some tables use abbreviations or symbols and supply a legend that defines what they mean.

DESIGNING TABLES

FIT CONTENT TO TABLES, AND FIT TABLES TO CONTENT

One of the biggest challenges of designing tables is fitting all the content appropriately into cells. Two factors affect how content fits into tables: (1) the space allowed by the page or screen design constrains the space available for the table from the outside, and (2) the contents of the table determine the space needed for the table from the inside.

Theoretically, tables could be designed with unlimited numbers of columns and rows, extending far beyond the field of view. Microsoft Excel, for example, allows users to create spreadsheets larger than one million rows and 16,000 columns—likely more than anyone can see at one time. When designing documents, however, we're usually limited to the visual space a single user can see at one time: a single page, a spread of pages, or a screen. And even though we can create large tables on-screen for users to scroll through, they can quickly get lost once the column heads scroll up beyond their view.

So the first consideration of table design involves taking into account the format and layout of the page or screen. Tables should fall within the grid design of the page or screen. For example, in a simple business document, a table should typically not extend beyond the main text area (or in other words, into the margins). This often causes problems because many tables contain lots of horizontal data elements—that is, cell data longer than it is tall.

Given page design space constraints, designers of paper documents sometimes rotate a table 90 degrees so the table can take advantage of the often greater height of the page—for example, on an $8^1/_2$" × 11" page, laying out the table so the rows run along the 11" axis. Doing so can allow some additional space, but it can also inconvenience users, who must physically turn the document to read the table and then turn it upright again. If you know a document will include a table that must be rotated this way, it's best to design the entire document around the table rather than inconvenience the user.

Just as the format of the page or screen constrains the design of the table from the outside, the data held in a table determines the table's dimensions from the inside. For example, if we have one cell that must contain more data than others, that cell dictates the width of the columns and the height of the rows. Figure 9.8 shows the compromises that are sometimes necessary when accommodating the data in the table to the surrounding format of the document.

CHOOSE APPROPRIATE TYPOGRAPHY FOR TABLES

Tables can hold a large amount of data in a relatively small space. Because this compression can cause legibility problems for users, it's important to choose simple and clear typefaces for tables.

For this reason, many designers use sans serif typefaces (see chapter 6 for more information on serif and sans serif typefaces). Serif faces work fine for text, where users' eyes move primarily horizontally, but in tables, users must read data both vertically and horizontally. Sans serif

Figure 9.8. Creating a table that fits the format and content. Table design is constrained from the outside by format and from the inside by content. This table needed to accommodate some long entries, making some columns very wide. The designers responded to this problem by using a wide landscape format for the entire document to give more space to the table. Inside the table, they also used a sans serif typeface in a very small font to maximize available space. But as a result, the type is difficult to read. *Source*: Columbia Aircraft, "Columbia 400."

	Equipment and Features	Lancair Columbia 400	Cirrus SR22	Beechcraft Baron G58	Beechcraft Bonanza G36	Mooney Bravo GX	Piper Mirage
Powerplant	Turbocharged Intercooled	●	N/A	N/A	N/A	●	●
	Horsepower	310 HP	310 HP	300 HP	300 HP	270 HP	350 HP
Electrical System	Dual Electrical System	●	●*	●*	●*	●*	●*
Avionics	Primary Flight Display with ADHARS and Flight Director (10.4")	●	O	●	●	●	●
	Multi Function Display (10.4")	●	●	●	●	●	●
	Coupled Autopilot with Altitude Preselect, Autotrim and GPSS	●	O	●	●	●	●
	GPS (Approach/IFR) Nav/Com with Glide Slope	●	●	●	●	●	●
	Second GPS (Approach/IFR) Nav/Com with Glide Slope	●	O	●	●	O	●
Equipment	Inflatable Door Seals	●	N/A	N/A	N/A	N/A	●
	Built In Oxygen	O	N/A	O	O	●	●
	Speed Brakes	O	N/A	N/A	N/A	●	N/A
	Air Conditioning	O	N/A	O	O	O	●
	Automatic Climate Control System	O	N/A	N/A	N/A	N/A	N/A
	Ice Protection	O	O	O	O	O	●
	Heated Prop	O	N/A	O	O	O	●
Certification Standard	Certification Standard	Utility	Normal	Normal	Normal	Normal	Normal
Performance	Max Cruise Speed	235 Kts @ FL250	185 Kts @ 8,000 Ft	203 Kts @ 5,000 Ft	176 Kts @ 6,000 Ft	220 Kts @ FL250**	213 Kts @ FL250
	Stall Speed	59 Kts	59 Kts**	75 Kts	59 Kts	59 Kts	59 Kts
	Range - Economy Cruise	1,301 NM/55% power (205 Kts) @ FL250	939 NM/55% power @ 17,000 Ft**	1,520 NM @ 8,000 Ft 194 Gal	914 NM @ 6,000 Ft 74 Gal	960 NM/55% power**	1,345 NM
	Range - Max Cruise	715 NM/ FL250	744 NM/75% @ 10,000 Ft**	1,025 NM @ 5,000 Ft 194 Gal	697 NM @ 8,000 Ft 74 Gal	750 NM/75% power**	990 NM**
	Service Ceiling	25,000 Ft	17,500 Ft	20,688 Ft	18,500 Ft	25,000 Ft	25,000 Ft
	Take off Distance Hard surface, gross weight, sea level	1,300 Ft	1,058 Ft**	2,300 Ft	1,913 Ft	1,080 Ft	1,090 Ft

Source for all data, except where noted otherwise, is the official website for each aircraft manufacturer
* Not all Dual Electrical Systems are created equal. The Columbia 350 and 400 each use completely redundant, dual alternator, dual bus systems capable of operating the aircraft with no power shedding in the case of partial failure. This is not true of all aircraft with Dual Electrical Systems. See www.flycolumbia.com for more information.
**Source Plane & Pilot Magazine Aircraft Specification Database

● = Standard, O = Optional, NA= Not Available

Original table

Primary Flight Display with ADHARS and Flight Director (10.4")
Multi Function Display (10.4")
Coupled Autopilot with Altitude Preselect, Autotrim and GPSS
GPS (Approach/IFR) Nav/Com with Glide Slope
Second GPS (Approach/IFR) Nav/Com with Glide Slope
Inflatable Door Seals
Built In Oxygen
Speed Brakes
Air Conditioning

Close-up of second column

●
●
●
●
O
O
O
O
O
Utility
235 Kts @ FL250
59 Kts
1,301 NM/55% power (205 Kts) @ FL250
715 NM/ FL250
25,000 Ft
1,300 Ft

Close-up of third column

typefaces lack the horizontal "feet" of serif faces, so they tend to look more upright and enable vertical scanning. Sans serifs also tend to look more open and clear in tight spaces than serif faces do. Finally, many sans serifs are narrower than their serif counterparts, which saves horizontal space in a table.

Consider figure 9.9, a table of lots for sale in a collectible postage stamp auction. We'll use this data in several examples to show different design techniques, evolving the table design as we go along. Here, notice how the sans serif version of the table reads more clearly than the serif version, particularly in the vertical axis.

Figure 9.9. Using sans serif typefaces in a table. Sans serif typefaces maximize available space and make it easier to scan vertically down columns. *Source*: Created by the author.

Lot #	Cat #	Condition	Comments	Price
1	137	S	A fine example	1740.00
2	149	XF/S	Clean & sharp	500.00
3	430	XF/S	A rich green, sharp perforations	185.00
4	88	S	A very pretty stamp; no gum skips	440.00
5	371	S	Post Office fresh	460.00
6	138	VF	Vertical pair	360.00
7	383	VF/XF	Small thin lower right corner	225.00
8	292	VG	Ex-Miller Collection	1150.00
9	27	XF	Good example	385.00
10	389	F/VF	Good centering for this issue	150.00

A table using a serif typeface

Lot #	Cat #	Condition	Comments	Price
1	137	S	A fine example	1740.00
2	149	XF/S	Clean & sharp	500.00
3	430	XF/S	A rich green, sharp perforations	185.00
4	88	S	A very pretty stamp; no gum skips	440.00
5	371	S	Post Office fresh	460.00
6	138	VF	Vertical pair	360.00
7	383	VF/XF	Small thin lower right corner	225.00
8	292	VG	Ex-Miller Collection	1150.00
9	27	XF	Good example	385.00
10	389	F/VF	Good centering for this issue	150.00

A table using a sans serif typeface

If column heads or row heads are important cues to the content of the table, they should present some typographical contrast. Use a larger size, a greater density, or a different typeface to provide a clear contrast between the heads and the cell content—but not so different that the heads don't seem to belong to the same table. The example in figure 9.10 does the job adequately by bolding the column heads.

Avoid the temptation to overuse all caps in tables. Tables are already rigid and square enough—there's little reason to emphasize that rigidity and squareness with all caps. Besides, the additional contrast between letter forms in lowercase or with initial caps will make scanning and reading the table easier (see figure 9.11). Lowercase letters also take less space, as you can see from the relative narrowness of the lowercase table.

Figure 9.10. Emphasizing column heads with boldface type. This version of the table in figure 9.9 emphasizes the column heads with boldface type. *Source*: Created by the author.

Lot #	Cat #	Condition	Comments	Price
1	137	S	A fine example	1740.00
2	149	XF/S	Clean & sharp	500.00
3	430	XF/S	A rich green, sharp perforations	185.00
4	88	S	A very pretty stamp; no gum skips	440.00

Figure 9.11. Avoid using all caps in tables. Text in all caps makes the cell data harder to distinguish and takes up more space than lowercase type (or lowercase with initial caps, as shown here). *Source*: Created by the author.

Cell data and column heads in all caps

LOT #	CAT #	CONDITION	COMMENTS	PRICE
1	137	S	A FINE EXAMPLE	1740.00
2	149	XF/S	CLEAN & SHARP	500.00
3	430	XF/S	A RICH GREEN, SHARP PERFORATIONS	185.00
4	88	S	A VERY PRETTY STAMP; NO GUM SKIPS	440.00

Cell data and column heads with initial caps

Lot #	Cat #	Condition	Comments	Price
1	137	S	A fine example	1740.00
2	149	XF/S	Clean & sharp	500.00
3	430	XF/S	A rich green, sharp perforations	185.00
4	88	S	A very pretty stamp; no gum skips	440.00

ORDER TABLES CONSISTENTLY AND LOGICALLY

Because tables organize data both horizontally and vertically, they offer a number of options for ordering data. For example, you can order a table numerically or alphabetically by the contents of any single row or column. You can also switch around the columns to help readers find the information they need in a logical progression. For example, in figures 9.9 through 9.11, the table is ordered by the lot number column, from Lot 1 to Lot 10. This makes sense if the designers intend users to refer to this table while the auction is under way, using it as a guide to what's coming up next.

But imagine if the table were designed as a price list to go in a catalog. In that case, it would make more sense to order the table's contents by the catalog number so users could read along in the catalog and refer to the table when necessary (figure 9.12). It would also make sense to change the order of the columns to place the ordering column leftmost.

USE ALIGNMENT TO INCREASE USABILITY

Tables offer many options for horizontal and vertical alignment of cell contents—in other words, where cell data "sits" in the cell.

Figure 9.12. Ordering a table. In this version, the auction table is ordered by catalog number (rather than by lot number as in the previous figures). This ordering would make more sense if the table appeared in a price list to go in a catalog so users could refer from the catalog to the price list easily. Notice that the order of columns has also changed: Cat # is leftmost, since it serves as an index to the table data. *Source*: Created by the author.

Cat #	Lot #	Condition	Comments	Price
27	9	XF	Good example	385
88	4	S	A very pretty stamp; no gum skips	440
137	1	S	A fine example	1740
138	6	VF	Vertical pair	360
149	2	XF/S	Clean & sharp	500
292	8	VG	Ex-Miller Collection	1150
371	5	S	Post Office fresh	460
383	7	VF/XF	Small thin lower right corner	225
389	10	F/VF	Good centering for this issue	150
430	3	XF/S	A rich green, sharp perforations	185

Vertical alignment. Vertically, different kinds of cell content call for different alignments, and inappropriate alignments can cause significant reductions in usability. For example, consider the table of stamp auction lots in figure 9.12. All of the columns are left-justified, but that justification might make reading some of the fields more difficult.

In figure 9.13, the changes we made to the vertical alignment let users scan the numerical columns more easily, particularly in the Lot # column, which creates an order for all the rows. Centering the Condition column emphasizes the condition code for each stamp. The Comments column contains only text, so a left alignment makes sense there, just as it would in a list. And the Price column, aligned at the decimal point, allows users to distinguish between expensive and inexpensive stamps very quickly.

Horizontal alignment. Horizontally, it's usually best for the cell data to sit with about an equal amount of space above and below the data, in relation to the cell boundaries. Some typefaces, however, may require some careful manipulation of cell data alignment to give the data a comfortable space in its cell. You want to provide enough space in the cell to accommodate caps, ascenders, and descenders: you wouldn't want anything to extend beyond the cell boundaries or make one row taller than the rest.

Figure 9.13. Aligning a table. The stamp catalog table has been revised with a more appropriate pattern of alignment to increase readability. *Source*: Created by the author.

Columns with numbers are right-aligned to emphasize numerical order.

Column with condition code is center-aligned.

Column with text is left-aligned.

Column with currency is aligned at decimal point.

Lot #	Cat #	Condition	Comments	Price
1	137	S	A fine example	1740.00
2	149	XF/S	Clean & sharp	500.00
3	430	XF/S	A rich green, sharp perforations	185.00
4	88	S	A very pretty stamp; no gum skips	440.00
5	371	S	Post Office fresh	460.00
6	138	VF	Vertical pair	360.00
7	383	VF/XF	Small thin lower right corner	225.00
8	292	VG	Ex-Miller Collection	1150.00
9	27	XF	Good example	385.00
10	389	F/VF	Good centering for this issue	150.00

As we already discussed, it's best to set cell contents in lowercase type. When doing so, provide roughly equal space between the baseline and the lower cell boundary as between the x-line and the upper cell boundary (see figure 9.14).

Whatever horizontal alignment you choose, be consistent. The table in figure 9.15, for example, would probably slow down and confuse

Figure 9.14. Using appropriate spacing within cells. Make sure to use balanced horizontal spacing between text and cell boundaries, accommodating caps, ascenders, and descenders. In most situations, try to space the x-line and the baseline equally from the top and bottom cell boundaries, respectively. *Source*: Created by the author.

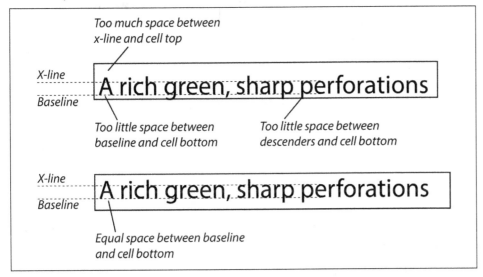

Figure 9.15. Use consistent horizontal alignment in a table. An inconsistent horizontal alignment in cells might confuse users about what row they're reading, particularly if you don't use cell borders. In this case, the prices are closer to the bottom of each cell, while all of the other information is closer to the top. *Source*: Created by the author.

Lot #	Cat #	Condition	Comments	Price
1	137	S	A fine example	1740.00
2	149	XF/S	Clean & sharp	500.00
3	430	XF/S	A rich green, sharp perforations	185.00
4	88	S	A very pretty stamp; no gum skips	440.00
5	371	S	Post Office fresh	460.00
6	138	VF	Vertical pair	360.00
7	383	VF/XF	Small thin lower right corner	225.00
8	292	VG	Ex-Miller Collection	1150.00
9	27	XF	Good example	385.00
10	389	F/VF	Good centering for this issue	150.00

users, especially because it doesn't use borders. Notice that the price data is aligned closer to the bottom of its cells, while the matching content in each row is aligned near the tops of the cells. As users read across the rows, it would be difficult to figure out which price goes with which lot.

USE PROXIMITY TO BUILD RELATIONSHIPS BETWEEN DATA

Designing effective tables requires paying attention to both horizontal and vertical distance between a variety of elements, including the cell contents, cell contents and borders, rows, columns, and column and row heads. Differences in proximity can radically change the way users approach tables because they affect users' perceptions of good figure between the design objects (i.e., the cell data) we put in the table. In figure 9.16, the distance between the columns and the close proximity of rows give the columns strong figure; the columns appear more prominent than the rows. This encourages users to scan down each column. In figure 9.17, moving the columns closer together and the rows farther apart makes the rows seem the stronger figure, encouraging users to scan across rows.

Either of these approaches might be appropriate. It depends on how you think (or have found out) that users actually want to use the table. Just make sure that you don't put so much space between the rows or columns that users don't recognize the data as a table or have trouble recognizing both rows and columns visually.

Figure 9.16. Creating strong vertical figures. The rows are closer together than the columns here, making the columns a strong figure and encouraging users to scan vertically down the columns. *Source*: Created by the author.

Lot #	Cat #	Condition	Comments	Price
1	137	S	A fine example	1740.00
2	149	XF/S	Clean & sharp	500.00
3	430	XF/S	A rich green, sharp perforations	185.00
4	88	S	A very pretty stamp; no gum skips	440.00
5	371	S	Post Office fresh	460.00
6	138	VF	Vertical pair	360.00
7	383	VF/XF	Small thin lower right corner	225.00
8	292	VG	Ex-Miller Collection	1150.00
9	27	XF	Good example	385.00
10	389	F/VF	Good centering for this issue	150.00

Figure 9.17. Creating strong horizontal figures. Closer horizontal proximity and more distance between rows might encourage more horizontal scanning. *Source*: Created by the author.

Lot #	Cat #	Condition	Comments	Price
1	137	S	A fine example	1740.00
2	149	XF/S	Clean & sharp	500.00
3	430	XF/S	A rich green, sharp perforations	185.00
4	88	S	A very pretty stamp; no gum skips	440.00
5	371	S	Post Office fresh	460.00
6	138	VF	Vertical pair	360.00
7	383	VF/XF	Small thin lower right corner	225.00
8	292	VG	Ex-Miller Collection	1150.00
9	27	XF	Good example	385.00
10	389	F/VF	Good centering for this issue	150.00

USE ENCLOSURE FOR CLARITY AND EMPHASIS

Tables are essentially a grid of rectangles that enclose data, so think carefully about using enclosure to create successful tables.

Borders. We often default to using borders to emphasize the enclosure of cells, but in many designs, borders can actually obscure the data. After all, a border is an additional positive element that users must decipher. And borders take up extra positive space in the design, requiring additional negative space to keep them from crowding the table's contents.

Sometimes tables work better with limited or selective use of positive borders, relying on the alignment of cell contents and the negative space between cells, rows, and columns to emphasize the visual structure of the grid. Besides, using borders only selectively gives additional opportunities to emphasize data or encourage the users to see the table in a certain way.

Consider the tables in figures 9.18 and 9.19. Both tables look cleaner and easier to scan than the previous versions of this table that included a full grid of borders. The single horizontal rule in figure 9.18 gives extra emphasis to the column heads. The single vertical rule in figure 9.19 emphasizes the Lot column, helping users locate the lot that is being auctioned. Of course, in comparison, this deemphasizes the other columns and column heads, making them clearly subsidiary to the lot number.

Figure 9.18. Emphasizing column heads with horizontal rule. The horizontal rule emphasizes the column heads. *Source*: Created by the author.

Lot #	Cat #	Condition	Comments	Price
1	137	S	A fine example	1740.00
2	149	XF/S	Clean & sharp	500.00
3	430	XF/S	A rich green, sharp perforations	185.00
4	88	S	A very pretty stamp; no gum skips	440.00
5	371	S	Post Office fresh	460.00
6	138	VF	Vertical pair	360.00
7	383	VF/XF	Small thin lower right corner	225.00
8	292	VG	Ex-Miller Collection	1150.00
9	27	XF	Good example	385.00
10	389	F/VF	Good centering for this issue	150.00

Figure 9.19. Using vertical rule to emphasize a column. The vertical rule emphasizes the lot number column. *Source*: Created by the author.

Lot #	Cat #	Condition	Comments	Price
1	137	S	A fine example	1740.00
2	149	XF/S	Clean & sharp	500.00
3	430	XF/S	A rich green, sharp perforations	185.00
4	88	S	A very pretty stamp; no gum skips	440.00
5	371	S	Post Office fresh	460.00
6	138	VF	Vertical pair	360.00
7	383	VF/XF	Small thin lower right corner	225.00
8	292	VG	Ex-Miller Collection	1150.00
9	27	XF	Good example	385.00
10	389	F/VF	Good centering for this issue	150.00

Shading. Reverse text or shading can also be used for enclosure. In figure 9.20, we shift the balance of emphasis back to the column heads by enclosing the column heads with a dense black field and using reverse text (for more on reverse text, see chapter 6). This brings the focus back to the columns as strong figures while making the Lot column a slightly less strong figure.

Shading alternate rows in a gray screen or in color makes the rows a strong figure, while still not overwhelming the stronger elements of reverse-text column heads and the bordered Lot column (figure 9.21). If you decide that the columns are more important to users than the rows, shade alternate columns instead. You can also use shading as a tool for emphasis.

Enclosure is another method that can be used to inset tables within tables, usually by merging cells. One of the problems with the example we've been using is the awkward *condition* code for each stamp, which could really use some clear explanation. If we use a legend to accompany the table, we can create a visual guide that might help users diagnose each stamp's condition more readily (see figure 9.22). This table creates a new row at the top with a single merged cell that contains the Condition column head and adds sub column heads for the condition labels S, XF, VF, F, and VG. It also uses a symbol to mark the condition of the stamp, rather than the more complex condition code, which is explained in a

Figure 9.20. Using reverse text for emphasis. The reverse text places emphasis on column heads. *Source*: Created by the author.

Lot #	Cat #	Condition	Comments	Price
1	137	S	A fine example	1740.00
2	149	XF/S	Clean & sharp	500.00
3	430	XF/S	A rich green, sharp perforations	185.00
4	88	S	A very pretty stamp; no gum skips	440.00
5	371	S	Post Office fresh	460.00
6	138	VF	Vertical pair	360.00
7	383	VF/XF	Small thin lower right corner	225.00
8	292	VG	Ex-Miller Collection	1150.00
9	27	XF	Good example	385.00
10	389	F/VF	Good centering for this issue	150.00

Figure 9.21. Using shading for emphasis. Shading can help distinguish between rows or columns. It can also be used for emphasis, drawing users' attention to the most rare and expensive stamps. *Source*: Created by the author.

Lot #	Cat #	Condition	Comments	Price
1	137	S	A fine example	1740.00
2	149	XF/S	Clean & sharp	500.00
3	430	XF/S	A rich green, sharp perforations	185.00
4	88	S	A very pretty stamp; no gum skips	440.00
5	371	S	Post Office fresh	460.00
6	138	VF	Vertical pair	360.00
7	383	VF/XF	Small thin lower right corner	225.00
8	292	VG	Ex-Miller Collection	1150.00
9	27	XF	Good example	385.00
10	389	F/VF	Good centering for this issue	150.00

Lot #	Cat #	Condition	Comments	Price
1	137	S	A fine example	1740.00
2	149	XF/S	Clean & sharp	500.00
3	430	XF/S	A rich green, sharp perforations	185.00
4	88	S	A very pretty stamp; no gum skips	440.00
5	371	S	Post Office fresh	460.00
6	138	VF	Vertical pair	360.00
7	383	VF/XF	Small thin lower right corner	225.00
8	292	VG	Ex-Miller Collection	1150.00
9	27	XF	Good example	385.00
10	389	F/VF	Good centering for this issue	150.00

Lot #	Cat #	Condition	Comments	Price
1	137	S	A fine example	1740.00
2	149	XF/S	Clean & sharp	500.00
3	430	XF/S	A rich green, sharp perforations	185.00
4	88	S	A very pretty stamp; no gum skips	440.00
5	371	S	Post Office fresh	460.00
6	138	VF	Vertical pair	360.00
7	383	VF/XF	Small thin lower right corner	225.00
8	292	VG	Ex-Miller Collection	1150.00
9	27	XF	Good example	385.00
10	389	F/VF	Good centering for this issue	150.00

legend at the left. It employs vertical, alternating shading to help readers scan down the Condition columns if they are looking for stamps of a particular condition. But because these vertical shadings make reading across the rows more difficult, we've added horizontal borders between the rows as well.

DESIGNING TABLES FOR THE SCREEN

Most of the features of tables we've discussed can work equally well on paper or on screens. However, the Hypertext Markup Language (HTML) behind websites has some constraints. HTML includes tags to specify the following features of tables:

- *Table:* width, height, alignment, and background color
- *Row:* height and background color
- *Cell:* width, height, alignment, cell spacing (the distance between cells), cell padding (the distance between a cell's content and its boundaries), and background color
- *Cell borders:* width

In addition, in HTML we can merge cells by using the *rowspan* attribute (to merge a cell with one or more below or above it) and the *colspan* attribute (to merge a cell with one or more to its right or left).

Figure 9.22. Combining several table design techniques. Here, the conditions of the stamps are provided in a legend, and an embedded table gives users a quick visual sense of the conditions of the stamps. *Source*: Created by the author.

Conditions
- **S:** Superb
- **XF:** Extra Fine
- **VF:** Very Fine
- **F:** Fine
- **VG:** Very Good

Lot #	Cat #	S	XF	VF	F	VG	Comments	Price
1	137	O					A fine example	1740.00
2	149	O	O				Clean & sharp	500.00
3	430	O	O				A rich green, sharp perforations	185.00
4	88	O					A very pretty stamp; no gum skips	440.00
5	371	O					Post Office fresh	460.00
6	138			O			Vertical pair	360.00
7	383		O	O			Small thin lower right corner	225.00
8	292					O	Ex-Miller Collection	1150.00
9	27		O				Good example	385.00
10	389			O	O		Good centering for this issue	150.00

Cascading Style Sheets (CSS) allow much more flexible designs of tables in web pages, including greater control over borders. But their implementation is somewhat more complex than the scope of this book allows. See a good CSS reference, such as the World Wide Web Consortium's online CSS description for more information on how to use CSS for table layouts.

Finally, dynamic websites (those using a database to hold their contents) can use sortable tables, which are convenient for users in many situations. Usually, users sort the table by clicking on a column head, which then re-sorts the rows according to the contents of that column. These tables are particularly common in online web mail applications, which often use tables to display the messages waiting to be read, with columns for sender, date, subject, and so on. They require computer scripting to work, however, which places them beyond the expertise of some document designers.

Avoid using tables to lay out content when designing for online spaces. While tables help organize data, online they create accessibility problems. Screen readers may encounter difficulty when attempting to parse content displayed in a table. Tables online also come with other problems:

- They can be difficult to maintain.
- They lack flexibility when rendering on multiple screens is desirable.
- They impact search engine optimization.
- They are invalid in HTML 4.01 (though not in HTML5) (see, e.g., Kyrnin, 2021).

So, as always, consider your audience, purpose, and context carefully when deciding if you really need to use a table in your online design.

FORMS

Not many of us get through more than a few days without encountering a form to fill out. In fact, the familiarity of forms sometimes breeds contempt. When a form is badly designed, asks ambiguous questions, or requires repetitive information, we often find it annoying and distracting to complete.

But forms hold an important place in the exchange of data from users to producers of documents. Forms, after all, are the medium through which users communicate important data to producers. Companies rely on forms to gain important information about their customers, whether users wish to order a product, register it, or complain about it. Governments rely on forms to gather information from citizens—anything from Selective Service registration to taxes. Hospitals use forms to gain vital information from patients, such as allergic reactions and medical histories. And most employers use forms to gather information from their employees.

As we said at the beginning of this chapter, the difficulty in designing forms lies in the fact that they have two sets of users: those who fill out the form and those who must gather the data from the filled-out form. This two-way relationship embodied in forms can make them complex to design. For the form to gather data accurately, it must work well from both perspectives.

Users who fill out forms must understand the form well enough to enter accurate and complete data. If the form's design deemphasizes an important data entry field, for example, users might skip it. If the relationship is unclear between the questions the form asks and the spaces it provides for answers, the users might provide unusable data. The design also must adequately instruct users in how to fill out the form. Bagin and Rose (1991) analyzed over 4,000 responses regarding bad form design and determined that bad design results in loss of benefits and services. Nearly half of the responses included complaints that forms were too complicated and that instructions were unclear. This line of research on form design (e.g., Zimmerman & Schulz, 2000) suggests that good design can help minimize both of these complaints. If you need to collect information—medical, financial, personal, or something else altogether—you want your users to be able to easily provide that information.

At the other end of the process, form processors—whether human or computerized—must be able to extract the data quickly and efficiently, usually so it can be entered into a database. For forms that will be read by human eyes, this means creating a design that a person can scan quickly and accurately. For computer-read paper forms, this means creating a design that computer scanners can read effectively and that optical character recognition (OCR) software can translate into editable text with high accuracy. Online forms require that each answer field must have a unique and accurate name by which a computer can identify its contents and populate the correct field in the corresponding database.

COMPONENTS OF FORMS

Designing good forms requires knowing a bit about their typical parts and the conventions they use for gathering information in data acquisition fields.

FORM AREAS

Most forms have the following four main parts (see figure 9.23):

- *Masthead:* The masthead can include the form's title, as well as a logo or the name of the sponsoring organization.

- *Instructions:* Users need to know why and how to fill out the form, as well as what to do with it when they're finished. A section at the beginning of the form typically provides these explanations and directions.

Figure 9.23. Typical parts of a form. Forms typically include four essential parts: a masthead, instructions, a data acquisition area, and a footer. *Source*: Created by the author.

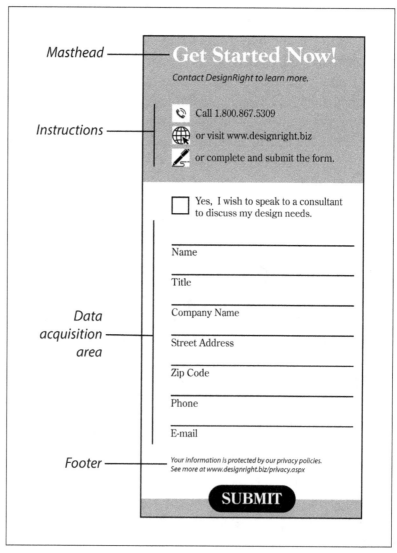

- *Data acquisition area:* This area typically makes up the bulk of the form and includes all the requests for information. It often ends with a request for a signature and date, which asks users to authorize their responses.

- *Footer:* The footer can include a variety of information. For the user, it might include final instructions about how to submit the form. For the organization, it might include a form number or version control information.

Not all forms require all four parts, and some parts might be abbreviated. For example, a simple magazine subscription card might have a very short masthead ("Subscribe to our magazine today!"), little in the way of introduction or instructions, and no footer. Still, users have seen so many of these forms that they usually have little trouble understanding that all they must do is fill out the data acquisition area and drop the form in the mail or click the Submit button.

DATA FIELDS

Forms gather data through a series of data fields in the data acquisition area. Each data field has two parts: a *prompt* and a *response.*

The prompt asks the user for information, and the user responds by supplying the information in a way the organization can process efficiently. Form designers must create prompts that encourage the appropriate response from users, as well as areas that constrain the response to the desired kind of information.

Although forms can look very complicated, only two kinds of data fields are usually sufficient to fulfill these needs:

- Alphanumeric entry fields

- Option fields

Alphanumeric Entry Fields

Alphanumeric entry fields prompt users to write a response in text or numbers within a designated space. That space might appear as a simple horizontal line, a box, or even a negative space that implies a box. The space allowed for response constrains the length of the response, but in general, users can enter or write whatever they choose in the field. These fields work best when gathering information that varies widely from person to person, such as address information or comments.

Forms designed for OCR processing might further constrain users with a box or implied box for each letter or number of the user's response so computer software can decipher the letters more easily. And computer forms (e.g., those on websites) can constrain the contents of alphanumeric fields by allowing users to enter only a certain number of characters in a field or requiring them to enter a valid email address.

Option Fields

While alphanumeric entry fields allow users to enter whatever text or numbers they choose, option fields restrict user responses to one or a few choices among a list of options. These fields allow organizations to gather information quickly on topics that are somewhat consistent from person to person, such as state of residence (one choice out of fifty), or if a person would like to opt-in for email advertisements (yes or no, one choice out of two).

Option fields increase the accuracy of response, but they might also exclude potential responses. For example, some organizations may still offer only two choices for gender, when we know that there are host of other options, and a lot of controversy has arisen over the options government forms provide to ask about a user's ethnicity or race. As a designer, you must be an advocate for the users of any document you design; think carefully about how the options you provide on a form might force incorrect or incomplete answers, or even potentially alienate users.

As a designer, you must be an advocate for the users of any document you design; think carefully about how the options you provide on a form might force incorrect or incomplete answers, or even potentially alienate users.

The most common form of option field is a checkbox. Checkboxes prompt users to enter a check mark, an X, or a tick mark in a square beside responses in a list of options. Checkboxes are common when there are several possible answers to one question.

For other questions, you may want to constrain the response to one option. One example is the ubiquitous machine-readable multiple-choice test form, which includes option fields as ovals to be filled in with a pencil. If the user fills in more than one oval, it typically invalidates that question.

For paper forms, you must inform users in the instructions whether they can make multiple responses or are restricted to one. HTML forms, however, can manage these responses automatically. A list of HTML checkboxes will allow users to mark more than one choice. But HTML uses radio button fields to constrain users to one response: if users change their mind and click on a different option than the one they chose initially, the form automatically deselects the first radio button and selects the new one. Online forms can also use drop-down menus, which similarly allow users to choose only one option from a list.

DESIGNING FORMS

Designing successful forms involves paying attention to at least these issues:

- Including clearly designed information about the form, such as its purpose, procedures, and organizational sponsor

- Grouping data fields visually

- Using alignment to encourage consistent and complete responses

- Designing response spaces that encourage accurate responses

INCLUDE CLEARLY DESIGNED INFORMATION ABOUT THE FORM

People don't often enjoy filling out forms, so they like to know *why* they have to do so and what they should do with the form after they fill it out. These rhetorical and instructional functions are usually carried at the top of the form. A clear visual design will draw users' attention to this area of the form and encourage them to fill out the form attentively and accurately.

The Sierra Club form in figure 9.24, for example, has a difficult rhetorical situation: it must convince users to become members of a voluntary organization and donate money to it. The form employs a distinctive image of wildlife on the left. Then it lists the benefits of participating along with a clear statement about the work the organization does, then offers a free gift for joining. All of the rhetorically argumentative elements appear on the left, and the eye tracks down the list because of their use of imagery and list elements—the large bag, for example, acts as a sort of visual anchor, pulling the eye down the page. Only after this rhetorical argument do the data fields of the form begin on the right, in the first screen (on the left) with option field selections for donation amounts, then in the next section (shown on the right) with form fields for personal information.

Different forms call for different approaches. The Internal Revenue Service (IRS), for example, doesn't spend much time persuading people to fill out *its* forms because the penalties for not doing so are pretty severe. But IRS forms gather very complex information, so they typically include

Figure 9.24. Design and forms. Visual design can draw attention to the *why* of the form. *Source*: Sierra Club, "Be a Champion for the Environment."

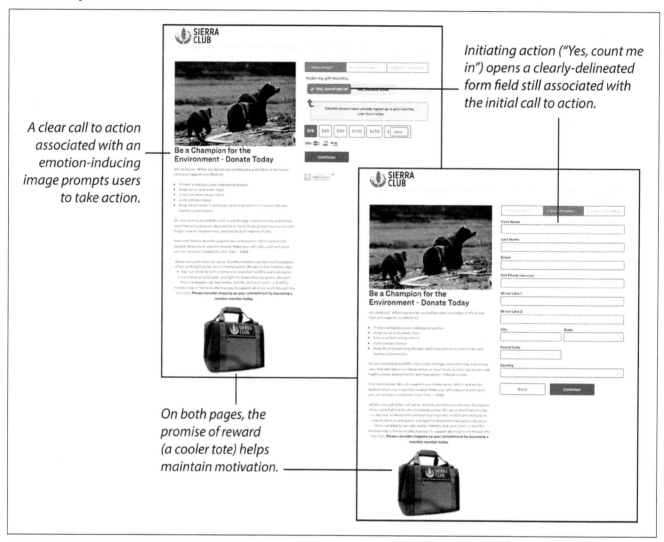

A clear call to action associated with an emotion-inducing image prompts users to take action.

Initiating action ("Yes, count me in") opens a clearly-delineated form field still associated with the initial call to action.

On both pages, the promise of reward (a cooler tote) helps maintain motivation.

more explanations and instructions about *how* to fill out the form and *what* to do with it. In fact, IRS forms come with separate instruction booklets that are much longer than the forms themselves.

GROUP DATA FIELDS VISUALLY

When forms ask for many different pieces of information, users may get tired and frustrated. And the more tired they get, the sloppier their answers to questions can be. Good form designs provide resting places for users by visually and logically grouping the information requested. For example, the US Postal Service (USPS) change of address form (figure 9.25) groups three different kinds of information:

Figure 9.25. Logically grouped fields. This form groups together different kinds of information in a logical way. *Source*: U.S. Postal Service, "Official Mail Forwarding Change of Address Order."

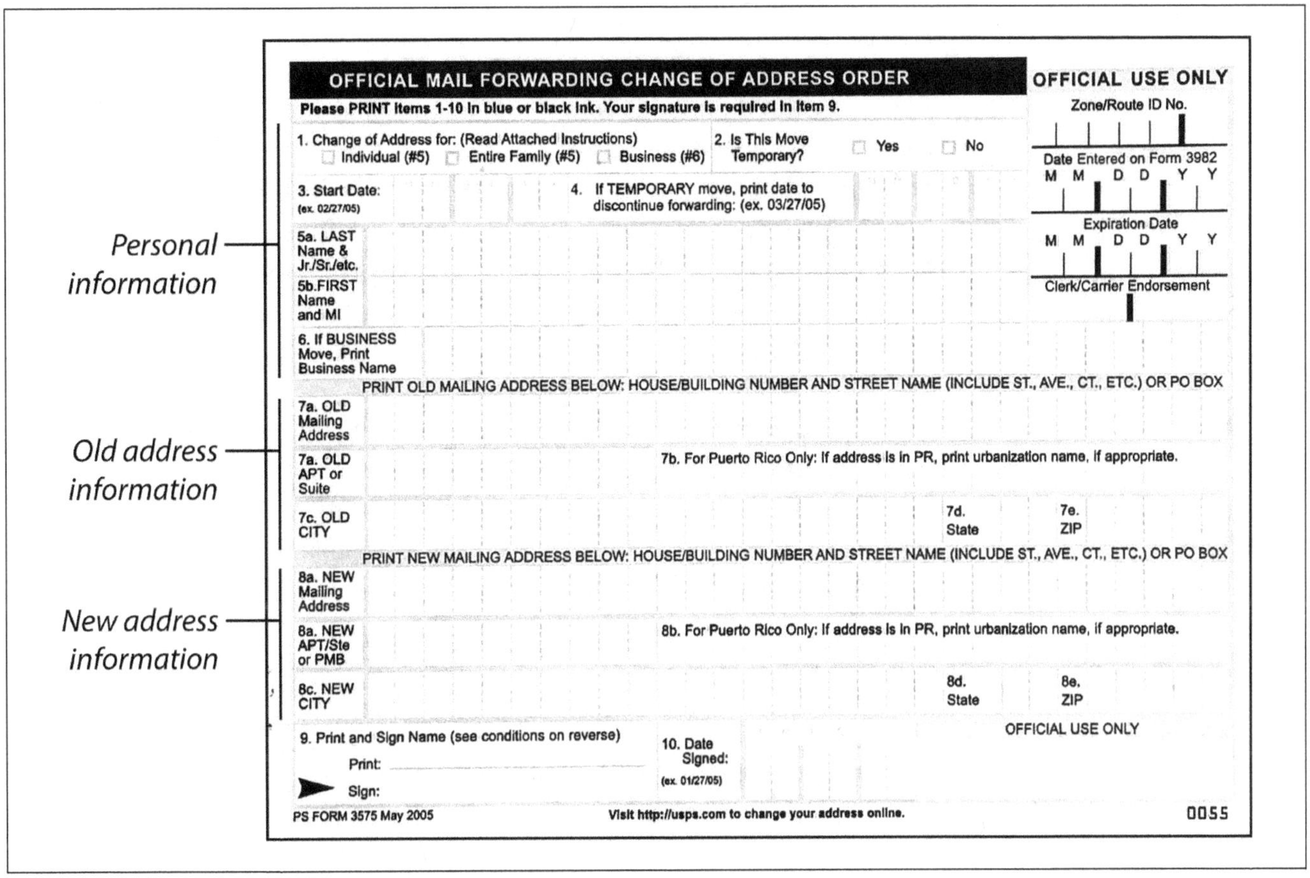

- Personal information (including fields for last name, first name, and business name)

- Information about the user's old address (including fields for the street address, apartment number, city, state, and zip)

- Information about the user's new address (including fields for the street address, apartment number, city, state, and zip)

This structure makes sense: to change your address, the USPS needs to know who you are, where you used to live, and where you're living now. But the form also emphasizes this structure visually by including a broader-than-usual border and a heading before the old mailing address section and the new mailing address section.

We can use a variety of techniques to visually emphasize the logical structure of the data. Simple descriptive or numbered headings, much as you might include in a straightforward business report, can do wonders. Alternately, we can use heavy rules (lines) between sections or enclose sections with borders, shading, or negative space.

USE ALIGNMENT TO ENCOURAGE CONSISTENT AND COMPLETE RESPONSES

One of the biggest challenges users face when filling out forms is simply seeing all the data fields and recognizing what they still have to fill out. Users often unintentionally skip areas because the fields are misaligned or scattered across the page or screen.

Aligned data fields make it very obvious when some fields have been filled out and others haven't. Aligned data fields also help organizations gather information efficiently from forms for the same reasons: it's easier to find information by running your eyes down an aligned set of data fields than having to look all over the place.

Figure 9.26 shows part of a form requesting information about employment. The data entry fields (all of them option fields) are arranged like text in paragraphs rather than fields on a form. As a result, it's difficult for users to see the options, and it's time-consuming for organizations to gather data from the forms after they've been filled out. Figure 9.27

Figure 9.26. Misaligned data fields. A form with misaligned data fields makes it difficult for users to see all of the options. *Source*: Created by the author.

What is the highest level of education you have achieved? ___ Primary school ___ Middle school ___ High school ___ Trade school ___ Associate's degree ___ Bachelor's degree ___ Professional degree ___ Doctorate

What best describes your current job? (Check all that apply.) ___ Unemployed ___ Part time ___ Full time ___ Hourly wages ___ Salary ___ Self-employed

Figure 9.27. A revision of the form in figure 9.26, with aligned data fields. Aligning the data fields makes the form easier to use. The users filling out the form can more easily see what options are available, and the users processing the form can quickly record what options have been marked. *Source*: Created by the author.

What is the highest level of education you have achieved?
☐ Primary school
☐ Middle school
☐ High school
☐ Trade school
☐ Associate's degree
☐ Bachelor's degree
☐ Professional degree
☐ Doctorate

What best describes your current job? (Check all that apply.)
☐ Unemployed
☐ Part time
☐ Full time
☐ Hourly wages
☐ Salary
☐ Self-employed

shows a much better design. Now the options are arranged vertically, so the person filling out the form can easily look down the list to see what applies. And the person reading the form can quickly see what items have been marked.

DESIGN RESPONSE SPACES TO ENCOURAGE ACCURATE RESPONSES

Users need a space in which to respond to prompts appropriately and conveniently. Some of the biggest hassles when filling out forms involve recognizing which response fields match with which prompts and having an adequate amount of space in which to respond fully. Use proximity and enclosure to create a relationship between the prompt and the response field and to create a usable space in which to respond. Figure 9.28 provides options for linking prompts to responses in alphanumeric forms. Any of these designs can work successfully, and each has its own advantages and disadvantages:

- *Placing the prompt on a response line* uses proximity to create a strong visual relationship between prompt and response. But it takes up a lot of room on the line.

- *Placing the prompt below the line* also creates a strong relationship between prompt and response, without taking up unnecessary room. But this design runs the risk that users

Figure 9.28. Arranging prompts and responses. Here are some options for creating relationships between the prompt and the response. *Source*: Created by the author.

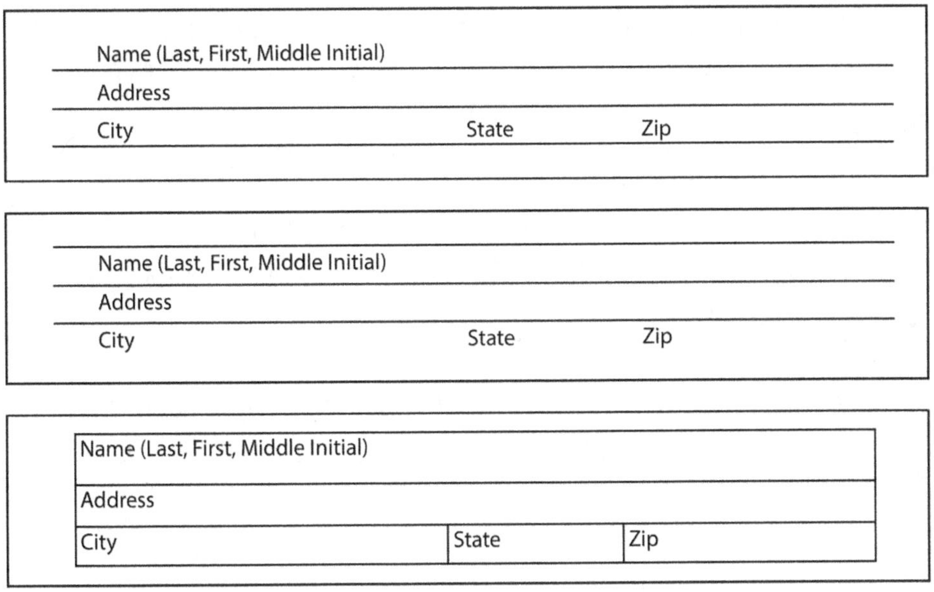

won't understand which prompt goes with which enclosed space: the user might enter a response in the space below the prompt rather than on the line above it.

- *"Hanging" the prompt from the top of a box* keeps it out of the way, and the box's borders enclose the prompt/response pair so strongly that few wouldn't know where to put a response. But it requires more vertical space (so users can write beneath the prompt), and it creates a boxy-looking form.

Regardless of what design you choose, be sure to provide users with adequate space in which to enter their response. Some personal names and city names, for example, can be remarkably long; you must design the form to accommodate the longest entry possible but still fit the form within the design of the page and the format of the document. Sometimes it helps to have a sense of what is possible: Mooselookmeguntic, Maine, and Kleinfeltersville, Pennsylvania, for example, both take of a good amount of space, and those are one-word names. The Village of Grosse Pointe Shores, A Michigan City, Michigan (the full phrase is its actual name) takes up even more.

As for option fields, most printed forms use checkboxes, although some might use a short blank line instead. The convention for checkboxes is remarkably strong, and most Western users will recognize the box as a place in which to make a mark.

TEST FORMS

Be sure to *test* your forms with real users before you send out the forms into the world. A frustrating, misleading, or inaccurate form can cause significant headaches to many people, including users who try to fill out the form, secondary users who try to use the resulting (usually faulty) data, and everyone else who has to mend bridges and clean up afterward.

Test your forms with real users before you send out the forms into the world.

Even a short usability test will give you valuable insights into how well users respond to your form, allowing you to solve problems *before* they get out of hand. For more information on conducting usability tests, see chapter 10.

DIGITAL FORMS

Paper forms have two distinct disadvantages. First, they rely on the clarity of users' handwriting, which can be illegible or idiosyncratic, too large or too small, or written in felt-tip markers, faint pencil, or even crayon. Handwriting can cause problems whether humans or OCR scanners process the forms. Second, paper forms must be processed by hand, either by the workers who are reading the form and physically entering data into a database or by workers who are sending the forms through sheet feeders on OCR scanners. These steps are inefficient and lead to significant errors in data capture.

Digital media alleviate these problems by making user responses more legible (to people and especially to computers) and by linking the form more directly and automatically to databases. The most common example of automated forms is a form on a website or in an application. The main advantage of digital forms is their automation of data collection; the main disadvantage is the necessary increase in expertise to create and integrate the forms with a website and with a database.

Digital forms add one significant new element to forms: a set of buttons for users to submit the form to the database or cancel their entries. The *submit button* takes the place of instructions on a paper form for submitting the form by hand, fax, or mail. The submit button also authorizes the form, taking the place of the signature on a paper form. For this reason, submit buttons almost always appear at the bottom of a form. It's also a good idea to include a *cancel button*, which resets the form and allows the users to begin again if they've made a mistake.

Most digital forms share many visual conventions with paper forms, and as a result, the same principles apply to their visual design. The complexity arises in making the automated links between the form fields and the database fields, which requires expertise in computer scripting languages, database design, and web design. Perhaps this is why many digital forms are often badly designed, relying on the basic visual default settings of HTML and common web browsers or default designs in app design programs.

In addition, some forms are designed to be filled out on a computer and then printed out for traditional submission. Adobe's Portable Document Format (PDF) allows for the creation of such forms, with alphanumeric fields in which a user can type and option entry fields on which a user can click to set a check mark or radio button. All other areas of the form (layout, prompts, introduction, instructions) are protected from user input. Microsoft Word provides a similar functionality.

Of course, this kind of digital form must still be processed just like traditional paper forms. But they can be very convenient in situations where only a few dozen or several hundred forms must be processed, or when the organization would like to distribute the form digitally.

EXERCISES

You've put your design thinking to the test in this chapter by thinking through how we collect and present information in lists, tables, and forms, and you've revisited a lot of the concepts you worked with earlier: similarity, enclosure, alignment, contrast, order, and proximity. Along the way you've thought a bit about the rhetorical implications of design. Now it's time to practice what you've learned. The exercises here should help!

1. Talk to a researcher at your university who uses survey forms as part of their scholarship. Consider asking these questions:

   o In what way can the design of the form affect responses?

   o What particular problems do you frequently see in form designs?

   o Have you ever had an experience where the design of the form interfered with or affected your results?

2. Find three forms on the web or in an app and critique their visual design. How do these forms share the conventions of print forms? How do they differ from print forms? Are there features common to print forms that are counterproductive for digital environments?

3. Choose a small employer or manufacturer in your community and imagine you work there as a personnel manager. You've used traditional paper job application forms for years. Given your situation and resources, would your company be best served by sticking with the traditional paper forms or shifting to an online application system?

   o What are the advantages and disadvantages of each approach to gathering data about prospective employees?

4. Think of something that people enjoy doing that could take the form of a list (local area concerts, movies, showings at art galleries, etc.). Once you have chosen a topic, how would you advise the advertising agency to list its options, and where would you put this list? Should the concerts or movies or gallery showings, for example, be listed alphabetically? By date? By artist or star? Should they be in print or online? As you work, think through the rhetorical implications of not only how you choose to list these events but where you list them and how the design choices you make influence different audience types.

REFERENCES AND FURTHER READING

Bagin, C. B., & Rose, A. M. (1991). Worst forms unearthed: Deciphering bureaucratic gobbledygook. *Modern Maturity*, *34*, 64–66.

Barnett, R. (2005). *Forms for people: Designing forms that people can use.* Business Forms Management Association.

Kyrnin, J. (2021, Sep. 30). *Why you should avoid tables for web page layouts.* ThoughtCo. https://www.thoughtco.com/dont-use-tables-for-layout-3468941

Meyer, J., Shamo, M. K., & Gopher, D. (1999). Information structure and the relative efficacy of tables and graphs. *Human Factors*, *41*(4), 570–587.

Neeley, K. A., & Alley, M. (2011, June). The humble history of the "bullet." In *2011 ASEE Annual Conference & Exposition* (pp. 22.1462.1–22.1462.14).

Zimmerman, B. B., & Schultz, J. R. (2000). A study of the effectiveness of information design principles applied to clinical research questionnaires. *Technical communication*, *47*(2), 177–194.

UNIT 3: PRACTICES

10

PROJECT MANAGEMENT

LEARNING OBJECTIVES

Upon completing this chapter, you will be able to:

- Discuss different models of project management

- Articulate different ways of working with, and communicating with, your design team

- Plan your client communication strategy

- Outline a general interview process

- Generate a general overview of a design process

- Document and report on your design process

∽

This chapter includes a scenario designed to give you a sense of how the various elements presented in this book work together during the design process. It follows a group of designers through an entire design project, from first contact with the client to presentation of deliverables. The scenario follows the general design process described in figure 10.1. The exercises in this chapter conclude with a series of questions that will ask you to think through the design process from several different angles.

Figure 10.1. General design process showing overlapping segments from initial client contact to product delivery. *Source*: Created by the author.

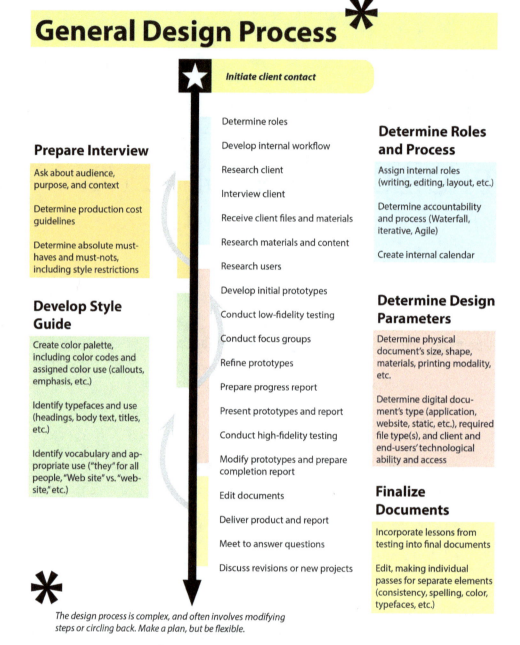

The team of DesignRight Solutions took on a design and documentation project for HouseHold Medical. Along with their manager, Rachel Hawkins, the team included Javier, Erin, Micah, Travis, Aisha, and Alex. Their project involved designing both a product card for a simple family first aid kit and a simple web page. The team faced typical technological and editorial challenges, but the most difficult part was coordinating the efforts of so many people.

The first thing they had to do was *get organized*. Working with the project stakeholders, the team conducted research and developed a plan. DesignRight's solution was to follow a fairly streamlined general design process that began with internal organization and ended with presentation of deliverables to their client. They could just as easily, however, have built a more complex PERT (program evaluation and review technique) chart organized by month. Figure 10.2, for example, shows how project

Figure 10.2. A PERT (program evaluation and review technique) chart. This PERT chart shows the project management of a document design project. Organization and planning are critical to the successful management of large projects. *Source*: Created by the author.

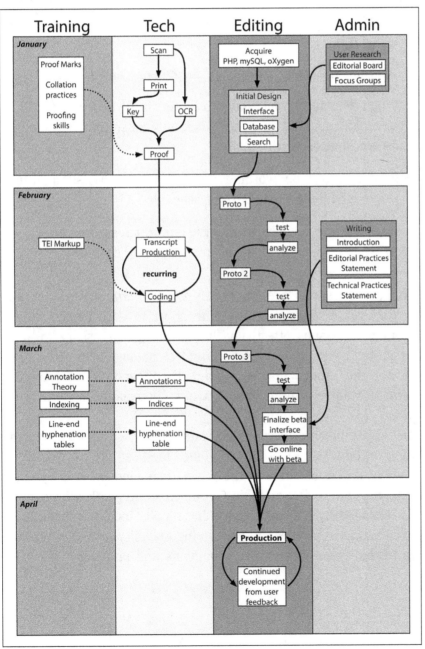

managers might divide their human resources into four teams (training, editing, technology, and administration) and how each team's tasks could be coordinated. (For more information on PERT charts, see chapter 7.) Regardless of how a team chooses to take on a project, the first step should always be getting organized.

Design projects are complex activities, requiring us to juggle a lot of variables. As a designer, not only do you need to think about the principles of design and make decisions about format, medium, color, and typography—some of the concepts you read about in units 1 and 2—but you also have to keep in mind big questions like these:

- Who are the clients? Why do they want us to create this document? What are their particular requirements, needs, and agendas?

- Who are the users? What are their needs and situations?

- When does the project have to be finished? How much will it cost? Or from another perspective, how much time and money do we have to work with?

- Who else will be working on this project? What are their responsibilities? What is their schedule?

- How will the design team track our progress on the project so we can know if we're on task and on schedule?

All of these questions are issues of *project management*. In practical terms, they center around how you're going to *do* a design project. To deal with these issues effectively, it's absolutely necessary to approach design projects with a good set of project management skills. Even if you're not the project manager yourself, understanding project management techniques will help you work successfully within a project team.

In this chapter we discuss two common project management models and then introduce a model that combines the best of both worlds. Along the way, we'll discuss some of the project management skills you'll need either to manage a design project yourself or to work within a project that is managed by someone else. These skills include scheduling projects, conducting user research, performing usability testing, working with design teams, and documenting your plans and progress.

Scenario Part 1: Getting Organized

Javier, Erin, Micah, Travis, Aisha, and Alex all work for DesignRight Solutions, a general-purpose design and documentation company managed by Rachel Hawkins. DesignRight is hired by HouseHold Medical to create documentation for a new product: a simple family first aid kit called MedPak. The kit will be priced affordably and is designed not only to be a useful household tool but to introduce potential buyers to HouseHold Medical's product line.

Rachel Hawkins, the manager, splits several of her employees into two teams. Javier, Erin, and Micah are in charge of any print-based design and documentation, and Travis, Aisha, and Alex are in charge of digital and online solutions. The team members are all involved in other projects with DesignRight Solutions as well, and so they know that getting themselves organized is a priority.

The first thing the two teams do is assign internal roles and work out both team accountability and overall project accountability. For the print team, they agree that Javier will be in charge of the bulk of the writing, Erin in charge of research and editing, and Micah in charge of design and layout, with all members contributing as needed to each other's work. The digital team agrees that Aisha will be in charge of writing, Travis in charge of research and editing, and Alex in charge design and layout, again with all members contributing as needed. As a group, the teams nominate Alex as the primary point of contact with their client.

The group agrees to follow an iterative model of design and agrees to share information and design protypes across the entire group to ensure that print and digital documents are consistent in color, typeface, and general aesthetics. Once the group is organized, they are ready to prepare for their initial meeting with HouseHold Medical.

PROJECT MANAGEMENT

Project management is the collective term for the techniques we use to *define*, *control*, and *assess* work on a project. Because many documents respond to specific situations and problems, it's easy to understand why they're often thought of as *design projects*. Design and project management go hand in hand—particularly when a design project is large enough to require the coordination of complex factors to bring a document to reality. The most important of these factors are people, time, and money.

- *People:* All document design projects involve clients, users, and a designer, but large projects typically include a design team whose members have various responsibilities for creating content, conducting research, and creating the design. Project management coordinates the efforts of a design team.

- *Time:* Projects take place in a finite period of time, from initial concept to final execution. Project management helps determine how much time is required to do the job, as well as how to break up the time and tasks into reasonable, manageable, and measurable units. Project management also estimates how much personnel time is necessary for a design team to complete a project.

- *Money:* Design projects take place in a real world where people and resources cost money. Project management helps to predict or estimate project costs. Even small projects that require little production cost might require considerable personnel costs.

Project management will help you evaluate and control the development of your design projects. Without conscious project management efforts, you risk creating unsuccessful finished documents.

WORKING WITH DESIGN TEAMS

As a document designer you might work alone on a project, particularly if it's a small-scale or a low-level design project. In those situations, you'll have to think and act quickly and assume all the roles of project management yourself. But more often you'll work with a *team* of people, all focused on creating a successful document.

This design team might include people with a variety of roles:

- Document designers

- Writers or content contributors

- Illustrators

- Editors

- Subject matter experts (SMEs)

The structure of the team might range from hierarchical to egalitarian, but typically all of these people will work under the direction of a project manager who coordinates the team's efforts and serves as the client's primary contact person. As the document designer, you might also be the project manager yourself, or you might work for a project manager. Each member of the design team is an important contributor to creating a successful document. Ideally, each is an expert in their own area and has valuable insights and vital skills to offer to the project.

MODELS OF PROJECT MANAGEMENT

Each project requires its own unique approach. But most project managers agree that there are two main schools of thought in project management. The *waterfall model* tries to control project development by creating a formal plan and set of specifications for the document before any drafting or visual design actually takes place. The plan and specifications govern the allocation of people, time, and money to the project. The *iterative model* of project management begins to create prototype documents at the beginning of the project and solicits user feedback to help control the direction of the project. In the sections that follow, you will see how each approach has its own strengths and weaknesses.

THE WATERFALL MODEL

In the waterfall model, project managers organize a project as a linear series of sequential phases from initial research and planning through delivering the finished product. This is called a *waterfall* because each stage is considered relatively unitary; like water cascading from one rock to another, the project completes one phase before going on to the next. The waterfall model organizes each project around the same general design process. JoAnn T. Hackos (1994) suggests that this process has five stages (pp. 28–29; see figure 10.3):

Figure 10.3. The waterfall model of project management. In the waterfall model of project management, the design process is mapped out linearly; each phase is completed before moving on to the next. *Source*: Created by the author.

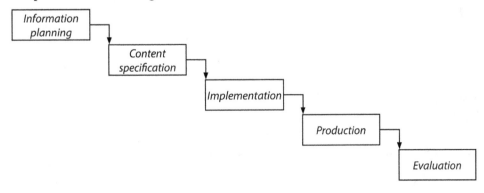

- *Information planning:* The project managers gather all possible information about the project and users and then write planning documents that describe the project in detail, create a project timeline, define the deliverables to be created, and assess what personnel will be needed.

- *Content specification:* The project managers create a document detailing the proposed contents for the deliverables proposed in the information planning stage.

- *Implementation:* The project team creates the deliverables specified in the content specification stage, taking them through several drafts.

- *Production:* The project managers see the deliverables through production, often through a professional printer.

- *Evaluation:* The project team looks back over the project to learn from its successes and mistakes.

This model gives considerable control to the project manager. The planning and specification stages allow for accurate estimations of time and costs. The staging, with its clear milestones and project documentation (e.g., the planning documents, the specifications document, and periodic progress reports), helps the project manager assess whether the project is on track, on time, and on budget. The extensive project documentation also promotes good communication between the project team and the client. And the evaluation at the end gives the project team an opportunity for reflection and learning so their next project will be more successful than the last. One common way to show progress checkpoints in the waterfall model is to create a *Gantt chart* (for more information on Gantt charts, see chapter 7). Gantt charts clearly block out proposed work in relation to proposed deadlines (see figure 10.4). They give people a quick overview of all the project elements and how those elements should occur in a given timeline.

Figure 10.4. Example Gantt chart often used in waterfall model planning. *Source*: Created by the author.

Project Element	Jan. 1 – 15	Jan 16 – 31	Feb. 1 – 15	Feb. 16 – 28	Mar. 1 – 15	Mar. 16 – 31	Apr. 1 – 15	Apr. 16 – 30
Plan	████	████						
Specify		▓▓▓	▓▓▓	▓▓▓				
Implement				░░░	░░░	░░░		
Produce							████	████

The waterfall model has some limitations, as JoAnn Hackos points out, particularly when it comes to meeting client and user needs accurately and quickly. Organizations that are less careful in their management of document design are likely to skimp on planning, but planning shortcuts often lead to a content specification that doesn't accurately respond to users' needs or the client's objectives. The result could be wasting a lot of time and money on something that doesn't work (1994, p. 40). Even with good planning, the lack of user input early in the process can lead to inadequate specifications. Waterfall models also assume a relatively long development process. Hackos uses the example of a six-month, 1,000-hour project. The length of this process means waterfall models can have trouble delivering a product quickly and responding to rapidly changing user needs.

THE ITERATIVE MODEL

If we look at how most design teams actually work, it becomes clear that they push forward a bit, look back to see where they've come from, then push forward a bit more. This gradually progressing series of actions and reflections, known as *iterations*, is often called *post-process* because it recognizes creative work as cyclical rather than linear (figure 10.5).

The iterative model takes advantage of the tendency to work in cycles, gradually progressing toward a desired goal. One such approach, *rapid application development (RAD)*, was first proposed by James Martin (1991) as an alternative to the waterfall model in software design.

Figure 10.5. The iterative model of project management. Each prototype is designed and tested. After the team analyzes the test, they feed the results back into the development of the next prototype. *Source*: Created by the author.

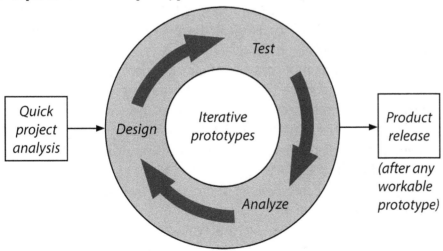

Rather than spending time on sequential stages of research, planning, and specifications, RAD jumps into creating prototypes of the product, which are then shown to prospective users or even released as introductory or *beta* versions of the product. The users' feedback then guides the development of the final product. Whereas the waterfall approach usually assumes that a project will take several months, RAD provides a product that meets most users' most urgent needs within sixty to ninety days. RAD uses a rigidly finite time frame for a development project: if the project isn't finished within that time frame, the designer eliminates some features rather than extend the schedule. The resulting product may be incomplete or imperfect, but it can be refined in later versions. Iterative approaches such as RAD are particularly popular in the open-source software movement, and some companies have taken iterative design as their primary project development model. Google, for example, often releases beta versions of its software years before rolling out a "final" version, which still undergoes continued development.

Similar to RAD, Agile is another iterative model of project management. Agile also came out of the software industry and arose from a group of developer's recognition that customer needs and expectations often change during software development, and project teams should be able to respond to evolving ideas over the life of a project. In 2001, a group of developers met and wrote the "Agile Manifesto" which argues that developers should privilege:

- Individuals and interactions over processes and tools

- Working software over comprehensive documentation

- Customer collaboration over contract negotiation

- Responding to change over following a plan (Beck et al., 2001)

As Rebekka Pope-Ruark (2015) notes, this focus on collaboration and interaction is vital to the training of those who will go on to work in industry. Agile iterative project management relies on projects moving forward one milestone at a time and allows for discovery and flexibility in any developmental process. While waterfall planning is theoretical, in that it supposes that nothing (or very little) will change over the course of a project, Agile iterative models ask developers to learn rapidly and adjust to changing technologies, needs, and expectations. This is often referred to as a "fail fast" approach. Your project will likely encounter setbacks, so work quickly to find those points of failure, learn from them, and move on.

Though there are many different ways to engage in iterative design practices, most iterative design projects typically start with a quick analysis of the rhetorical and practical situation surrounding the project and then develop a series of rough and incomplete prototype documents to show to users. These prototypes might start with sketches on paper (also known as *paper prototyping, rapid prototyping,* or *rapid paper prototyping*), conceptual models, or mock-ups of page or screen designs, but the goal is to produce better prototypes with each iteration.

For each prototype, the designers gather user feedback to guide the development of the next prototype. Users might be asked merely to comment on certain aspects of a prototype, either individually or in a *focus group*. Or they might be asked to do basic tasks with the prototype, in a process known as *usability testing* (for more information, see "Usability Testing: Getting User Feedback on Prototypes" in this chapter).

This process might involve omitting features that the design team had planned to include. For example, the designer might think that color would be a nice addition to a document but find through testing a black-and-white prototype that color isn't necessary to fulfill users' immediate needs. The designer could then decide whether to wait until a later version of the document to include color. Or in a website design project, the designer might initially plan to include a web forum for users but decide that there isn't time to implement this feature within the initial project timeline.

One advantage of iterative design is that it responds directly to users' needs by involving users throughout the design process—an approach also known as *participatory design (PD)* or *joint application design (JAD)*. Involving users throughout the design process means that you can respond quickly to users' needs and develop products that users actually want to (and can) use. Rather than working through an entire linear development process, you can create a "good enough" version—referred to in most Agile terminology as a *minimum viable product*—rapidly, and then use the feedback gained from that version to guide the development of better versions. This rapid response gets needed information into users' hands sooner.

On the downside, iterative design can lead to dynamic or even chaotic design projects because the target is constantly moving as you learn from testing each prototype. Iterative design also relies on creative, innovative, and flexible design teams; people who like a more linear approach often feel uncomfortable with the uncertainties of responding dynamically to user input. Management can also feel out of control with iterative design, since the model doesn't rely on the strict scheduling, milestones, and reporting of the waterfall model. Iterative design also makes estimating costs and deliverables more difficult because it is hard to predict users' responses to a prototype. Finally, by jumping into the design process without much planning, iterative design projects sometimes wallow around in uncertainty long enough for the project to lose momentum.

Scenario Part 2: Meeting the Client

In their first week of preparations, the DesignRight Solutions group determines their internal roles and accountability process and also begins research into their client. Erin and Travis take the lead on this part, and together they put together an overall profile of HouseHold Medical.

From the company's web presence, Erin and Travis learn that HouseHold Medical is a newly formed company whose goal is to provide personal medical equipment at an affordable price. It has a distinctly modern persona, with medical kits offered in easy-to-carry bags that can fit into (or be attached to) backpacks, though the general purpose appears to be ensuring that families have ready access to general first aid materials, from bandages to burn creams. Erin and Travis determine that the product DesignRight has been hired to work with is not yet available but has been designed as an entry-level product for potential buyers. Currently, HouseHold Medical's product line is fairly expensive, and it looks like the company wants to become more accessible to more people.

Travis and Erin go through HouseHold Medical's social media content and determine that the company has a fairly well-developed presence, with a message that shows commitment to issues of social justice and equity and a desire for people from varying income levels and educational backgrounds to be able to use its products. From what they can tell, HouseHold Medical's primary point of sale is online through its website, and it drives traffic there from social media promotions and online ads.

Travis and Erin share this information with their group, and together they come up with a list of interview questions that cover each of their areas of focus. Their list includes the following:

- We know that you want us to design both digital and print content for your new product, MedPak. What materials do you think are crucial for physical documentation, and what materials crucial for digital documentation?

- Who do you think the target audience for this new product will be, and how will they find out about it?

- Who generally buys your products, and where are they most often used?

- What are your primary goals for this product—what do you want it to do?

- Do you have a company logo that we need to use?

- Is there any particular color scheme or branding colors that you would like to see?

- Do you have a specific typeface used by your company?

- Is there anything in particular you would like to see incorporated into our designs?

- Is there anything in particular that you know you definitely do not want to see used in our designs?

- Do you have any printing or distribution restrictions? For example, can we use color on the print documents, or design them to use glossy paper? Do digital documents need to be in any particular format that you are aware of?

- We generally complete work of this scope within two to three months, and we'll probably have protypes ready within a month or so. Does that time frame work for you to schedule another meeting?

- Since none of us are medical professionals, we will likely have questions along the way as we develop these materials. If we email you questions, what kind of turnaround should we expect for responses?

Since their manager, Rachel, has already worked out financial details, they don't ask questions about pay, but instead focus on audience, purpose, context, and time-related variables.

At the initial meeting they all introduce themselves and establish that, going forward, Alex will be the primary point of contact between Design-Right and HouseHold Medical. They learn that HouseHold Medical wants a fairly simple design package: a physical product card to be included in the MedPak family first aid kit and a web page that fits in with HouseHold Medical's current branding that uniquely situates MedPak as an entry-level introduction to household first aid.

HouseHold Medical asks that the web page be more of an image and information site—a digital brochure—that will be easy to share over social media and serve as a landing page for new potential customers. Most of HouseHold Medical's current buyers spend over $150 at a time, but the company has received many requests for more affordable kits, so this MedPak product will open up a new demographic. The company hopes that these will also become common gifts, thereby further broadening their reach.

HouseHold Medical lets them know that its products appear to be mostly used at or close to home, despite their portability, but that some users take the products camping and hiking. Research suggests that the kits are often used at night.

The company provides a logo, a simple set of linked *H*'s with a rooftop (figure 10.6), and asks for familiar health-kit related colors (red, black, white, and blue), though it's open to options. The group is told that they are not to use any actual photographs of people using the product in the physical product card, but that images of actual people are fine for the web page. Also, product images are on the company's standard marketing page, so it would be preferable if the simple informative page focused more on use, description, and testimonials, rather than yet another image of a first aid kit.

Figure 10.6. HouseHold Medical's logo. *Source*: Created by the author.

They are told that there are no printing or digital distribution restrictions, and that they might generally expect a twelve-to-twenty-four-hour turnaround on any emailed queries. HouseHold Medical provides the group with three MedPaks so that they can get a feel for the contents.

Before leaving, DesignRight team members clarify with HouseHold Medical the expectations for deliverables and any reports. HouseHold Medical lets them know that there is no need a final formal written product report along with the deliverables, but does request a brief presentational overview of the design and testing process.

The group thanks HouseHold Medical reps for their time, returns to DesignRight's studio, and gets to work.

PLANNED ITERATIONS: A MIXED APPROACH

The waterfall model and the iterative model are seemingly opposing approaches to project management. The waterfall model is controlled, methodical, and consistent, whereas the iterative model is dynamic, creative, and changeable. You may work in organizations that favor either of these developmental models, or another altogether. In Agile, for example, one of the most common approaches is the Scrum model, which is managed by a Product Owner, who determines *what* needs to get done, in concert with a Scrum Master, who determines *how* work actually gets done. Teams are often fixed at seven members plus or minus two (five to nine, depending), and projects are scoped out in two- to four-week sprints with daily meetings and a final, fixed deadline for the minimal viable product. Teams agree to Team Norms as social contracts that dictate internal processes, from meeting conduct to how to offer critiques, and projects are guided by what are often referred to a "guardrails" in place in order to keep projects in line. The language may seem odd at first, but Agile Scrum project management is increasingly common in the workplace.

Though waterfall and iterative models seem opposed, another model of project management is the *planned rapid document development (PRDD)* model (figure 10.7), which retains some of the staging of the waterfall model while placing iterative prototyping within a finite development stage. It also retains the project documentation common in the waterfall model, thus promoting clear communication between the project team and the client.

PRDD includes five stages. Each stage is accompanied by a series of either formal or informal reports to document the design process and ensure clear communication between the design team and the client:

- *Stage 1: Research.* The project team meets with the client for a kickoff meeting about the client's agendas and expectations for the project, writing up the results in an informal *kickoff meeting report*. The team then conducts user research to build a clear understanding of the project and uses this research to write a formal *project analysis report*.

- *Stage 2: Design planning.* In this relatively short stage, the design team writes a planning document, specifying for the client and the team the iterations planned for stage 3, including a schedule for usability testing.

Figure 10.7. The planned rapid document development (PRDD) model for project management. PRDD combines the approaches of the waterfall and iterative methods to project management. Formal reports are in bold; informal reports or team documentation are in italics. *Source*: Created by the author.

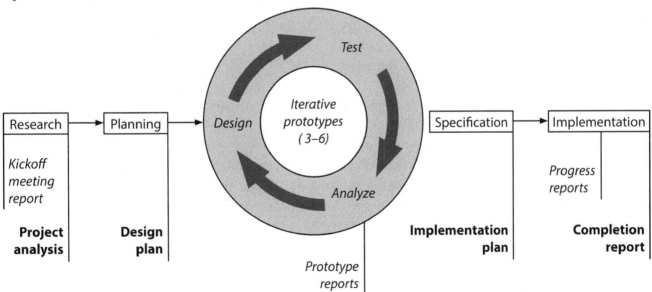

- *Stage 3: Iterative prototyping.* Within a strictly bounded time frame, the project team creates a series of three to six prototypes and conducts focus groups and usability tests to get user feedback. The goal of these iterations is to build toward a firmer idea of the document's design and contents. After each prototype cycle, the team writes an informal *prototype report* to codify its findings and recommend directions for further development.

- *Stage 4: Specification.* The team finalizes the document's design and contents, writing a formal *implementation plan* to guide stage 5.

- *Stage 5: Implementation.* The team implements the design finalized in stage 3. This stage includes a formal *completion report* as an opportunity to document the lessons learned in the project so they can be applied to future projects or future developments of the same project.

The PRDD approach retains some of the advantages of the waterfall model while gaining some of the advantages of the iterative model. It breaks down projects into reasonable stages with clear project deliverables: a set of reports that help the client and the design team understand where you're heading. It also sets aside a specific stage for research at the beginning of the project and provides a major product rollout at the end.

At the same time, PRDD involves users early in the process, through both user research and iterative prototyping, and it allows the design team to create a workable document rapidly. The process can also reiterate as often as necessary or as time allows, returning to the prototyping stage to improve the document in future versions. If time runs short, any workable iteration can be released as an introductory beta version, with further development to follow.

Of course, this is only a general model. Most projects will have some constraints or dynamics that might make you modify the model more toward iterative design or toward the waterfall model. If you're working on a small project, for example, you might consider replacing progress reports and testing reports with a project blog that serves the same purpose less formally, or use one of any number of online management tools such as Discord, Slack, or GitHub. If you're working on a larger project, you might include more formal documentation, such as reports for each of your user research activities in stage 1. Overall, try to balance organization with flexibility; sometimes you must change your plans to meet new conditions.

In the sections that follow, we'll discuss how to conduct, complete, and document each stage of PRDD.

STAGE 1: RESEARCH

The goals of the research stage are to understand the user and client situations and to determine what kinds of documents might meet user needs and fulfill client agendas. To do this, you will need to conduct some field research to find out as much about the user and client as possible within a reasonably small time frame.

Remember that the primary purpose of this stage is to get a relatively quick sense of the rhetorical situation of the project before moving into prototyping. Don't get stuck on this stage, but spend enough time here to get a good sense of the users' situations. In a typical sixty-day project, you should spend about a week conducting initial research into clients and users.

CLIENT RESEARCH

Many design projects start with the client. Clients frequently approach designers with an idea—to create a brochure, a website, an informational sign, a kiosk application, a newsletter, or some other kind of document. Your first job as a designer, then, should be to talk to clients about their agendas and expectations for the design project.

The project kickoff meeting. Ideally, you should sit down face-to-face with your client for a project kickoff meeting to talk about the client's agendas and expectations. The meeting should include all members of the initial project team, as well as all stakeholders in the client's organization. If you can't meet in person, try a conference call or a web conference—Zoom, GoToMeeting, CISCO Webex, Microsoft Teams, and more allow for screen and file sharing in addition to video and voice capabilities. As a last resort, an exchange of emails can suffice—although at these beginning stages of the project, email can take a lot longer than face-to-face meetings and possibly lead to misunderstandings.

The main purpose of the initial meeting is for you to find out what the client wants and how the client envisions the project. During the meeting, encourage the client to be as specific as possible about their ideas and ask careful questions to get a clear picture of the project. The kickoff meeting should include at least the following:

- Introductions of personnel and stakeholders

- Client's initial project description, including a description of anticipated users

- Client agendas

- Client expectations and project scope

Good interview skills are important here. Be prepared to ask questions, and try to draw your client out and get them talking about the life of the design you will be working on. This will give you a sense of how to best meet their needs, and the needs of the end-users. At the meeting, it often helps to repeat what the client has said to make sure you understand what they are thinking. In the DesignRight scenario, for example, one question in the initial meeting might have been "So, if we understand you correctly, you want a product card that shows the contents of the first aid kit and introduces people to basic first aid materials. Is that right?" Take careful notes. It's easy to hold a successful meeting and then quickly forget what happened in it!

If you are working independently, or finances have not yet been established, make sure that your first meeting also covers finances and financial expectations, including pay schedule, rates of pay, and client's ability to pay for any necessary developmental expenditures (if that is not already built into your company fee structure).

Client agendas. An *agenda* is what the client wants to *do* or to have *happen* as a result of the design project. Agendas are usually broader than the document itself. For example, a client might ask you to design a brochure, but to do that effectively you'll need to understand what the client wants the brochure to do, which could be many things. The client could have any of the following goals for the brochure:

- To promote a product or service

- To convey information to users

- To answer frequently asked questions, decreasing customer service demands

- To encourage users to contact the organization for more information

Ask specific questions of the client to understand the agendas as completely as possible. The answers might even lead you to suggest that another kind of document might be more suitable than the document initially envisioned. Keep a list of the agendas you identify in your discussion with the client. It might be useful here for you to think about what Hassenzahl and Roto (2007) describe as "do goals" and "be goals." "Do goals" support what the end-user of a product needs to actually be able to do: find the appropriate bandage material in a first aid kit, for example, or search a web page. "Be goals," however, are the result of who we want to be as the users of a product. How does using a product make us feel? User-centered design advocate Donald Norman argues throughout the book *Emotional Design* (2004) that "attractive things work better" and "be goals" help us think through that argument. As Hassenzahl and Roto argue, "Beauty in products . . . may be viewed as an unnecessary luxury. But imagine using an ugly product every day" (2007, p. 10).

Remember that, in most cases, your client is not a professional document designer—you are. It is your job to listen to their needs and expectations and offer suggestions and alternatives that might better suit their needs than their original idea. After all, they hired *you* to do what they cannot. A client might ask for a website, for example, when what they need is a product card. Take the extended project example in this textbook, for example. HouseHold Medical needs information to be readily available for those using their family first aid kit. Since there is no way of knowing the exact circumstances where someone might need first aid instructions, it makes sense to provide an overview both online and on a physical document in the kit. The kit might need to be used while on a hike, or in the event of a disaster where the power might be out.

Level of design. Discuss the level of design the client thinks will be appropriate for the document. As we discussed in chapter 1, some documents are more important to the client's overall mission than others. You wouldn't want to spend the client's money unwisely by overdesigning a document that's not really that important to the client's mission. On the other hand, you wouldn't want to waste time creating a document with too low a level of design—something that might embarrass the client in front of an important group of users or fail to meet users' needs and expectations.

To figure out the appropriate level of design, first discuss the client's sense of the document's potential users. The client often interacts with users regularly and has a good intuition of their situations and needs. The client will also have a good idea of how important the users are to the client's mission. If every user is very important and must be impressed with the document, then a higher level of design is warranted. If it's not particularly important to impress the users, a lower level of design might do the job just as well.

After discussing users, take your initial list of the client's agendas and work with the client to *prioritize* the list by dividing the items into three categories:

- *Mission-critical:* The document must do these things and will fail if it doesn't.

- *Optimal:* The document will work better if it does these things.

- *Optional:* The document will probably still work okay without these things, but they'd be nice to have.

These prioritized agendas will help you fine-tune the level of design for the project.

Client expectations. It's also a good idea to ask your clients explicitly what their expectations are, both of you and of the design project. Most clients have a variety of assumptions about what you will do for them in the project. These expectations often arise from the following issues:

- How long the project will take

- How much the project will cost

- How many people will be involved on the project team

- How much the client needs to be involved

- How and how much you'll communicate with the client about your progress

- How the project will be broken into stages

- What the deliverables will be—what you'll give the client at the end of the project

Some of the client's expectations about these issues might make a lot of sense; others might lie beyond what you can fulfill.

Making the client's *implicit* expectations *explicit* allows you to negotiate between what you can provide and what the client expects. It also gives you an opportunity to communicate clearly with your client about expectations. If the client knows what to expect from you, the client is less likely to be disappointed or frustrated by your work.

During this part of the communication and negotiation process, it is also important for you to make clear your own expectations, particularly those regarding client communications. If you know that your group will be using an iterative approach, for example, you might want to come to a shared agreement over how long you might expect to wait for a response to an email query. You may need to ask the client for information along the way: logo files, images, copyright permissions, written copy, and more. Make sure that your group has at least opened these expectations as a line of communication from the outset.

The kickoff meeting report. As soon as possible after your kickoff meeting, write up your notes in an informal report to the client. For small projects, even an email to the client will suffice. Reiterate the information you gathered in the meeting, and ask the client to review the document for accuracy. This process will highlight more questions that need to be asked or areas that need clarification. This report also gives clients the opportunity to "take back" any ideas that they decide don't mesh with their agenda. If the project is going to change, it's best that it change during these early stages rather than after you've invested your time and effort.

USER RESEARCH

The client's sense of users' situations and needs is a useful starting point for understanding prospective users. But clients' intuitions about users might not always be accurate. User research tries to develop a more solid picture of users and their needs within their own contexts. Design teams conduct user research primarily by observing users working in their own contexts and by asking those users targeted questions.

Finding users to research. One of the challenges of user research is finding users and getting their permission to observe or talk to them. Many potential users are happy to talk to you about their situations, including what kinds of documents might make their lives more convenient or their work more effective. Your client might be able to point you toward potential users to observe. The client's customers might be happy to talk with you because they already have at least some interest in improving the client's products or services. If the project anticipates creating internal documents for the client's employees, you should have little difficulty finding people to talk to. For design projects at your college or university, you might ask friends or members of student organizations to help.

Members of the general public are the most difficult kind of users to find and observe. If your project anticipates a public user group, you might need to advertise for participants. Regardless of the context, try to find participants who are as close as possible to the anticipated users in demographics, background, and interests.

Use research and ethics. Of course, it wouldn't be ethical to observe people without their knowledge and explicit permission. Before engaging in any user research, ask potential participants to sign a statement that gives you permission to talk to them, observe their work, and use the results in your design project. You will especially need written permission if you intend to record participants on audio or video. Naturally, you should avoid asking users to do anything they might find uncomfortable, unethical, or coercive.

Before conducting any user research activity, talk with the participants for a few minutes about the design project and about the activity in which they'll be participating. Let them know what to expect, as well as how long your observation and the ensuing discussion will take. Tell the participants how the data you gain will be used—particularly any video or audio recordings you make. Assure them that throughout the research activity, they are free to ask questions, decline to respond, stop participating, or even leave if they feel uncomfortable. After conducting user research, it's customary to reward each participant with a small gift to thank them for their participation. A gift certificate to a local restaurant or coffeehouse, a nice pen, or a coupon for goods or services at the client's business can all work well. This step is less necessary if the participants are employees of your client or if you're working on a student design project for a college course, but even a few home-baked cookies will go a long way toward making your participants happy and cooperative. In all cases, thank participants for their time and effort in helping you design a better document.

User research can involve a variety of techniques, including site visits, shadowing, task analysis, individual interviews, and focus groups.

Design Tip: A Checklist for Conducting User Research and Usability Testing

Before the research activity or test:

- Obtain the participants' permission in writing.

- Explain the purpose of the activity, and describe what you'll ask users to do.

- Assure users that their participation is voluntary and that they can stop participating whenever they like.

During the research activity or test:

- Observe carefully.

- Take notes.

- Consider recording users with audio or video (but only with their prior permission).

After the research activity or test:

- Thank the users for their participation.

- Consider providing the users with a small gift.

Site visits and shadowing. Site visits involve actually going to the environment where the anticipated document will be used and observing user activities in that environment. Site visits can help you develop a general sense of the environment in which your document design will be used. You may be allowed to walk around by yourself, but it's more likely that you'll have a guide or host to show you around the site and introduce you to people. Take careful notes, and, if possible, get permission to take some digital photos to remind you of what you saw.

While you're at the user site, it can be a good idea to get permission to follow a specific user around for an hour or two to see how and where they work. This technique is known as *shadowing.* Shadowing helps you build an in-depth picture of a single user's activities and attitudes toward the situation the document you're designing will address. While shadowing, avoid interrupting the participant's work or getting in the way. Instead, focus mostly on observing the participant and taking notes about what questions to ask later. At the end of the visit, talk with the participant about what you observed and ask any questions you might have.

Another user site activity is known as *bodystorming.* As described by Brian K. Smith, bodystorming asks designers to physically engage with space and place as a way to performatively envision how the work they create will appear in the wild. He writes:

> Like many other design methods, bodystorming includes observing learners in the context of activity and documenting their actions as a form of needs and/or requirements analyses. It differs by asking designers to leave their offices and act out potential solutions to problems in the real world. Bodystorming involves being physically present in environments where interventions will occur and assuming the roles of participants in those contexts to understand how design solutions are affected by existing relationships between peers, artifacts, prior knowledge and other environmental constraints. (2014, p. 72)

Bodystorming asks designers to move around to consider everything from lighting to traffic to ambient background noise (and more) in their designs.

Task analysis. Task analysis is a more intense kind of shadowing that focuses on watching a participant perform a specific task. It's particularly common in design projects where the expected deliverable is a set of instructions or a procedure. If your goal is to help other users perform the task successfully, you need to watch a user who already knows how to do the task. The more focused your observation, the more successful you will be in accurately reconstructing the instructions or procedure.

In some situations, you can have users perform the task in controlled conditions, such as a usability lab. But it's often best to watch users perform the task where they normally would. Try to get their permission to record them as they perform the task. This kind of recording isn't absolutely necessary for a successful task analysis, but it is helpful to be able to review the task repeatedly afterward. Even if you are able to record, take careful and precise notes about your observations.

To conduct a task analysis, first ask the participant to complete the task on their own. Then ask them to repeat the task again and narrate what they are doing as they do it—a technique known as the *talk-aloud protocol*. A variant of this process is known as *think-aloud protocol*. In talk-aloud, a participant narrates their actions. In think-aloud, a participant narrates their actions along with their reasoning behind the actions. Both approaches have their uses with some practitioners arguing that talk-aloud protocol helps researchers assess *what* a participant does, while think-aloud emphasizes *why* a participant does what they do, perhaps offering insight into seemingly subjective individual experiences. Finally, ask if you can do the task yourself under the participant's direction. Take notes about your impressions during the task—particularly at points where you hesitated or got lost. Compare these repetitions of the task. You may find that the participant breaks down the procedure differently in narration than in any previous descriptions of the task. Or you might find in doing the task yourself that the participant has internalized steps they don't even consciously recognize anymore.

Interviews and focus groups. Interviews and focus groups are the best ways to gain insight into users' needs, situations, and attitudes. An interview is usually conducted with only one other participant, whereas a focus group might have five to eight participants. As in task analysis, video or audio recording is helpful.

Interviews and focus groups should not be free-form conversations, however. Start by creating a relatively small set of discussion questions to ask the participants; about a half-dozen main questions are all you can hope to cover in an hour-long session. That means you must choose and sharpen your questions carefully before the session begins. Focus on the issues that are most important for you to understand about users and their situations. If you have short, basic questions to ask (e.g., demographic information), consider preparing a survey or form for participants to fill out *before* the session.

Keep your questions open-ended and non-leading. Rather than asking "Do you have a positive impression of this organization?," you might ask "What impressions do you have of this organization?" Also avoid leading participants into giving the answers you want to hear. The question "What do you like about doing this task?" would probably make participants focus on what they like, even if they might not like doing the task very much. Asking "How do you feel about doing this task?" will give you a much better sense of the participants' honest attitudes, positive or negative.

Keep in mind, however, that people don't always tell the truth in interviews and focus groups. Individual interviews are basically structured conversations with a stranger; participants sometimes wish to present themselves and their situations in a different light than what might be objectively true. Focus groups also involve group dynamics. Sometimes a focus group can be swayed by the opinions and attitudes of a single dominant or vocal member. At other times, peer pressure can make individuals tend to conform to the group, even when their opinions might differ markedly from those of the group.

WRITING THE PROJECT ANALYSIS REPORT

After you have conducted all of the client and user research you have time to do, write a project analysis report. This report should be addressed to the client and describe the project in terms of the findings of your research. The goal of the report is to create a good mental picture of the project, for both the client and the project team. The project analysis report also helps you to communicate with your client, explaining how you intend to proceed and asking for client feedback on your progress.

The project description is typically broken down into the following segments:

- *Introduction:* Introduce the report with a section that explains what this report will cover.

- *Initial project description:* Describe the project as the client described it to you, including the client's expectations.

- *Client goals and agendas:* Summarize the client's goals and agendas for the project, including the mission-critical, optimal, and optional objectives. (These first two sections can arise from the kickoff meeting report, incorporating any subsequent client feedback.)

- *User research:* Summarize the methodologies and results of your user research activities.

- *Initial deliverables recommendation:* Given what you have discovered about the client and the users, make an initial recommendation of what kind and scope of document or documents will best fulfill client agendas and user needs. You might find it useful to include sketches or wireframes (low-fidelity mock-ups of a user interface) here.

Of course, you should adapt this general outline to fit your own project. Submit a copy of the report to your client and ask for feedback.

Scenario Part 3: Initial Prototyping

Back at DesignRight's studio, Erin and Travis begin researching HouseHold Medical's client base more thoroughly and start to compile a list of people they might reach out to for some low-fidelity usability testing. As they do, however, they realize that most of their team fits into the demographic HouseHold Medical seems to be working toward as an audience, and propose to save themselves time and their client money by doing simple in-house testing rather than full-blown usability testing. They contact HouseHold Medical about this potential change of plans, and company reps agree, noting that the products are already well-tested and they are fine with largely following expert-based design advice. The company has its own team of site managers and designers and can make minor alterations to existing designs on an as-needed basis.

Micah, Alex, Javier, and Aisha unpack the simple medical kits and make a list of contents. The packs include:

- HouseHold Medical's standard first aid guide

- A flashlight

- A pair of scissors

- Tweezers

- A thermometer

- A tube of antibacterial cream

- A pack of aspirin

- Two small self-stick bandages

- Two large self-stick bandages

- A roll of gauze

- A pack of sterile gauze pads

- A roll of medical tape

- A rolled elastic bandage

- A chemical cold pack

- Two alcohol disinfectant wipes

- Two tubes of sterile saline

- A set of sterile gloves

After talking through the materials, the groups agree to start with simple, hand-sketched prototypes to make sure their designs are on the right track before they invest too much time. They talk through potential design ideas, then adjourn, leaving Micah, as they often do, to rough out sketches of both the site and product card based on their conversations.

"It's missing a few things, isn't it," Javier says.

"Yep," responds Travis, "but I think we have to work with what we got."

"We do," agrees Aisha.

"Can we change the logo?" Javier asks. "I hate the bump that sticks out on the right side of the double-*H*."

"Afraid not," replies Aisha. "We're stuck with it since that's what they use on everything."

On Monday, both teams meet together to go over the sketches.

Micah starts with the product card (see figure 10.8).

Figure 10.8. Micah's initial sketch (a low-fidelity prototype) and notes for the product card. *Source*: Created by the author.

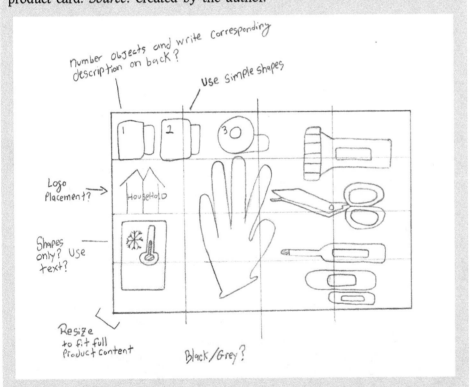

Micah follows DesignRight's standard introduction-to-product policy, which means offering only contextualizing information before showing the image, then letting the group talk through the design. The group follows this model because they know that they won't be present when actual users engage with their designs.

The group critiques the sketch, agreeing that they like the simple images but that the logo should be placed in the upper left (or bottom right), corner, the thermometer needs to show a temperature on it for increased context, more medical imagery (e.g., cross symbols) should be present, and the gauze rolls look a bit too much like toilet paper. The group then discusses how text will work on the card, ultimately deciding to get rid of the numbering system and instead use a translucent image on the reverse side with textual overlay.

The group then moves on to discuss Micah's simple web page sketch (figure 10.9).

Figure 10.9. Sketch for simple web page. *Source*: Created by the author.

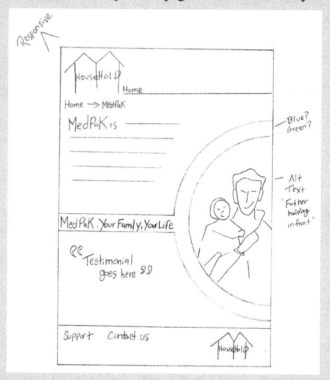

Again, the group agrees that they like the overall feel of the design, but that a few modifications will be necessary. Alex, as the group's primary web designer, notes that creating body text that maintains consistent spacing with a rounded image on differently sized screens will be difficult. Alex notes that the team's design should start with mobile technologies, as it's always easier to build for small spaces and then scale up, rather than build for large spaces and then scale down. To that end, Alex suggests using a simple division that places a pull quote immediately after the body text. Otherwise, however, they think they can get the site to approximate the image.

After seeing the initial sketches, the group puts together a basic style guide, agreeing on a primary palette of black and gray, with color used only for emphasis on the web page, and American Red Cross Red (hex code #ED1B2E, CMYK of 0, 88.6, 80.6, and 7.1) used for emphasis on the product card. For the typeface on the product card, the group decides on

Minion Pro, as its simple serifs allow for easy reading and won't detract from the simplicity of the image designs.

For the web page, the group decides on Times New Roman, agreeing that though the typeface has often been decried as overly plain, its serifs and widespread incorporation on devices means that the page will be easy-to-read and maintain its appearance no matter where it is viewed. Finally, they decide that most of their work will draw from language on HouseHold Medical's website, but they clarify that they will be writing "first aid," not "first-aid," anytime the phrase appears. They then separate to work on their individual components.

STAGE 2: DESIGN PLANNING

Stage 2 probably won't take very long because it simply involves writing a planning document to guide the iterative design in stage 3. For shorter projects, this planning document might even be appended to the project analysis from stage 1.

The goal of this document should be to develop a workable schedule and tentative plan for the design iterations you'll create in stage 3. The number of design iterations will naturally depend on the level of design and time available for the project. Mission-critical and complex projects will probably benefit from more iterations so you can explore different approaches to the design, while less-critical or simpler projects might need fewer.

As much as possible, you'll also want to plan for what *kind* of iterations and testing you'll perform. Part of iterative design is that you will need to use each iteration to guide the development of the next, but you should have a general idea of what sequence of iterations you're planning to follow. Remember that you must plan sufficient time for usability testing of each iteration, as well as for analyzing and reporting the results.

Because this planning document explains what actions you'll take throughout the rest of the project, it's considered somewhat binding. Make sure that you promise only what you can deliver. Submit the planning document to your client and ask for feedback—if you haven't already, now is the time to negotiate a shared idea of how the project should move forward.

STAGE 3: ITERATIVE PROTOTYPING

As we said before, the objectives of iterative prototyping are to get something into users' hands quickly and then use their feedback to improve the product in future iterations. In PRDD, you'll move through three to six prototypes, advancing toward the completion of the document with each iteration. The number of prototype iterations you choose to go through will depend on a variety of factors, including the complexity of the project and the amount of time you have to release a workable document to users.

In stage 2 you should make a tentative plan for the number and type of prototypes you'll create in the project analysis report, but you can decide how many are necessary at any point during the prototyping process. If you run out of time after coming up with a prototype that does a basic, good enough job, you have the option of releasing it as a beta version. You can always develop this prototype later in more polished versions.

CREATING PROTOTYPES

Typically, iterative design moves from rough prototypes in early iterations to more finished ones later on. These prototypes are referred to in terms of their *fidelity*, or truth to the final intentions of the project, from *low-fidelity (lo-fi)* to *high-fidelity (hi-fi)* (figure 10.10).

Figure 10.10. The fidelity of prototypes. The fidelity of prototypes can be measured vertically, from rough to polished, and horizontally, from incomplete to complete. *Source*: Created by the author.

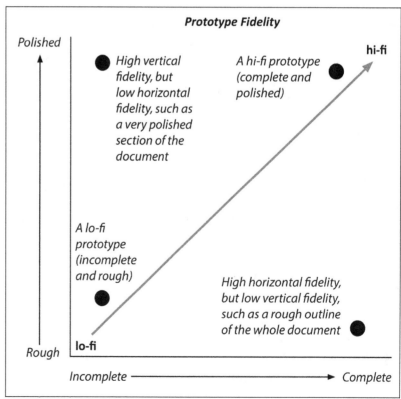

Fidelity can be measured vertically or horizontally, depending on *completeness* and *polish*. A *vertical prototype* is polished but not complete. It shows only a single working portion of the document. Examples of a vertical prototype would include a single section of a paper document that includes actual content or a single web page with all of the links activated (but with little or no content on the pages the links lead to). A vertical prototype works, as far as it goes—we can use the parts that the prototype includes but none of the other parts of the document.

A *horizontal prototype* is more complete but less polished than a vertical prototype. An outline or sketch of a document's entire structure is probably the most horizontal of prototypes, including all of the parts but no details. A highly finished page layout sketch with no real content would also be considered horizontal because it focuses on global design issues (the page design that will be applied to the whole document) but no actual content. A horizontal prototype doesn't actually work, but it gives a good comprehensive picture of the entire document.

Both kinds of prototypes can be important in different iterations, depending on the goals of your project. If you need to iron out the whole structure of the document you're designing, horizontal iterations will help you meet that goal more quickly than a polished, vertical prototype. But if you need to get a mission-critical part of the document working right away, a vertical prototype might take priority.

Don't underestimate the value of lo-fi prototyping in either the horizontal or vertical dimension. Hi-fi prototypes might look more complete and polished, but they take longer to create than lo-fi prototypes, which can be created quickly and repeatedly. Lo-fi prototypes are particularly useful for defining general concepts, such as the overall structure of the document and basic page layout concepts—figures 10.8, 10.9, and 10.11 are good examples of lo-fi prototypes. Even a very lo-fi prototype can be incredibly useful early in a project; sometimes a simple sketch on paper or on-screen will help you develop a global vision of the project's direction. Early lo-fi prototypes can also help you gain valuable user feedback quickly, which you can roll into the development of the next, higher-fidelity prototype. For a complex website, for example, you might consider rapid paper prototyping the entire site using sticky notes on a large wall. You could even usability test paths through the document by asking users to move from one "page" to the next through written "links." This would allow you to easily and cheaply draw out various layouts without ever having to code a single page—a particularly valuable approach if you are paying an outside individual or company for actual site coding.

Ideally, a design project should work toward both horizontal and vertical fidelity—that is, a complete *and* polished version of the document. But remember that the most important goal of iterative design is to fulfill as many of the client's agendas as possible as early as possible, working from the mission-critical agendas first. You can always go back to fill in less important details and polish the design in subsequent iterations, and you should bear in mind that secondary documents might draw from those earlier mission-critical documents in everything from typeface decisions and color palettes to chunks of written text.

In the next three sections, we'll introduce some more specific kinds of prototypes that you may want to develop in your design projects. In general, you'll find three types of prototypes most useful. Moving from lo-fi to hi-fi, these types include conceptual prototypes, design sketches, and operational prototypes.

Conceptual prototypes. Conceptual prototypes are typically very lo-fi explorations of potential concepts, such as the structure or navigational features of the document. Because they don't try to show anything close to a finished design, they're usually more horizontal than vertical.

One common early-conceptual prototype technique is *card-sorting*. The goal of card-sorting is to get users to help you organize a document in an order that makes sense to them rather than just to you. To conduct a card-sorting exercise, first brainstorm a series of potential concepts, topics, or sections the document might include. Write those concepts on index cards, one concept per card. Then ask a potential user to sort the cards into groups and place them within whatever order makes most sense. Finally, interview the user to ask why they categorized and ordered the cards as they did, taking particular note of what logical method they used to organize the cards. Because this takes only a few minutes, you can repeat the exercise with several users in a couple of hours.

Working from the results of a card-sorting exercise or from other ideas, you might create a *conceptual diagram* of the document you can use to structure an interview or focus group discussion. A conceptual diagram doesn't look anything like the finished document; rather, it's an abstracted version of a potential document. Figure 10.11 shows a conceptual diagram used in the design of chapter 4 of this book.

A conceptual diagram can also be a flowchart that shows a user how to go through a task using the document. Potential users can then give feedback on whether they would use the document to perform the task that way or some other way. In an even more conceptual mode, you might ask potential users to help you draw a conceptual diagram that models the structure or use of the document. For example, you might ask users to help you draw a site map that shows the structure of a website.

Design Tip: Version Control

As you develop sequential prototypes or drafts in a document design project, you must keep track of successive versions—a practice known as *version control*. Version control is particularly important when you're working with a design team on multiple iterations of prototypes. Otherwise, as several people work on the same project, it becomes easy to lose prototypes, overwrite prototypes, or develop competing prototypes. Some companies with large design teams even invest in version control software to help them keep track of versions.

The simplest way to keep track of version control is to develop a consistent naming convention for your prototypes. For example, you might simply number the prototypes sequentially (1, 2, 3). If you're working with more than a few versions, however, you might want to have major and minor revision numbers (A1, A2; B1, B2, B3; etc.). Software revision numbers (0.5, 0.9, 1.0, 1.1, 2.0, etc.) can also work well. Numbers under 1.0 are usually considered beta versions, making it easy to see whether the version you're working with has already been fully released.

For paper prototypes, simply mark the version number on the back in pencil. For digital prototypes or textual drafts, add the version numbers to the beginning of the filename. You might also want to add the date: for example, 01_InformationPamphlet_5-5-2025.indd, 02_Information Pamphlet_5-10-2025.indd, 03_InformationPamphlet_5-17-2025.indd, and so on.

Make sure everyone on your team understands and uses the naming conventions you've established for your design projects.

Design sketches. Unlike conceptual prototypes, which focus on refining the structure or navigation of a document, design sketches help you refine the visual design of the document. As such, design sketches are often more vertical than horizontal.

Lo-fi design sketches are often completed with those old standbys, pencil and paper. For example, thinking about the design of this book resulted in the design sketches in figure 10.12.

Pencil and paper are cheap and fast, so they are a good way to try out a variety of designs quickly. You might want to use relatively large sheets of paper, subdivided into areas with about the same aspect ratio as your planned page or spread format. That way, you can see multiple sketch iterations on the same sheet for easy comparison. Once you've created a few strong ideas with paper and pencil, you can branch out to colored pencils, crayons, or markers for more complete design sketches. If you're working collaboratively, a whiteboard or paper easel can be a great tool for sketching out design concepts. A bulletin board can be

Figure 10.11. A conceptual, low-fi prototype used in the development of this book. Conceptual prototypes are a good way to get a lot of ideas on paper quickly. Don't worry about making them too neat! *Source*: Created by the author.

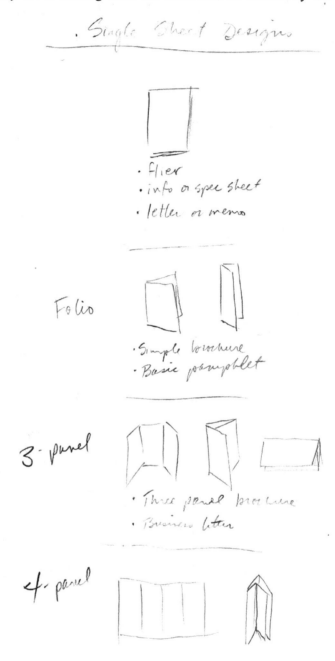

a useful place to tack up different iterations with pushpins so you can compare them all at once.

If you are collaborating at a distance, sketching is still a great way to try a lot of designs quickly. There are many available technologies available that allow you to create virtual whiteboards and share sketches.

Figure 10.12. Early design sketches for the first edition of this book. These design sketches represent initial ideas about the cover and layout of this book. As with conceptual prototypes, speed is more important than accuracy. Get all your ideas down as quickly as possible, and then refine those ideas with further sketches. *Source*: Created by the author.

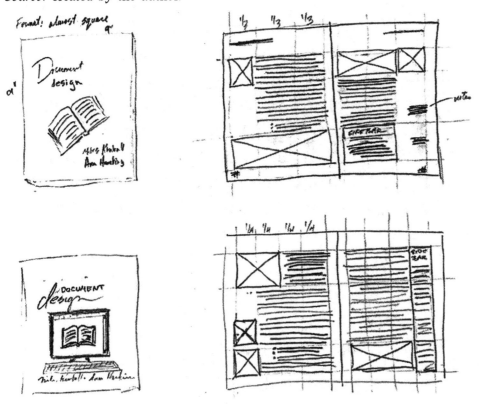

The trick in any initial sketching is not to let technology get in the way of the design—find the simplest tools you can use easily, and use them to sketch and share rough drafts. Even the drawing tools in Microsoft Word can serve this purpose well.

Use the sketching process to try out as many ideas as possible. Work quickly and creatively—there's little sense in making the sketches beautiful or perfect, since you'll discard most of them anyway. Once you see concepts or ideas you think work well, start creating more polished sketches that have a higher level of fidelity so you can show them to potential users and get some feedback.

After you've gone through a series of rough design sketches, you might want to create something a bit more complete, like a full-page outline or wireframe. This will be more time-consuming, but the higher level of fidelity moves your forward in the design process. Figure 10.13 shows a page outline of the layout for this book.

Figure 10.13. A digital page layout sketch for *Document Design*. This vertical prototype was created with a page layout program, so it's a relatively polished iteration of the design. The design evolved over time, but a vertical prototype like this was useful for getting feedback on more specific design ideas. *Source*: Created by the author.

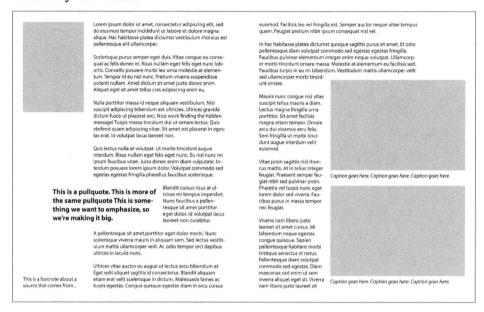

To create these more fully-formed electronic sketches, start with whatever software is fast, easy, and available. Your goal is to test your ideas and get feedback, so at this stage, don't worry about getting just the right image or text. Placeholder text (*lorem ipsum*) and placeholder images can stand in for real content for the time being.

For later prototypes with more vertical fidelity, you might want to create design sketches in the program that you intend to use to produce the final version of the document. You also might want to start importing real content, both images and text.

Operational prototypes. Operational prototypes have a relatively high level of horizontal fidelity (completeness), although they might have less vertical fidelity (polish). The goal of creating an operational prototype is to determine how well the prototype fulfills user needs. Typically, that means that the prototype will need to have all of the features to fulfill mission-critical agendas. (Of course, if one of the mission-critical agendas is that the document should look very polished, then an operational prototype might have a high level of vertical fidelity as well.)

An operational prototype is particularly suited to usability testing, since it should have enough functionality for users to do something with it. Because they're relatively complete, operational prototypes are also close to being releasable, at least in beta versions.

HEURISTIC ANALYSIS: GETTING EXPERT FEEDBACK ON PROTOTYPES

Once you have completed a prototype, you'll need to get some feedback on it so you can develop the next iteration. The two most common ways to do this are heuristic analysis and usability testing.

We've already discussed usability testing in limited terms as a way to get feedback from users. But for some prototypes, you might choose to rely on expert feedback rather than on user feedback. This technique is usually called *heuristic analysis* because it asks someone who knows a lot about design or the project—an expert, the client, or sometimes even a design team member—to evaluate the prototype in terms of how well it fits a *heuristic*, or an established set of principles. The evaluator analyzes the prototype in terms of these principles, noting where and how it falls short. For example, an expert evaluator might note that a prototype lacks clear alignment of page elements, leading to problems with the coherence of the design.

For a document design project, those principles should probably include our standard principles of design:

- Similarity

- Enclosure

- Alignment

- Contrast

- Order

- Proximity

For website design, you might consider using Jakob Nielsen's ten usability heuristics (2020), widely available online. These heuristics include making sure that users always know what is going on; design makes sense to real users; unintended actions can be readily fixed; designs are consistent; design is employed to reduce the likelihood of errors; design aids recognition; design supports use of shortcuts by experienced users; irrelevant information is avoided; error messages make sense; and any necessary documentation is usable.

Heuristic analysis is an organized way to get a quick sense of whether a prototype is on track. Rather than simply asking for an opinion about the prototype, you're asking *how well* the prototype meets generally accepted principles of design. The limitation of heuristic analysis is that although experts know general principles very well, they might not understand what it's like to be a real user of the document. Only users can give you that perspective.

Scenario Part 4: Prototyping

Javier, Erin, and Micah get to work on the product card. Micah uses Adobe Illustrator to design a series of vector-based graphic images based on the original sketch. Javier and Erin work together to research and write the copy that will accompany the translucent versions of the images on the back of the card.

The written copy they develop is as follows:

- Flashlight: A flashlight is valuable for when the power goes out, when you are outdoors, or when you need to be able to see a wound more clearly.

- Bandage scissors: Scissors are important for cutting bandages or tape, or for cutting clothing away from a wound.

- Tweezers: Tweezers help you remove splinters, or grasp small objects.

- Thermometer: A significant change in your temperature could indicate illness. The average body temperature is at or around 98.6°F (37°C).

- Antibacterial cream: Use this cream on wounds to prevent infection and kill bacteria.

- Aspirin: Aspirin can be used as a pain reliever, or given to adults with chest pain. Always call for medical help if you suspect heart problems.

- Self-stick bandages: Put these bandages on small wounds to help keep them clean and prevent infection.

- Sterile gauze pads: Gauze pads help cover wounds larger than self-stick bandages can handle. Use rolled gauze and either tape or the elastic bandage to keep in place.

- Rolled gauze: Use rolled gauze to help secure gauze pads or other materials in place around a wound.

- Medical tape: This tape is pressure-sensitive and is used to hold bandages in place over a wound.

- Rolled elastic bandage: Use to help hold pads in place, to help you splint a body part, or wrap tightly to restrict blood flow to a large wound.

- Chemical cold pack: A cold pack can help you alleviate pain and reduce swelling.

- Disinfectant wipes: Use these wipes to clean and help sterilize wounds.

- Sterile saline: Use sterile saline to wash out wounds or remove irritants from your eyes.

- Sterile gloves: These non-latex sterile gloves will protect you from bodily fluids.

Micah saves two files in Illustrator, one with no transparency and the other with the opacity set to 10 percent. The images are then linked to InDesign, where they are set onto a 5" × 7" card with images only on the front, and the translucent images plus descriptive text set in Minon Pro at 8 points in bold, with item types in italics on the back (the images are placed and linked so that any edits to the original files later carry over to the InDesign file). Each of the text blocks are placed over their associated images and aligned with surrounding blocks of text (see figures 10.14 and 10.15).

Figure 10.14. Front of the MedPak product card. *Source*: Created by the author.

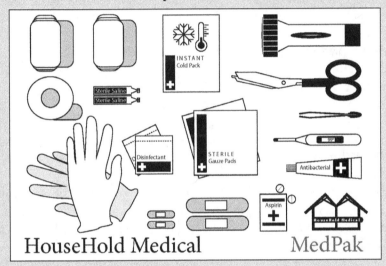

Figure 10.15. Reverse of the MedPak product card. *Source*: Created by the author.

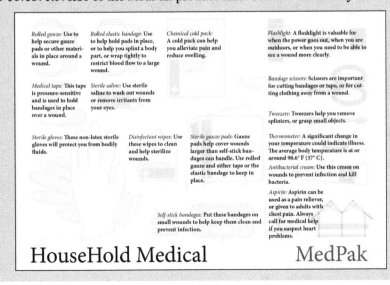

While Javier, Erin, and Micah work on the product card, Travis, Aisha, and Alex work on the simple web page. Aisha works with Micah's original sketch to develop a single, reactive page that works on multiple screen types and sizes, and Travis and Aisha work on the written copy for the page. While they are writing, Alex uses lorem ipsum as dummy text to see how the page will look on-screen.

For the image, Aisha decides to use something from Creative Commons. She wants to reflect Micah's original design, and so she looks for images that will fit with their themes for HouseHold Medical's MedPak product along with the family-centric ideas they have been working with. She searches Creative Commons for work that can be used commercially and adapted as needed, ultimately deciding that Iván Erre Jota's "Father with Child. . . . II" suits their needs perfectly. She crops the rectangular image in Photoshop into a circle, then adds a ring of blue based on the blue in the father's shirt (hex #1fa9c9). When she adds the image to the site she designed in DreamWeaver, she adds the alternative (alt) text, "Image of a father and child. "Father with Child . . . II" by *Iván Erre Jota* is licensed under CC BY-SA 2.0."

As she works, Travis and Aisha write two simple paragraphs of text for the site:

MedPak is your first line of defense for everyday injuries at home. From minor cuts and scrapes to burns and bumps, MedPak helps protect your precious family from infection, and can help you keep things in hand while you seek more advanced medical treatment.

Every family needs a MedPak. The kit includes a light, bandages, a cold pack, and even sterile saline for rinsing wounds or cleaning out eyes. Make sure your family has the best first line of defense. MedPak. For your family.

After looking at Alex's design at several screen sizes, they decide to do away with Micah's original tagline, "MedPak, Your Family, Your Life," and instead add a testimonial quote. They reach out to HouseHold Medical, and, as promised, by the next day receive the following:

Medpak saved my little girl's life. We had one of the early test kits at home, and she fell and cut herself badly. I was able to bandage her leg and get her to the hospital. Without it, I don't know what we would have done. —Jorge Ramirez

Since the site is, ultimately, designed to sell MedPaks, Alex adds a "Buy Now!" link under the testimonial pull quote, ensuring that the link is a different color and size from the rest of the body text, and that it is underlined, as all of these elements aid accessibility.

Once everything is put together, the group decides that they are pleased with the outcome (figure 10.16).

Figure 10.16. MedPak simple web page image. *Source*: "Father with Child . . . II" by *Iván Erre Jota*, CC BY-SA 2.0.

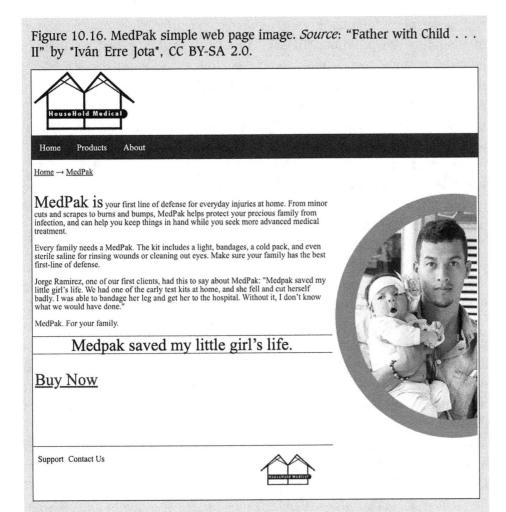

USABILITY TESTING: GETTING USER FEEDBACK ON PROTOTYPES

After you've completed a prototype, you can use it to get feedback from real, or at least potential, users. You can use some of the techniques we've already discussed for user research, including focus groups and individual interviews. But the primary approach to getting feedback on prototypes is *usability testing*.

Usability testing might sound intimidating, but the general concept is simple: *to find out how usable a document is, ask some people to use it*. If you observe their use during a structured task carefully, you can learn a lot about how to improve your next prototype.

Some corporations and universities have invested in expensive usability testing laboratories, but you can get good results less formally with a set of techniques known as *discount usability testing*, pioneered by usability guru Jakob Nielsen. Discount usability testing usually involves nothing more than a prototype (on paper or on the screen, as the project demands), a participant, and some means of recording the participant's actions and comments, such as video or audio recording. Discount usability testing doesn't require many participants. As Nielsen and Landauer reported in their 1993 paper, the first few participants will find most of the problems with a prototype; later participants will find fewer and smaller problems, decreasing the return on investment in the testing process. Nielsen and Landauer suggest that the optimal number of participants is three to five people. Because users often find the same errors, they noted that most usability tests can stop after testing five users, as upward of 85 percent of the potential problems will have been reported, and to continue testing would require more and more people with greatly decreased return on investment.

Usability testing techniques. The most common techniques used in discount usability testing are scenario testing and the talk-aloud or think-aloud protocols. *Scenario testing* involves presenting participants with a prototype and a few small tasks to perform. For example, you might test a prototype website by asking the participant to order a product, find the customer service phone number, and find the answer to a frequently asked question. You might test a paper document by asking the participant to find a particular section, find the definition of an important term, or follow a set of instructions in the document.

The think-aloud protocol involves asking participants to narrate what they are thinking and doing as they perform the requested tasks. This technique brings the participant's mental processes out into the open, where they can be observed, recorded, and discussed. Using the think-aloud protocol, you can readily detect when users are confused, when they hesitate, or where they get stuck in the task. This information is invaluable for making sure that users can perform tasks successfully. Remember the difference between talk-aloud and think-aloud: In *talk-aloud*, a participant might narrate that "I am clicking the blue, underlined link." In *think-aloud*, the participant is encouraged to offer explanations, so they might say, "I am clicking the blue-underlined link because it really stands out against the background, and the underlining tells me it is clickable."

When conducting think-aloud and talk-aloud protocols, it is important that you consider your participants' backgrounds—their culture and language. A participant engaging in think-aloud protocols whose first language is not English, for example, may have a difficult time expressing their thoughts as they go because they must first translate their thoughts into English—a process that could take them out of the moment in performing whatever tasks you have asked of them. Remember that you are not assessing the participant, but the process.

Fidelity and usability testing. You can perform usability testing on lo-fi or hi-fi prototypes, as well as on vertical or horizontal prototypes. Lo-fi testing is fast and cheap, and it can give you some great feedback early in the design process. With hi-fi prototypes, however, you will probably be testing something closer to a draft of a document that might be near a beta release. You can even do usability testing on a document or product that has already been released, gaining feedback on how to improve the next version of the document.

Creating hi-fi prototypes takes more time, but it can result in higher-quality feedback in usability testing, since participants are using something closer to the final version of the document. For example, with a hi-fi prototype, you can time how long it takes users to complete the tasks. This information can give you a precise sense of whether one iteration of your design improves over a previous iteration: if the participants can perform the task more quickly, you can conclude that the new design is more usable.

Planning and conducting usability tests. Just as you did with user research, be sure to get permission in writing from usability testing participants. Explain to the participants that they are not required to participate and that they can stop at any time. (See box "Design Tip: A Checklist for Conducting User Research and Usability Testing" in this chapter.)

When planning the scenarios for a usability test, think about two factors: the client's mission-critical agendas and previous user feedback. As you'll recall, the client's mission-critical agendas are what the document *must* do to be successful. These agendas can usually lead directly into the scenarios for prototype usability testing. For example, if one of the mission-critical agendas of a software manual is that users must be able to install the software successfully, then one scenario should probably ask users to install the software using a hi-fi vertical prototype of the document section that includes the installation instructions.

Feedback from previous usability tests or user research can also help guide your development of scenarios. If you found in your user research that most users say they prefer a pocket-sized format for the document you're designing, you might ask the usability test participants to carry the prototype document around and use it for a while. If they stick it in their pockets, your earlier research was accurate; if they don't, you might consider changing the format. This is actually a form of what we referred to earlier as *bodystorming*. Jeff Hawkins, for example, one of the founders of Palm Computing, famously carried around a block of wood to stand in for the portable device his company was designing. He would interact with the device, pretending to enter data and use it as if it were a real device. In so doing, he was able to performatively assess what their future portable device would need to be able to do (and look like).

To keep usability testing consistent between multiple participants, it's usually best to create a written plan, known as a *script*. The script should specify what the observer should say to introduce the test, what the participant will be asked to do, what the test is for, and how long the test will take. It should also include the scenarios, written out either for the participants to read themselves or for the observer to read to the participant. This written script helps to keep the usability test consistent for all participants, giving you more reliable results.

Your script, if usability testing several participants at once in a lab setting, might begin like this:

> Hello, I'm [Name], thank you for agreeing to participate in this study. Today, we will be asking you to complete a series of tasks on [what you are testing] and then participate in a short focus group to talk about your experiences with [what you are testing]. As you go through the process, please remember that we are testing [the product], not you. We need to see if and where you have problems using [the product], so please do not be ashamed or embarrassed if you have difficulty along the way. That's actually what we are looking for!
>
> (Have participants sit down at the computers.)

And so on. Your script should include management cues that will keep the experience consistent across different users and different times. You want data collected from participants at eleven o'clock in the morning on Monday to be comparable against data collected from participants tested at three o'clock in the afternoon the following Wednesday.

While conducting the usability test, be sure to ask participants to talk, or think, aloud. You might even have to remind them while the test is in progress, since many people will forget while they're concentrating on the task. Let the participants know that if they get stuck on a task, they can ask you for a hint. But try not to take over the task and give the participant direct instructions, because the goal is to see how the prototype works on its own. At the end of the usability testing, be sure to thank participants for their help.

WRITING PROTOTYPE REPORTS

For each phase of prototype usability testing, write an informal report addressed to the client that summarizes the test and your findings. Use your best judgment to determine what counts as a *phase*. If you conducted several quick usability tests of a succession of paper prototypes in one afternoon, you might want to write one report on the results rather than multiple reports. For short or fast-moving projects, you might want to write your reports even more informally on a project blog or wiki.

Prototype reports should include at least the following sections:

- *Introduction:* Explain the purpose of the report, record the date of the usability test, and summarize your general recommendations.

- *Methodology:* Describe how you set up the usability test, including how you chose participants and determined the scenarios.

- *Results:* Summarize the results for the usability test.

- *Discussion:* Analyze the results and discuss their significance for further development of the prototype.

- *Recommendations:* Recommend what changes the testing indicates the team should make to the project's direction.

- *Appendices:* Include materials such as the test script, transcripts, and other raw data.

The goal of these reports is to record and communicate the results of the usability test and to recommend what actions to take in the next phase of prototype development. It's more important to write a concise report quickly and move on to the next phase rather than to write an extended report and hold up the iterative design process.

STAGE 4: SPECIFICATION

At the end of the prototype stage, you should have a good sense of what kind of document you'll be producing, including its medium, format, organization, page layout, typography, navigation, and paratextual features (tables of contents, lists of figures, indexes, appendices, glossaries, etc.). In the specification stage, you'll codify these features into a guide you'll use in the implementation stage. This guide, called the *implementation plan*, should be addressed primarily to the project team, with the client as a secondary audience. You'll need the client's feedback on and approval of your implementation plan, but the plan's main purpose is to guide the team's efforts as you bring the document to a releasable state.

Not all projects are big enough in scope to require a formal implementation plan. But all projects, no matter how small, require at least some informal specification to help you keep design features consistent throughout the design. Larger projects—particularly those that require multiple documents that will feature a consistent design—need a more formal guide that can extend across all of the deliverables. Large projects are also likely to involve more people on the project team, and those people will need the implementation plan so they can work in concert to create a unified design. The implementation plan usually includes two main parts: a style sheet (or template, for smaller projects) and an implementation schedule.

CREATING A STYLE SHEET OR TEMPLATE

To keep the design consistent throughout multiple-document projects, you'll want to create a style sheet that specifies the design of all the page elements, the page layouts, the colors of the document, and even the color system to be used. A style sheet might include the following information:

- Heading 1: Eras Bold ITC 16pt, 1p0 before, 0p2 after

- Heading 2: Eras Demi ITC 12pt, 1p0 before, 0p2 after

- Body text: Ingra 10pt, first-line indent 0p3

- Header: running, Eras Light ITC 10pt

- Color scheme: PMS 289; PMS 158

The style sheet should also include a labeled diagram of the page layout, including information about margins, columns, gutters, and the position of headers and footers. For website design projects, the style sheet will probably be formatted as a CSS (Cascading Style Sheet) file, which you can incorporate directly in the design.

Alternately, you may prefer to create a template instead of a style sheet. A template is simply a page layout file (for a website, these might include HTML, PHP, Javascript, and/or CSS files, or full Bootstrap or WordPress templates, among others) that include all of the preceding elements actually put into action. You can set up the page layout, including columns, margins, and gutters, as well as create paragraph or character styles for the different textual elements. In other words, the template serves as an empty shell for the design. After creating the template, you can simply Save As to create a new version into which to port content. Templates are particularly useful for design projects that have several deliverables with consistent designs, such as a project creating a series of pamphlets or a newsletter that will come out periodically.

CREATING AN IMPLEMENTATION SCHEDULE

The implementation plan should also include a schedule that specifies when each part of the design project will be completed. For example, it might include separate milestones for copy (text), copyediting, art, and layout. It should also specify when the document will be ready for production. If the deliverable is a paper document that will be printed professionally, be sure to include the print shop's milestones as well, including proofs (see chapter 11). If the deliverable is an electronic document such as a website, specify when the site will go live on the internet.

STAGE 5: IMPLEMENTATION

The best way to track and communicate your progress is to write an informal *progress report*. The purpose of a progress report isn't just to say "Everything's going fine; here's what we've done so far." Instead, a progress report measures progress against the implementation plan. For example, if the plan specified that the graphics for the project were to be completed by November 15, a progress report should measure whether the graphics were in fact completed on time. If they weren't, the progress report will need to discuss what went wrong and how the team plans to get back on schedule.

Write your progress reports with the client as the primary audience. Communicating regularly about your progress will help both you and the client understand where the project stands and what to expect as the project continues.

Progress reports are only one way to document and keep track of progress, though. There are many different online and application-based project management tools, and your group may find it useful to work with something as simple as a shared calendar; manage project process, files, and code through something like GitHub (an online platform for collaboration and version control in software design); or choose any one of many other options, both paid and free. On smaller or less formally managed projects, you might consider keeping a running blog of your progress rather than writing separate progress reports. Online tools may allow you to broaden your concept of audience for progress information to include users, who might be very interested in how your project is going and when they can expect to see it complete or usable.

Scenario Part 5: Deliverables

DesignRight Solutions contacts HouseHold Medical and schedules a meeting to introduce the deliverables. They meet, and DesignRight presents HouseHold Medical with the design files, image files, and site code needed build the web page into their existing site code.

During the presentation, Alex, as primary contact and the voice of the design group, walks HouseHold Medical reps through their design choices. They talk about the group's informal, in-house usability testing, DesignRight's justifications for the design and layout of the product card, and the image choice and responsive design-based coding of the page.

Finally, Alex thanks HouseHold Medical for hiring DesignRight Solutions for its design needs and reminds reps that they have thirty days to request significant design revisions under their existing contract. Alex again provides contact emails for the design team and the team manger, Rachel Hawkins, and thanks them for their time.

WRITING THE COMPLETION REPORT

Finally, design projects often end with a *completion report* addressed to the client. The completion report sums up the entire project and makes recommendations for future development of the deliverables. It should also describe the lessons learned through the project. These lessons are important organizational wisdom that can affect the development of new design projects. Completion reports are often formal, written documents but can also take the form of presentations or memos.

If you're completing a student design project, the completion report is also a very good opportunity to absorb and reflect on everything you've learned while doing the project. Most of us learn a lot by doing, but if you can express what you have learned, you're more likely to take those lessons on to new experiences and greater successes in the future.

EXERCISES

You just thought your way through an entire design project from start to finish. You've considered different models of project management and thought through ways you might communicate with your design team.

You've thought about client work in fairly complex ways and worked through interviews, overviews, and documentation. Now it's time to practice. These exercises should help you operationalize many of the aspects of design you just read about.

1. Imagine that you're working on a project to design a new guidebook for international students at your university. The client—your university's international student center—envisions a brochure or handbook that includes information international students might not know about the school and community, such as how to open a bank account, where to get an ID card, how to use the local bus service, and so forth. The goal is to make international students' entry into the academic and local community go smoothly. What user research would you conduct for this project? What questions would you need to answer?

2. For a project that you're working on, create a PERT chart like the one in this chapter. Using your understanding of the mission-critical elements of the project, be sure to distinguish between critical paths (absolutely essential steps) and optional paths. For example, you can use red lines for critical paths and dotted lines for optional paths. You don't need any particular software to design something like this—you could easily make one in Word or PowerPoint using the drawing features.

3. For a project that you're working on, keep a journal that describes your impressions and thoughts as you go through the project. At the end of the project, look back on your journal and reflect on what you learned.

4. Form a group with a few other people. Together, talk through the following questions related to the scenario. When you are done, write up a brief memo outlining your answers.

   o Do you think DesignRight Solutions created deliverables that HouseHold Medical will be able to use? Why or why not?

   o DesignRight had several people working on what could be seen as a fairly simple project. What is the value of working with complex teams on a project like this?

   o What additional steps would you have taken in the design process if you were in charge of a design group?

o If HouseHold Medical had requested a more formal usability testing process, what might that process have looked like? Put together your own presentation of DesignRight's materials to HouseHold Medical. What do you think are the most important elements to focus on in this presentation and why?

REFERENCES AND FURTHER READING

Barnum, C. M. (2001). *Usability testing and research*. Longman.

Beck, K., Beedle, M., Van Bennekum, A., Cockburn, A., Cunningham, W., Fowler, M., . . . Thomas, D. (2001). *Manifesto for Agile software development*. http://www.agilemanifesto.org

Cervone, H. F. (2011). Understanding agile project management methods using Scrum. *OCLC Systems and Services*, *27*(1), 18–22. https://doi.org/10.1108/10650751111106528

Hackos, J. T. (1994). *Managing your documentation projects*. John Wiley and Sons.

Hassenzahl, M., & Roto, V. (2007). Being and doing politics. *Interfaces*, *72*, 10–12. https://doi.org/10.2307/j.ctt2005rdp.5

Kuniavsky, M. (2003). *Observing the user experience: A practitioner's guide to user research*. Morgan Kaufmann.

Lauren, B. (2018). *Communicating project management: A participatory rhetoric for development teams*. Routledge.

Martin, J. (1991). *Rapid application development*. Macmillan.

Nielsen, J. (1994). *Guerrilla HCI: Using discount usability engineering to penetrate the intimidation barrier*. Nielsen Norman Group. https://www.nngroup.com/articles/guerrilla-hci/

Nielsen, J. (1994/2020). *10 usability heuristics for user interface design*. https://www.nngroup.com/articles/ten-usability-heuristics/

Nielsen, J., & Landauer, T. K. (1993, May). A mathematical model of the finding of usability problems. In *Proceedings of the INTERACT '93 and CHI '93 conference on human factors in computing systems* (pp. 206–213).

Nielsen, J., Clemmensen, T., & Yssing, C. (2002). Getting access to what goes on in people's heads? Reflections on the think-aloud technique. In *Proceedings of the second Nordic conference on human-computer interaction* (pp. 101–110).

Norman, D. (2004). *Emotional design: Why we love (or hate) everyday things*. Basic Books.

Pope-Ruark, R. (2015). Introducing agile project management strategies in technical and professional communication courses. *Journal of Business and Technical Communication*, *29*(1), 112–133.

Smith, B. K. (2014). Bodystorming mobile learning experiences. *TechTrends*, *58*(1), 71–76. https://doi.org/10.1007/s11528-013-0723-4

11

PRODUCTION

LEARNING OBJECTIVES

Upon completing this chapter, you will be able to:

- Discuss pros and cons of the six main types of printing technologies

- Evaluate the types of ink and paper needed for a print project

- Articulate the communication needs of both your client and print shop and explain their workflow

~

Up until this point, we have primarily discussed what happens *before* a document is produced—the process of researching, planning, and designing that ends with a completed prototype. The final stage, production, can be complicated, particularly for designs that will be printed by a commercial print shop like the one in figure 11.1. Production includes the steps we need to take before we deliver a design to a print shop or prepare it for printing ourselves—a process known as *prepress*.

Whether you're preparing a document for simple photocopying or sending it to a full-service print shop, you will design more successful documents if you understand production processes. Printing is a complex industrial process, and understanding that process will help you anticipate what the printer needs to produce your design. Understanding print production will also help you communicate effectively with printers, your partners in moving from a design to a real product.

Figure 11.1. One machine in a large commercial print shop. Printing is an industrial process that requires special skills and expensive equipment, such as the six-color offset lithographic Heidelberg press shown here. Knowing a little about printing processes will help you communicate successfully with your commercial printer. *Source*: Photo by the author.

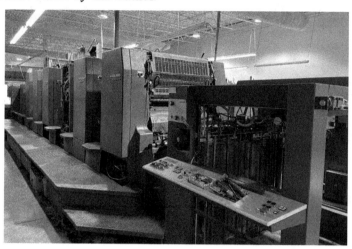

Of course, production applies primarily to printed documents, since electronic documents are essentially "produced" when the design is complete and made available to users. So in this chapter, we'll focus on common printing technologies. Then we'll discuss some of the processes that happen after printing, such as folding, gathering, trimming, and binding. Finally, we'll cover some of the things you'll need to think about when communicating and working with professional printers.

PRINTING TECHNOLOGIES

As a designer, you should know about six main kinds of printing technologies:

- Intaglio
- Relief
- Lithography
- Reprography
- Inkjet
- Commercial digital printing

We'll define each of these technologies, emphasizing those you are most likely to encounter as a document designer. This information should help you make good decisions about which printing technologies are best for your project. Although the sections in this chapter provide the details, refer to table 11.1 for a quick overview of the uses, strengths, and weaknesses of each of these technologies.

Table 11.1. Printing Technologies: Uses, Advantages, Disadvantages

Printing Technology	Common Uses Today	Advantages	Disadvantages
Intaglio	Printing currency, stock certificates, and similar official documents	Beautiful and difficult to duplicate	Requires highly skilled artists Plates wear out quickly
Relief	Printing packaging materials (flexography)	Useful on flexible materials like plastics and cardboards	Imprecise Pooling at edges of inked areas
Lithography	Printing high-quality, complex color documents such as books, brochures, pamphlets, and reports	Inexpensive in large print runs Increasingly automated Flexible output in nearly any size and format	Industrial, requiring expensive machines and highly skilled personnel Expensive in short print runs Hard to fix mistakes
Reprography	Printing short runs of single-sided documents in standard paper sizes	Inexpensive in short print runs Fast desktop access Easy to fix mistakes Within control of designers	Expensive in long print runs Lower quality than lithography, especially in terms of color and images
Inkjet	Color proofing Quick printing on low-level designs Large-format printing (signs)	Relatively inexpensive equipment Commonly available	Very expensive consumables, especially for process color printing Often poor print quality, especially for process printing
Commercial digital printing	Print-on-demand (POD)	Very fast Small print runs	Very expensive Lower quality than traditional offset lithography

PRECURSORS: INTAGLIO AND RELIEF PRINTING

The primary difficulty in printing is how to get ink onto the paper where we want it, while keeping it out of the areas where we don't want it. Intaglio and relief printing do so with two complementary strategies: inking a *depressed* surface or inking a *raised* surface (figure 11.2).

Intaglio and relief and printing are rarely used today, but being familiar with the transmission from one technology to the next will help you understand some of the terms your printer uses and the options they provide.

Intaglio printing (also known as *gravure*) uses *depressions* in a metal plate to hold the ink that will later transfer to the paper. These depressions are created either mechanically, by *engraving* the plate, or chemically, by *etching* parts of the plate with acids. Both approaches leave the part to be printed lower than the surface of the plate. Ink is then rolled onto the plate and the excess wiped away, leaving ink only in the depressions. When the paper is pressed onto the plate with high pressure, the ink transfers from the depressions to the paper.

Intaglio printing requires highly skilled artists to engrave or etch the plates, and the plates are difficult or even impossible to correct if there are any mistakes. For a long time intaglio printing was the primary method for printing illustrations and maps. Today, it's mostly used to print currency and stock certificates because the highly artful nature of intaglio plate engraving makes it hard for counterfeiters to copy the plate.

Taking the opposite approach, *relief printing* uses a *raised* surface on which the ink can be spread, usually with a roller. Paper is then pressed onto this raised surface, transferring the ink to the paper. The earliest

Figure 11.2 Intaglio, relief, and planographic printing methods in profile. Intaglio printing inks a lowered surface. Relief printing inks a raised surface. Planographic printing places the inked and un-inked areas on the same plane. *Source*: Created by the author.

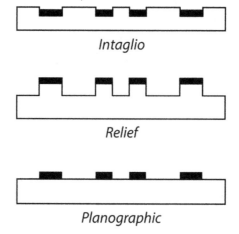

relief printing began in China in the ninth century with carved wooden blocks. Movable type, developed by Bi Shēng, using a glue and clay amalgamation similar to porcelain, saw its first use in China in the eleventh century. Other forms of movable type were developed over time, with cast-metal type developed in Korea in the thirteenth century. But the most influential application of relief printing appeared in the fifteenth century in Europe with Johannes Gutenberg's development of movable type for *letterpress* printing, the primary technology for printing until the 1950s.

In letterpress printing—the primary relief printing technology from Gutenberg to the middle of the twentieth century—each letter of the typeface was molded in reverse on a small block of metal, with the area to be printed raised to take the ink. In a process called *composition*, a worker would choose pieces of type one at a time from subdivided drawers called *cases*. The lower case held the small letters, also known as *miniscules*, and the upper case held the capitals (*majuscules*) and punctuation marks—eventually, small letters became known as *lowercase* letters, and capitals as *uppercase*. The compositor arranged the individual pieces into lines separated by strips of lead, from which we get the term *leading*—the space between lines of type. After completing a whole page of type, the compositor would tighten it into a frame using blocks and wedges. The printer placed this frame in the press and inked the raised surfaces of the type. They then used a printing press to press paper onto the framed page of type to receive the impression of the ink. At the end of the press run, workers took the type out of the frame and separated, cleaned, and distributed it back to the cases, one piece at a time.

This process became more efficient after the 1890s with the introduction of machines such as the Monotype, which molded individual pieces of type on demand and composed them into text, and the Linotype, which molded a whole line of type in one piece. But even with these advances, letterpress printing was labor-intensive and expensive. As a result, it fell out of mainstream use between the 1950s and the 1970s with the development of photolithography (which we discuss in the next section). But because people used letterpress printing for five centuries, the conventions developed by letterpress printers have become very closely associated with all printed documents, such as justified type, leading, and typefaces. In addition, many of the typefaces we use for documents today are based on Renaissance typefaces designed for letterpress printing.

LITHOGRAPHY

Intaglio and relief printing have a significant disadvantage: it's hard to create the depressed or raised surfaces needed to separate the inked areas from the un-inked areas. Planographic printing, however, places both the inked and un-inked surfaces on the same plane, neither raised nor lowered (see figure 11.2).

The most significant planographic method is *lithography*, invented by Alois Senefelder in 1798. To get the inked and the un-inked surfaces on the same plane, lithography relies on the principle that oil and water don't mix.

Early lithography (which means "stone-writing") started with a smooth, flat stone. A printer or artist drew on the stone with an oily crayon (*crayon* comes from the French for "chalk pencil"), and the stone was wetted with a mixture of water and gum arabic (a thickening agent). The stone was then inked with oily lithographic ink, which would stick to the oily crayon marks but not to the water-soaked areas of the stone. Paper was then pressed on the stone at moderate pressure, and the ink was transferred to the paper. This early lithography was used increasingly throughout the nineteenth century to produce illustrated books, maps, and inexpensive art for decorating the home.

PHOTOLITHOGRAPHY

The manual process of lithography became more efficient with the development of *photolithography*, which came into common use in the 1950s.

Traditional photolithography begins with *camera-ready copy*, which requires treating type and images separately. For type, computerized typesetting machines are used to create the prototype blocks of text on paper at high resolution. For images, photographic prints are photographed again through a fine, grid-like filter called a *screen*. The screen allows light through its holes and blocks light with its lines. The result is a *halftone* image of the imposition, made up of tiny dots that blend together to simulate the photo (if you recall, we discussed this same process in regard to *halftone separations* in chapter 8). Until around 1980, the halftones and typeset blocks of text were then pasted with wax or glue onto a paper or clear plastic substrate in the proper arrangement for the page design. This assembled camera-ready copy was photographed once again to create a negative of the whole page or imposition. Finally, a positive plate of the image was created by shining a strong light through the negative onto the *plate*, a thin, flexible plate of metal coated with light-sensitive chemicals. The areas of the plate where the light shone through the negative underwent a chemical reaction that made them receptive to lithographic ink. The areas where the light was blocked remained unexposed; these areas were then wetted in the lithographic process to keep them from receiving ink.

Fortunately, this manual assembly or *paste-up* process is now managed digitally to make the plate. But basic lithographic printing happens in much the same way now as it did in the last century. At the press, the developed plate is attached to a roller called the *plate cylinder*. As the plate cylinder rotates, the press inks and waters the plate. The exposed portions of the plate receive the ink and repel the water, while the unexposed areas receive the water and repel the ink. Finally, a sheet of paper is rolled through the press between the plate cylinder and the *impression cylinder* (a drum on which the plate is mounted), and the ink transfers to the paper, creating a positive image of the page (figure 11.3).

OFFSET LITHOGRAPHY

One disadvantage of this basic lithographic printing process is that the plates wear out quickly because they come into direct contact with the relatively abrasive surface of the paper. So around 1900, printers developed *offset lithography*, which protects the plate and makes it last longer. Almost all lithographic printing done today uses offset lithography.

Offset lithography puts a rubber-covered *blanket cylinder* between the plate cylinder and the impression cylinder. The inked plate cylinder

Figure 11.3. Side view of lithographic and offset lithographic presses. The offset press introduces a blanket cylinder between the plate and the impression cylinder to make the plate last longer. The rubber blanket picks up the ink from the plate and presses it on the paper, keeping the abrasive paper away from the plate itself. *Source*: Created by the author.

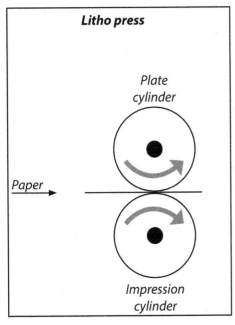
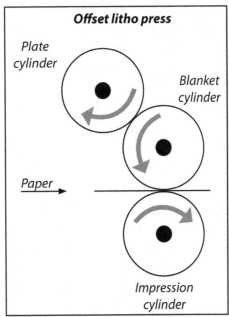

presses against the blanket cylinder, which takes up the ink from the plate and then presses the ink onto the paper. Because the rubber blanket is nonabrasive, the plate lasts much longer—usually enough to print tens of thousands of impressions.

Motorized offset presses can print hundreds of sheets a minute this way, using a series of vacuum and roller feeds to pick up sheets, send them through the press, and collect them at the other end. Even faster are *web presses*, which take up the paper from a continuous roll rather than individual sheets. The paper from this continuous roll is trimmed to final size after printing. Newspapers and high-press-run magazines are typical users of web printing.

FOUR-COLOR PROCESS LITHOGRAPHY

Offset photolithography also lends itself easily to *process color* printing, which was invented in the eighteenth century but came into effective use in the 1920s (see Griffiths, 1996). Process color is used for documents that need the fullest possible range of colors, such as magazines, annual reports, and promotional materials. It involves making and printing from four plates for each impression—one each for cyan, magenta, yellow, and black, abbreviated as *CMYK*. (See chapter 8 for a full discussion of CMYK and color management.)

Process color printing requires four plates of the page, also called *color separations*. Until recent years, these plates were created photo-lithographically using a series of camera filters (in addition to the screen for halftones). Each of these filters strains out one of the three additive primary colors, red, green, and blue (RGB), allowing the other two to combine and produce the subtractive color plates in cyan, magenta, and yellow (CMY). Finally, a grayscale filter produces the key plate (the K in CMYK), so called because it's usually the first plate printed, and the other three are matched up or *registered* to it. The key plate adds saturation, clarity, and darker shadows than CMY can produce on their own.

The resulting four plates are then printed one at a time onto each sheet of the print job, producing what looks like a full-color image. Many offset lithographic printing presses now also include more than one printing assembly, so a sheet can be printed in all four-color impressions at one pass (see figure 11.1 at the beginning of this chapter—each of those towers you see is a separate printing assembly for just that purpose). Even more complex presses, called *perfecting presses* or just *perfecters*, can print multiple colors on both sides of the sheet simultaneously. In larger print shops, it's not uncommon to find perfecters that are able to print six colors (CMYK plus two spot colors) on one side and four colors on the other side, all in one pass of the paper through the press. (For more information about spot color ink, see "Spot and Process Inks" in this chapter.)

FURTHER REFINEMENTS IN LITHOGRAPHY: COMPUTER-TO-PLATE AND DIRECT IMAGING

Newer developments in lithography bypass the need to create and photograph camera-ready copy to make a plate. Today, the separation process typically takes place inside page layout or imaging software, where each separation is called a *channel*. Each channel can be manipulated and adjusted digitally, giving considerable control over the output.

With *computer-to-plate (CTP)* or *digital imaging (DI)*, process color lithography is automated even further. CTP uses a machine called a *plate processor* to create the lithographic plate directly from a digital file rather than through photography. The plate can then be printed normally in a traditional offset lithographic press. DI extends this development by combining the plate processor with the press itself. Using DI, the press receives the digital file at one end and outputs printed material at the other, reducing manual intervention even further.

Both of these processes shorten the distance between the designer and the printed document by making the production process more streamlined. In essence, they allow a tighter integration between the digital files that designers submit to print shops and the resulting product that comes out of the press.

ADVANTAGES AND DISADVANTAGES OF LITHOGRAPHY

The greatest advantages of offset lithography are high quality and cheap per-unit cost for long print runs. The biggest single expense in doing a lithographic job is typically making the plates; once you've gone to that expense, offset lithography is relatively cheap per copy. As a result, the higher the volume of copies printed, the cheaper the price per copy becomes. If the plate costs $200 to make, then the first copy made with the plate will cost about $250—the cost of the plate plus a bit more because the printer will waste some sheets of paper in setting up the press (a process called *make-ready*). But if we print a second copy, the plate and setup costs will be spread over twice as many copies, cutting the price per copy in half, to just a few cents over $125 (even one sheet of paper costs a bit). Print four copies and the unit price goes to a little more than $62.50, and so on. Over thousands of copies, the cost of plate-making and make-ready is amortized until it's a very small part of the per-unit cost. Conversely, for short-run print jobs (anything fewer than several hundred copies), the expense of creating the plates makes offset relatively expensive per copy.

The biggest disadvantage of lithography also has to do with plates—namely, that *mistakes on lithographic plates cannot be corrected*. To correct a mistake after a plate is made, the print shop must make the correction in the camera-ready or digital copy and make a new plate, which doubles the plate cost. Other disadvantages of lithography arise from the highly mechanized nature of the process, which requires expensive equipment and highly trained personnel.

REPROGRAPHY

Because printing technologies like lithography are so complex, innovators have long searched for ways to bring printing into the hands of less-skilled people. The Xerox corporation introduced a technology in the mid-twentieth century that allowed people to reproduce documents using a relatively simple machine that anyone could operate. The machine used a process called *reprography*, also commonly known as *photocopying*. Later, reprography was adapted to mesh with computers in *laser printing*. Reprography is a tremendously useful technology for document production—particularly with the kind of ephemeral documents many of us create each day.

THE REPROGRAPHIC PROCESS

The most common form of reprography is photocopying. Photocopiers use the principle that electrical charges attract their opposites. Here's how it works:

1. You place an original document on the photocopier's impression window and activate the machine.

2. A bright light scans across the original. Where the original is black (e.g., text or images), no light is reflected. Where the paper is white, light is reflected onto a positively charged rotating drum.

3. This reflected light changes the places on the drum where you *don't* want ink to a neutral charge, leaving a positive charge only on the areas where you *do* want ink.

4. The drum picks up a powdered, negatively charged plastic ink called *toner*. Because opposites attract, the negatively

charged ink sticks only in the areas where the drum is still positively charged.

5. The photocopier picks up a sheet of blank copy paper and gives it a positive charge before sending it to the drum.

6. The negatively charged ink image on the drum transfers to the positively charged paper.

7. A heating element called the *fuser* melts the ink onto the paper.

This last step explains why paper always leaves a photocopier feeling warm.

Laser printers work very much like photocopiers. But instead of shining a bright light on an original, laser printers use a laser to change the charge on the drum. The laser is digitally controlled by the printer's computer processor to change the drum's charge to positive only in the areas where we want images or text to appear. The rest of the printing process works just like steps 4 to 7 in the photocopying process.

Color laser printers have recently become more common as their prices have fallen. As you might guess, color laser printers work by applying the four process-color inks (CMYK) to each page. They transfer the powdered ink to the page and fuse it there exactly the same way as a monochrome laser printer does—just with four repetitions using four separate drums.

ADVANTAGES AND DISADVANTAGES OF REPROGRAPHY

Reprography has significant advantages for document designers. The equipment is inexpensive and common enough that companies, organizations, and individuals can afford it. After half a century of development, the technology has become very easy to use. For short-run documents—from one to a few hundred copies—reprography gives control and flexibility to document designers, allowing us to create prototypes or finished documents from our desktops cheaply and on demand, rather than having to wait for a print shop to run an offset job.

Mistakes in the document are also easy to catch and fix with reprography; you can print drafts cheaply and make corrections at any time. Print shops or specialized copy centers benefit from reprography as well, using the technology to print short document runs cheaply and quickly. At these facilities, reprographically produced documents can also be folded, stitched, and bound just like more traditional offset documents. Color reprography even allows printers to produce runs of full-color documents on demand (see "Evolving Printing Technologies" in this chapter).

However, reprography does have two significant disadvantages when compared to lithography: lower quality and higher per-copy cost. Reprography handles line-drawn images like logos and diagrams well, but it reproduces photographs and other halftone images poorly. The inks used in reprography can be dense and dark under the best conditions, but as anyone who has eked out a few too many printouts from a laser printer cartridge knows, it's common to get pale or streaky darks with reprography. Quality problems can also arise when laser printers and photocopiers are poorly maintained. Photocopies often have stray gray streaks on the white areas, which are caused by worn-out drums, inconsistent toner, or smudges and dust on the imposition glass.

In long print runs, reprography also costs more per copy than offset printing. As we discussed before, with offset lithography an economics of scale kicks in as print runs get longer; each copy made is cheaper than the last. But with reprography, every copy costs exactly the same, no matter how many you make. So while lithography is often cheaper for long print runs, reprography wins out for short print runs. There's also a point, usually somewhere around several hundred copies, where reprography and lithography cost about the same per copy. At that point, other factors such as quality and time come into play in the decision about which technology to use.

INKJET PRINTING

The kind of printing you're probably most familiar with is performed by the printer sitting next to your computer. If you're like most computer users, that printer is probably an *inkjet*. Inkjets have become so common that they're often included when you buy a computer. And there are some good reasons to remember the humble inkjet in your document design work.

In some ways, inkjets are the simplest of modern printing technologies; they work by directly squirting tiny drops of ink onto the paper. More specifically, inkjet printers work by propelling drops of ink through a series of microscopic holes in the print head. The print head moves across the page in sequential rows, spraying out precise, computer-timed pulses of ink to create the printed image.

This technology also lends itself conveniently to four-color printing. Inkjet printers can print CMYK inks through four heads (or more) simultaneously. Some inkjet printers use a single combined ink cartridge with connected reservoirs for the four ink colors; others use four separate ink reservoirs and heads, so you can replace each color as it runs out. Others use a tank system that can include not only CMY and K, but other inks, such as red and gray.

ADVANTAGES AND DISADVANTAGES OF INKJET PRINTING

Inkjet printers are cheap and ubiquitous; you can readily find them in many stores. They've made color printing available to nearly everyone who has a computer, and they can also be useful for quick and dirty proofing of color print jobs.

But the per-unit printing cost of inkjet printing is high, primarily because of two factors: ink and paper. Inkjets go through ink cartridges quickly, and the cartridges aren't always cheap. Even on tank-based systems, ink can be expensive. On lower-quality machines, a single new set of cartridges may cost more than the printer itself! And if you're printing in color, you'll need even more ink as the process colors are overprinted on top of each other. With cheap inkjet printers, it's common to get only forty or so pages of four-color printing from a single set of CMYK cartridges. Though many inkjet systems have moved to digital cartridge monitoring, your inks may still run out at the most inconvenient time—like right in the middle of a page, leaving the rest of the printed pages with three process colors instead of four, or midway through a series print run.

Paper also plays a role in inkjet print quality and expense. Although inkjets can disperse precisely metered drops of ink onto the paper, the quality of the paper determines whether the ink will stay where the printer put it. If the paper is porous or rough, it can allow the ink to bleed beyond its original spray area, effectively decreasing the resolution of the output. If you use a higher-quality and more expensive coated paper to avoid bleeding, you might run into problems with smearing: instead of soaking into the paper, the ink sits wet on the surface until it dries, just waiting for anyone or anything to touch it.

Finally, although manufacturers have made great strides recently in inkjet output quality, inkjet printers often provide an unmistakably streaky, pale, and poorly registered output. This comes from the sequential passes of the print head; as the printer wears out, the print head doesn't move as consistently as it should. Streaks and pale output can also arise from ink clogging as the print head becomes contaminated with gunky, dried ink.

Higher-end inkjet printers are capable of excellent production value. And for many projects, they can give you a quick and relatively cheap sense of how your design might look when printed offset, but the output of inkjets usually differs remarkably from the output of process color offset lithography. If you want more accurate color proofing, or even at-home production quality equipment, you might want to invest in a more expensive dye-sublimation printer or color laser printer.

Inkjets can also be used for final copies of ephemeral documents—low-design-level work. Sometimes a quick flyer printed on an inkjet is just the ticket for advertising an event. But be aware of the conditions in which the document will be used: inkjet inks are water-soluble, so documents printed with inkjets don't stand up well in humid or wet environments.

NEW PRINTING TECHNOLOGIES

As printing technologies have advanced in the past several years, a gradual coalescence of computers and printing presses has developed. In some ways, the goal of new printing technologies has been to use computers to automate the printing process so much that the industrial process of printing will move toward the convenience and flexibility of desktop printers—just on a larger scale and with higher quality. Broadly speaking, these new technologies can be called *commercial digital printing*, but there are several different technologies currently under development.

Digital offset printing, for example, entirely eliminates the need to create a separate plate for each page. One iteration of this approach, developed by Hewlett-Packard, uses a *Photo Imaging Plate (PIP)* on a drum in place of the traditional metal lithographic plate (Hewlett-Packard, 2016, p. 6). In a sense, the PIP combines the electrically charged impression drum of reprography with the mechanical processes of offset printing. The PIP can be altered with a laser to accept or reject an electrically charged liquid ink on the fly. The negatively charged liquid ink (much like the electrically charged reprographic toner) sticks to the PIP only where we want an image. The resulting ink image is then transferred to a blanket cylinder and then to the paper, just like in traditional offset lithography.

The big advantage of this technology is that the PIP can change on the fly, even potentially with each impression. This allows each copy of the document to have different content—for example, we can insert a personalized name or address label for each copy that goes through the press. Because digital printing bypasses plate creation entirely, it's remarkably flexible and fast: a digital file goes in one end of the press, and finished sheets come out the other end.

Other similar approaches set aside the lithographic process entirely, using industrial-capacity laser printers or color laser printers to produce finished documents. The resulting *commercial reprographic printing* provides quick response and flexibility when producing small-run documents. The documents are produced faster and less expensively than those output from desktop laser printers or office photocopiers—and they usually have a higher quality.

This flexibility enables *print on demand (POD)*. Strictly speaking, POD is an approach to printing rather than a particular print technology. Since the input of digital printing is simply a digital file, it's possible to print very small print runs of documents as they're needed, just as you do with a desktop printer. The print shop can keep digital files for different documents on hand and then print out more copies when necessary—even if it's only one more copy.

POD can still be more expensive than traditional commercial printing, but the costs are decreasing as digital printing technology becomes more efficient. Additionally, the ability to print only the materials needed at a fixed price can make small print runs more economical—if you only need twelve copies of a book, the slightly higher per-book price might far outweigh the cost of cheaper-per-book print runs that require the printing of hundreds, or even thousands, of copies.

PAPER

Printing technologies are only one part of document production. The paper on which a job is printed is an important and often-overlooked element in itself. In this section, we'll examine the different types and characteristics of paper used in printing.

TYPES OF PAPER

Document designers enjoy a myriad of papers to choose from today—enough to make specifying paper for a print job a bewildering experience. Because we're always on the lookout for a new paper to make a document look distinctive or interesting, designers drive this variety by encouraging paper manufacturers to develop new paper products.

Despite this variety, printers make one broad but useful distinction in the types of paper they use for document production: text stock and cover stock. *Text stock* (book paper) is designed to hold printed text and images. Text stock can come in many colors, but it's typically a relatively thin paper that takes ink well with minimal bleed-through (if the paper is to be printed on both sides). It can be as simple as standard photocopy paper or as complex as the specialty papers used for printing high-design documents like glossy sales brochures and annual reports.

Cover stock is typically a thick, stiff, hard-surfaced paper that folds well and holds up to handling and abrasion. Cover stock can also come in a wide variety of finishes, textures, and colors. As the name suggests, its most common use is in bindings that protect text stock.

Card stock is yet another distinction that we can classify as falling between text stock and cover stock. Card stock is thicker than text stock, but not typically surfaced like cover stock, so it is used, as the name implies, for postcards, folders, bookmarks, reply cards, and so on.

The thickness and quality of paper stock can have a significant effect on users' perceptions of the document and the client it represents. A book covered with paper or card stock, instead of cover stock, for example, might not carry quite the same gravitas as one that fits our more common conceptions of a glossy, slick-feeling cover. At the same time, a pamphlet or zine meant for cheap printing and rapid distribution might feel overly formal if cover stock is used instead of paper stock throughout. As always, think through audience, purpose, and context when choosing your paper types.

CHARACTERISTICS OF PAPER

Text stock, cover stock, and card stock are pretty broad distinctions, so this section introduces some more specific terminology used to specify papers so you can accurately describe what you are looking for. Given the great variety of papers available, it's worthwhile to get samples before specifying a job—especially for high-design projects, which are more likely to require unusual papers. Printers also try not to keep large stocks of paper on hand, so they might need to order unusual stock from a wholesale paper supplier, which can take a day or two. Plan accordingly, and talk to your printer about paper early in the production process.

Paper for printing has several interacting factors that can affect specifications for print jobs:

- Size

- Grain

- Weight

- Thickness

- Opacity

- Finish

- Brightness

- Color

SIZE

The width and height of raw paper—the paper the printer must work from before it has been printed, trimmed, folded, or bound—can affect or even determine what sizes and shapes of documents we can design. Knowing the standard sizes of these sheets as they come from paper manufacturers will help you design documents to match these dimensions, resulting in jobs with less waste and accordingly lower costs.

Most document printing uses sheet-fed presses, which take up sheets of paper one at a time and send them through the print heads. Paper sheet sizes are specified by width and height—for example, 28" × 34". Either before or after printing, sheets can be trimmed on the edges, and larger sheets can be split into smaller sheets (see "Trimming and Bleeds" in this chapter).

In the United States, paper sheet stock comes in dozens of standard sizes. The most common US press sheet size used for offset printing typical documents such as manuals, pamphlets, brochures, and reports is probably 25" × 38", but this can differ significantly depending on the type of paper, the paper supplier, and the press capacity (some presses can accommodate bigger sheets than others). If this seems large, remember that commercial printers often print multiple page impressions at a time—a 25" × 38" press sheet allows for eight 8½" × 11" impressions to be made simultaneously (see figure 11.4).

Figure 11.4. Using large press sheets to print multiple copies of a document can save money and time by decreasing press run times. This sheet could be eight 8½" × 11" flyers, or eight brochure fronts, or something else entirely. The more efficiently you use your paper and press, the more money you save. *Source*: Created by the author.

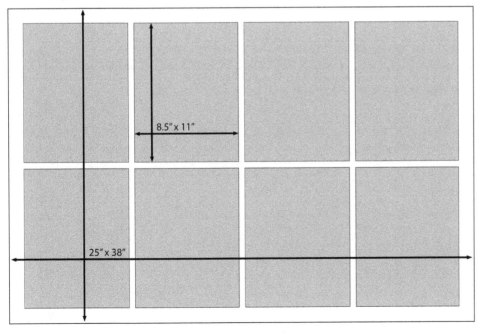

European Union countries use paper sizes based on International Standards Organization (ISO) standards, specified in millimeters. Especially common ISO sheet sizes are in the A-series standard, based on an A0 (A-zero) sheet (see figure 11.5):

- A0: 841 mm × 1189 mm ($33^1/_8$" × $46^{13}/_{16}$")

- A1: 594 mm × 841 mm ($23^3/_8$" × $33^1/_8$"), half of an A0 sheet

- A2: 420 mm × 594 mm ($16^1/_2$" × $23^3/_8$"), half of an A1 sheet, or one-quarter of an A0 sheet

- A3: 297 mm × 420 mm ($11^{11}/_{16}$" × $16^1/_2$"), half of an A2 sheet, or one-eighth of an A0 sheet

- A4: 210 mm × 297 mm ($8^1/_4$" × $11^{11}/_{16}$"), half of an A3 sheet, or one-sixteenth of an A0 sheet

An A4 sheet, which is commonly used for business correspondence documents in Europe, is near the size of a US $8^1/_2$" × 11" sheet—just about one-quarter inch narrower and one-quarter inch taller.

Figure 11.5. European paper sizes, based on the International Standards Organization (ISO) A0 standard. Each size sheet is half the size of the next larger sheet: A1 is half of A0, A2 is half of A1, and so on. *Source*: Created by the author.

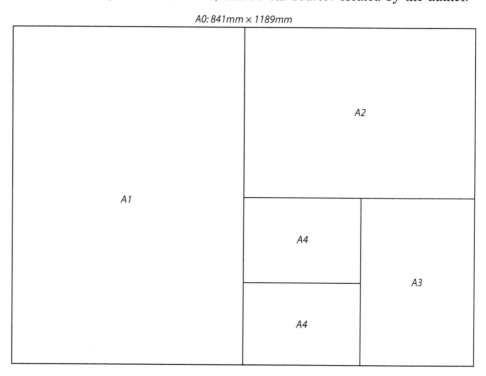

GRAIN

Paper is made of fibrous materials such as cotton, flax, or wood pulp. In the manufacturing process for most papers, these fibers line up and interlock in one direction, known as the *grain*. Paper folds and tears more easily with the grain (in line with the fibers) than against the grain (across the fibers). Accordingly, printers take grain into account for any document that will be folded or trimmed into smaller pages.

Paper is usually cut so the grain runs in the long dimension of the sheet. Two different conventions can give you clues about which direction the grain runs in a sheet. Most commonly, the second dimension expressed in a paper size specification is usually the direction of grain, so 28" × 34" indicates that the grain runs along the long dimension, and 34" × 28" indicates that grain runs in the short dimension. As a less ambiguous alternative to this convention, sometimes the grain direction is underlined (28" × 34").

Design Tip: Determining Paper Grain

You can recognize the grain in a sheet of paper in two ways:

- Fold the sheet gently in half without creasing it, first across its width and then across its height. Compare how stiff it is each way. The stiffer direction is against the grain because you're folding the fibers back on themselves.

- Tear a sheet. Paper tears straighter with the grain than against the grain.

WEIGHT

Two measurements of weight are common in printing: basis weight and grammage. The *basis weight* of paper is measured according to the weight in pounds of one ream (500 sheets) of the paper in its full sheet size as delivered by the manufacturer (i.e., before any trimming at the press). So calling a paper "50 lb. paper" means that one ream in its full sheet size weighs around fifty pounds. Basis weight, however, varies according to the paper size. A single sheet of 50 lb. paper in 25" × 38" would be a much "heavier" paper than a sheet of 50 lb. paper in 54" ×

77", since in the larger sheet size, the same weight would be stretched over a greater area.

Because basis weight can be ambiguous, paper manufacturers increasingly use *grammage* as a more consistent measurement of weight. Grammage is the paper's actual weight in grams per square meter (g/m2), even if the paper isn't available in sheets that size.

THICKNESS

Paper thickness is usually measured in thousandths of an inch. Thickness can affect both weight and opacity. But it's still an independent factor to consider, especially when specifying paper for a job that will involve folding because thicker papers fold less easily than thinner papers. Thicker papers *can* have greater strength in terms of stiffness and resistance to tearing than thinner papers, but this isn't always the case. Paper towels are very thick, but weak and easily torn, while the thin papers used in reference works such as dictionaries can be quite strong.

OPACITY

Opacity is a subjective measurement of the amount of light blocked by the paper. Thickness, weight, and finish can have an effect on how opaque a paper appears.

Opacity can be an especially important concern in designs that are printed on both sides of the paper or that require a lot of ink, which might show through a less opaque paper.

FINISH

Paper stock can come in a wide variety of *finishes*, which refers to the texture, smoothness, and hardness of the paper surface. The most common finish specification is whether the paper is coated or uncoated. *Uncoated papers* typically have a matte, rough, absorbent surface. Standard copy paper is uncoated, as is the paper used for most books. *Coated papers* are covered with a layer of material such as clay, so they are shiny, smooth, and nonabsorbent. Coated papers are specified as either *C1S* (coated one side) or *C2S* (coated two sides). Coated papers are often used for process printing because they reflect light well and make colors seem more vivid; magazines and "glossy" final reports typically use coated paper. However, coated papers typically cost more than uncoated papers.

More about . . . Paper Finishes

Specialty finishes in today's papers, such as *vellum*, *linen*, *laid*, and *wove*, originally referred to the material or the manufacturing process for making the sheets. Vellum, also called *parchment*, was actually a prepared sheet of animal hide rather than paper; it typically offered a shiny surface and came in thick, individual leaves. Linen is made from fibers of the flax plant. Most papers between the Middle Ages and 1840 were made of linen. Today's vellum or linen papers are usually made from wood or cotton pulp pressed against a mold to create a surface texture similar to the original.

Laid and *wove* also refer to the surface texture created by the mold used in papermaking. In most papers from the Middle Ages to the 1820s, papermaking molds were a grid of narrowly spaced horizontal wires held together at intervals by thicker vertical wires called "chains." The paper made in these molds was left with a pattern of narrowly spaced lines that typify laid paper. These patterns are artificially pressed into papers today to make them look old-fashioned.

In the late eighteenth century, high-end papermakers developed a smoother finish by placing a fine woven mesh over the wire and chains to create a relatively smooth "wove" surface. Almost all papers manufactured today are wove papers.

Coated and uncoated papers also take ink differently. Ink soaks into uncoated papers, which can lead to fuzzy impositions and bleed-through to the other side of the sheet. Conversely, ink tends to sit on top of coated papers. This keeps impositions crisp and reduces bleed-through but can lead to smudging. Inks can also have different apparent colors when printed on coated or uncoated papers. A spot color ink on an uncoated paper might look significantly different from the same ink on a coated paper, since the coating changes the reflectivity of the paper (see "Ink" in this chapter).

One compromise between coated and uncoated papers is to use *calendered* papers, which are uncoated but mechanically smoothed or *burnished*. Like coating, calendering hardens the surface of the sheet, which cuts down on bleed-through and allows sharper impressions.

Paper can also have a variety of specialty textures, such as *vellum*, *laid*, *wove*, or *linen*. These finishes replicate the surfaces of older sheet-making technologies, so they're often used for designs trying to invoke an antique or old-fashioned feel. These qualities and terms are not standardized, so one paper manufacturer's laid paper can have a significantly different texture than another's.

BRIGHTNESS

A related characteristic to color is brightness, which refers to the reflectivity of the paper. White papers tend to reflect the most light, but some white papers are brighter than others. High-visibility colors (sometimes called *neon*) can also have high brightness—as much as or more than some white papers.

Bright papers provide the greatest figure-ground contrast between paper and black text, but they can be difficult to read from because the reflected light can fatigue the reader's eyes.

COLOR

Paper comes in a wide variety of colors and consistencies. Even the standard "white" paper we usually see used for business documents can come in a variety of shades or tints of white. Paper pulp can be bleached to remove the natural color of the fibers or colored with dyes to add color. In addition, sometimes paper manufacturers add visual features to the paper, such as bits of unbleached or dyed fiber, other plant materials, plastics, or preprinted patterns.

Paper color can significantly affect the design and printing of documents. Because the CMY inks used in process printing are translucent, process printing on a colored or even off-white paper can alter the intended color. Highly figured, colored, or decoratively patterned papers can also lead to insufficient figure-ground contrast between the paper and printed text. Some highly decorated patterned papers, such as marbled papers, are not meant to be used for printing at all—just for bindings or covers.

INK

When it comes to moving from design to print, ink is just as important as paper. Lithographic inks in particular are worth knowing something about because there are so many to choose from.

More about . . . Ink

Without ink, there is no printing, and our modern inks give us truly staggering printing abilities. Even a low-end inkjet printer can replicate an impressive array of colors, and the inks that professional printers use allow us near endless possibilities. But the inks we use today are quite different from those used hundreds, even thousands, of years ago.

In his *Printing Ink: A History, With a Treatise on Modern Methods of Manufacture and Use*, published in 1926, Frank B. Wiborg wrote, "So rapid are the changes that take place in the art of printing and so much is demanded of ink, paper and other agencies that is novel and complicated, that a technical book five years old may be obsolete today" (p. xv). This holds true today. But some of our greatest histories of ink come from, well, our history, and one of the most thorough short lectures on the subject is Charles C. Pines's "The Story of Ink," reproduced in *The American Journal of Police Science* in 1931.

In his lecture, Pines notes that our modern word *ink* likely evolved from the Latin "encaustum," which referred to a burnt purple-red ink used by Roman emperors (p. 291). Ink itself, though, is older (Wiborg sets its invention at around AD 220). In general, ink consists of a pigment and a vehicle—a dye or other coloring agent in oil, fat, or something else. Some of our oldest inks consist solely of soot and oil. Inks can be taken directly from nature as well—squid ink is quite powerful, and Pines describes it as "the most lasting ink of all natural substances" (p. 294).

Inks have been—and are—derived from charcoal, mollusks (e.g., Tyrian purple), tannins (iron gall inks), chemical processes such as aniline dye (also known as coal-tar inks), and more. He even describes a "river of ink" in Algeria where a river carrying heavy iron deposits meets a river carrying tannin—containing gallic acid—from a peat swamp. Though there may or may not be an actual river of ink you can sample for your own writing (it's been contested for decades), you could certainly try writing a note to someone in onion juice or milk instead, which can then only be revealed through heating. Our natural world is full of compounds we can use for design.

Printer's inks, however, are where our sense of modern production and ink making coincide. As Pines writes, "Lampblack, a form of carbon, made by burning rosin, turpentine, pitch, and petroleum oils, was used in the first printing inks until 1864," at which time a more efficient carbon—carbon black—was introduced. And if you're wondering how efficient the inks used in the 1860s were, he adds, "Nine pounds of ink containing one pound of carbon black and eight pounds of oil and other materials will print ninety copies of a three hundred page octavobook" (p. 299), so roughly 27,000 pages of text. By comparison, in 2019, *PC Magazine* noted that 11 mL of ink would print roughly 220 8½" × 11" pages. So you would need about 123 11-mL cartridges to print—generally—that same 27,000 pages, not accounting for the differences in page size and paper composition. Ink weighs about 1 gram per mL, so 11 mL of ink is about 0.0242508 lb. That's about 3 lb. of modern ink needed to accomplish a comparable print run to Pines's 9 lb. Our modern inks are not only more flexible, they're lighter and easier to dispense and disperse, which gives us a lot more creativity in the how we design our printing technologies. After all, some of the original handpresses were so large and heavy that they had to be bolted to building frames to stand up to the pressures generated by the people running them, while our modern printers sit quietly and often unobtrusively on our desktops.

SPOT AND PROCESS INKS

Process inks are standardized translucent inks in cyan, magenta, and yellow, along with a key ink in a dense, opaque color—usually black. *Spot inks*, however, come in thousands of hues, tints, and shades, including metallic, pastel, and high-visibility inks for high-design or specialty print jobs.

As we mentioned in chapter 8, spot inks are usually specified using the Pantone Matching System (PMS). The most common reference tool for this system is a Pantone chip set, a series of sample color chips printed on strips of cover stock punched at one end and mounted on a ring so different chips can be held next to each other for comparison. The color chips are labeled with a Pantone number, usually in three or four digits. The chips are also printed on both uncoated and coated paper, since the paper finish can change the apparent color of the ink.

Design Tip: Up-to-Date Chip Sets

Always refer to a current chip set for final ink specifications. Chip sets lose their accuracy as they fade with time, so they're stamped with an expiration date. Using a current chip set will help you specify spot inks accurately and consistently.

New chip sets can be expensive, so having an old chip set for desktop reference can be very useful. Sometimes a printer will give you an expired chip set—if you ask nicely!

TOTAL INK

Four-color process printing overprints the same sheet with several different colors of ink—even more so if combined with additional spot colors. This overprinting risks adding so much ink to the page that the ink cannot adhere to the paper. In other words, there's a limit to the *total ink* that can be laid on a page without causing problems. One such problem is *picking*, where the ink laid down by one print plate (e.g., cyan) sticks to the ink of subsequent plates, ruining the print job. Too much total ink usually occurs in places where three or four inks are overprinted in relatively equal quantities—in other words, in areas that are meant to be printed in gray.

But if the viewer is going to see gray anyway, why not just use black ink screened to gray? This idea is the basis of two techniques that address the problem of too much total ink: *gray component replacement (GCR)* and *under color removal (UCR)*. GCR replaces CMY with screened black in areas where CMY would be overprinted in equal strengths. In other words, GCR replaces three overprinted inks with one ink, decreasing the ink density on the page. Rather than completely replacing CMY with screened black in gray areas, UCR simply removes one process ink and adds black to make up part of the difference. UCR retains a subtle hue in grays and shadows, while still reducing total ink. Both GCR and UCR also have the advantage of increasing the apparent saturation of images, particularly photographs. You can specify GCR and UCR settings in Photoshop, InDesign, and similar programs.

VARNISHING

Some projects require paper with special durability—for example, on the cover of a document that will get a lot of abuse. In those cases, you can specify *varnishing*, a clear lacquer or plastic coating that's added to the printed sheet like a final layer of ink.

Varnishing prevents smudging and makes the sheet less sensitive to abrasion or wear. The varnishing material can delaminate or peel away from the paper, however, so its adhesion to the substrate paper and ink must be taken into account. It also adds to the cost of the print job.

FROM DESIGN TO DOCUMENT

In this section, we'll discuss factors you need to understand about how printers get from document designs to finished documents. First, we'll discuss how items are arranged for printing (through imposition) and the issues one must consider when planning for printing (like numbers of pages and signature foldings). Then we'll move to how printed sheets are folded and assembled into documents.

PLANNING FOR PRINTING: IMPOSITION AND SIGNATURES

While we usually encounter finished documents in single pages printed on what seem like individual sheets of paper, most documents are printed on much larger sheets of paper, with several pages arranged on both sides of each sheet. To make this work, the printer must arrange the pages so that they end up in the correct order when the sheet is folded and trimmed.

This arrangement of pages on the front and back of a sheet is called *imposition*. After printing, the large sheets are folded to make up *signatures* (an old term from the early years of printing when each folded unit was marked, or "signed," with a letter of the alphabet to indicate its place in the assembled book). Each signature, when trimmed, produces a group of individual *leaves*, and each leaf has two pages: the front page, which is known as the *recto* of the leaf, and the back page, which is called the *verso*. The most common signature foldings are *folio*, *quarto*, and *octavo* (see figure 11.6).

- A *folio* is folded only *once*—for example, for a short four-page/two-leaf newsletter. The outside pages 1 and 4 will be printed on one side of the sheet, and the inside pages 2 and 3 will be printed on the inside.

- A *quarto* is folded *twice*—once in half to make a folio, and then in half again across the first fold. A quarto can carry eight pages and four leaves, twice that of a folio. Pages 1, 4, 5, and 8 are printed on one side of the sheet, and 2, 3, 6, and 7 on the other.

- An *octavo* is folded *three times*, by folding a quarto in half again across the second fold. An octavo can hold sixteen pages on eight leaves. For example, in the most common octavo impositions, pages 1, 4, 5, 8, 9, 12, 13, and 16 are printed on one side, and pages 2, 3, 6, 7, 10, 11, 14, and 15 are printed on the other.

Figure 11.6. Folio, quarto, and octavo impositions. Sheets printed in these impositions are folded and trimmed to create final pages. *Source*: Created by the author.

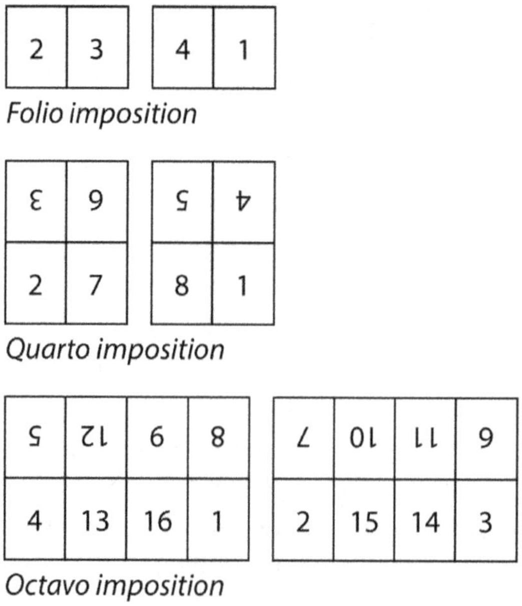

Folio imposition

Quarto imposition

Octavo imposition

Although people often think of folios as large documents and octavos as small ones, these terms have nothing to do with the size of the finished document. Depending on the size of the paper you start with, you can design a small folio or a large octavo: these terms refer only to how the sheet is folded.

You probably won't need to specify impositions and signatures yourself; the print shop will take care of this for you. But knowing about impositions and signatures is important because it affects how many pages can fit in a document most efficiently.

Signatures gather pages in four-, eight-, or sixteen-page groups, so most documents are designed to be printed in some combination or multiples of those numbers. For example, a forty-eight-page document can be printed in three sixteen-page octavo signatures ($3 \times 16 = 48$) or six eight-page quarto signatures ($6 \times 8 = 48$). Combinations of impositions are possible as well: a fifty-two-page document can be printed in three octavo signatures plus one folio signature, although the additional odd signature would increase the cost of printing.

More about . . . Design in Fours, Eights, or Sixteens

Design documents so they will be paginated in multiples of four, eight, or sixteen to maximize efficient production and reduce printing costs.

Printing documents that don't match up easily with signatures is more difficult and expensive. For example, it's complicated to print and bind a forty-nine-page document. With forty-seven pages, you could print a three-signature forty-eight-page octavo and leave a blank page—perhaps the verso of the first leaf. But a forty-nine-page document would require printing at least fifty-two pages and leaving three blank pages, which could look awkward and cost more. The only other alternative is to *tip-in* a single leaf that is printed on just one side and glue its edge into the spine somewhere in the document or to leave a tag on the spine side where it can be sewn in with the rest of the signatures. But tipping-in is labor-intensive and expensive. It's better to just design pages in multiples of fours, eights, or sixteens from the beginning.

FINISHING

The final steps in the printing process are called *finishing*. These include folding, scoring, perforating, gathering, trimming, and binding (also known as stitching), and they all usually happen at one stage in the press facility. Today, it's increasingly common for one machine to perform all of these processes automatically, accepting printed sheets at one end and producing nearly finished documents at the other end. You should specify how you want your document to be finished when discussing the print job with the printer.

SCORING AND PERFORATING

Scoring and perforating are sheet processes that can be useful in some applications. Scoring puts a crease in a sheet without actually folding it. The cover of a perfect-bound (soft-cover) book or report is often scored about a quarter of an inch from the spine, making the cover easier to bend without breaking the spine. Perforating allows readers to tear out a sheet or part of a sheet easily. It's often used to encourage users to tear out, fill out, and return a sheet or card, such as a survey, subscription card, or RSVP.

FOLDING AND GATHERING

We've already discussed folding as it applies to large printed sheets that are folded into signatures. Most print shops have machines that fold and *gather* signatures automatically, placing them in the appropriate order for the final document.

TRIMMING AND BLEEDS

Trimming is performed with an electrical or hydraulic shear called a guillotine. Most guillotines can cut through between 500 and 1,000 sheets at one stroke, depending on the paper stock. Paper might be trimmed before printing to set it up for the press, but whole documents can also be trimmed after printing and before or after binding.

Because a signature results from one large sheet being folded to make a gathering of leaves, in most cases, the folds must be trimmed to open up the leaves. Trimming allows the leaves to separate but remain attached at the spine. A folio requires no cutting because its leaves are joined only at the one spine fold. A quarto is trimmed only on the *top-edge*. An octavo is trimmed on both the top-edge and the *fore-edge* (the edge opposite the spine). Although only these trims are necessary to release the leaves, sometimes the document will be trimmed on all edges except the spine, simply to square up the signatures before binding.

A final trimming issue arises when we want to print sheets to the edge of the page—for example, the cover of an annual report, a four-color flyer, or a quick reference card. This kind of printing is called a *bleed*. Printing presses require a small amount of paper they can grab, called a *gripper edge*, to send the sheet through the press. Presses can't print on the gripper edge, which is trimmed away at the end of the printing process (figure 11.7).

As a result, printers must print bleeds on an oversized sheet and then trim the excess to make the sheet appear to have been printed on its entire surface. For example, printing an 8½" × 11" flyer might require designing for a 9" × 11½" sheet, leaving a narrow margin around the sheet that will be trimmed off so the ink will appear to go to the edge of the sheet. In these cases, designers use page layout software to place trim marks outside the printed area to indicate where the paper should be trimmed.

As we discussed in chapter 4, you can also design for special edge trimming in more complex patterns or shapes—anything from tabs on the fore-edge to rounded corners. This kind of trimming is usually called *die cutting* because it requires the use of a cutting head called a *die* to make the special trim.

Figure 11.7. Gripper edge and trim marks. The printing press grabs the gripper edge to pull the sheet through the press. The gripper edge is later trimmed off at the trim marks to leave the ink bleeding to the edge of the page. *Source*: Created by the author.

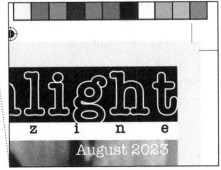

Detail of trim marks and bleed

BINDING

After folding, signatures are bound (again, you might want to look back at chapter 4). A single quarto or octavo signature might be *stitched* (stapled) at the spine to make a booklet, or multiple signatures can be glued or *sewn* together into a *codex* (a book). Documents printed in multiple signatures are typically bound rather than stitched. Although new binding technologies are constantly being developed, the most common book bindings today are *case* and *perfect*:

- *Case bindings:* Signatures are sewn or glued together at the spine, trimmed, and then pasted into a hard-cover case made up of the front and back covers and the spine.

- *Perfect bindings:* Single sheets or signatures are glued into a soft-cover, square-backed binding made of cover stock. Perfect bindings are trimmed after the cover is attached.

In addition, a variety of techniques can bind individual sheets (rather than signatures of multiple pages). The most common are the following:

- *Wire bindings:* Individual sheets are drilled with small holes.

- *Comb bindings:* Individual sheets are punched with a row of rectangular slots near the spine edge. With a special tool, the plastic comb is held open and its teeth are inserted into the slots. When released from the tool, the comb closes and holds the sheets.

There are also many proprietary binding technologies you might consider; discuss options with your printer.

PREPARING DESIGNS FOR THE PRESS

COMMUNICATING WITH PRINTERS

Document designers sometimes get anxious when it is time to hand over a document design to the printer. Up to this point, you've enjoyed a lot of control over the design process. But once your design is in the print shop, the control shifts to a team of people with different priorities and pressures. Printers want to be your partner in creating a great document; one of their goals is to make you happy so you will bring more work to them. But it's easy to allow misunderstandings and mixed signals to undercut the smoothness of the experience or the quality of the finished product.

The key to avoiding problems is to communicate *early*, *frequently*, and *clearly* with your printer:

- *Communicate early:* Contact the print shop several weeks before you intend to deliver a job to them. They have other jobs to do, and yours must fit into their production schedule. The more complex the job, the earlier you should talk to the printer.

- *Communicate frequently:* Both before and during the printing process, check in with the print shop regularly—first to keep them apprised of your progress in getting the project into their hands on schedule and then to confirm and follow up on your specifications about the job.

- *Communicate clearly:* Communicate what you want *in writing*—email is often fine—and ask for a written quote that specifies everything the printer will do. Be as specific as possible in describing what you want; printers can give you only what you ask for. Also try to use correct printing terminology. It's easy to dismiss this terminology as jargon, but think of it as a specialized vocabulary that allows for precise communication about a complex process.

Your printer has as much at stake in doing a good job as you do, so don't be afraid to ask questions if you don't understand something or need clarification. Printers are usually very happy to take the time to talk to you, since clarifying misunderstandings early saves time and money. If your printer won't talk to you or answer your questions, find another printer.

More about . . . Printing Specification Checklist

Communicate the following information (as applicable) to the print shop so they can prepare your estimate:

- Your contact information
- Requested delivery dates
- Quantity of finished pieces
- Finished size
- Folds
- Number of inks on each side
- Types of paper for cover and interior
- Bleeds
- Finishing and binding specifications
- Mailing or delivery needs
- Billing information

GETTING ESTIMATES

After you've communicated the specifics of the print job to the print shop, they will give you a formal estimate of the printing costs. This estimate will be based primarily on the specifics of the job, such as the number and type of pieces to be printed, the number of colors, finishing, binding, paper, and inks. If you have time and some flexibility, you can get competitive estimates from several print shops.

The estimate will usually include a charge for *make-ready*. As we discussed earlier, this term describes the waste sheets that the printer must use to set up the press before printing the actual run of your document. Some jobs require more make-ready than others, but setting up the press can increase the amount of paper the job requires by 10 to 15 percent.

If you aren't happy with the estimate, negotiate changes as soon as possible. You may want to change some of the specifications of the job to keep costs down. Ask the print shop for advice on ways to economize. They can tell you exactly where the unexpected costs are coming from.

DELIVERING YOUR DESIGN TO THE PRINTER

This may sound very obvious, but make sure the materials you deliver to be printed are as correct and accurate as possible. After a certain point, mistakes in print jobs cannot be fixed without considerable added expense. If mistakes do arise and they are your fault, be prepared to pay extra to have them fixed. Sometimes an entire print run must be discarded and reprinted from scratch because of embarrassing mistakes or legal reasons. This effectively doubles the cost of printing.

Most print shops can give you detailed specifications for how they want you to deliver your design. These commonly include the page layout file formats the print shop will accept, the storage media they can access, and how to manage associated image files and fonts.

FILE FORMATS

Print shops often prefer to receive page layouts as Adobe Acrobat files (PDFs), since these are common file types that can carry all of the page description and fonts. PDFs work particularly seamlessly with digital printing technologies. The print shop can make changes or adjustments to a PDF including correcting misspellings, replacing pages, and more, but it is your responsibility to deliver a working design. Remember, you are the designer, not your print shop (unless you've contracted them as such) so the responsibility of design—for better or for worse—is on you.

If you're not comfortable with that level of responsibility, ask if you can deliver your print jobs to the print shop as an application file. Most print shops readily accept a variety of page layout file formats, including those of the page layout software industry leaders, Adobe and Quark. For Quark, the file extensions are typically either QXD (QuarkXPress Document) or QXP (QuarkXPress Project). For Adobe, the file extensions can be IND (InDesign), FM (FrameMaker), PSD (Photoshop), AI (Illustrator), and more. Always check with your print shop to see what technologies they can work with. If there are small problems with the design, such as out-of-gamut colors or missing fonts, the print shop will be able to correct the problem (although big problems will probably incur an additional fee).

Few print shops like to work with home computer applications, such as Microsoft Publisher or Microsoft Word. These applications create severe limitations for preparing the design for printing. If at all possible, avoid using these applications for jobs that will be professionally printed.

LINKED AND EMBEDDED OBJECTS

Print shops will need not only your page layout files but also all of the associated files you used to make it, including image files and sometimes computer font files. The two ways to give the print shop all the files they need to create your document are linking and embedding.

With *embedding*, you actually insert the images in the page layout file. This makes a neat and tidy package, but it can create monstrously huge files. With *linking*, you simply insert a reference to the image files in the page layout file. When the page layout program renders the design on-screen, it retrieves these separate files and shows them where indicated. The advantage is that the page layout file stays much smaller. The disadvantage is that if the linked files are on another hard drive or computer, the page layout program can't access them.

Whether you link or embed, give the print shop all of the image files you used in the design, just in case the printer needs them to make a last-minute repair or modification. You will usually want to provide the files in TIFF or high-quality JPG format. Again, though, find out what your printer needs.

Be sure that you have used the CMYK mode in preparing your images, not RGB! The printer can replicate only CMYK colors; any RGB colors you specify might fall outside of the CMYK gamut. (See chapter 8 for a discussion of color gamut.)

It's also a good idea to send along any unusual computer font files. Print shops usually have large libraries of font files, so they don't always need you to supply the font file with your print job, and most can download any proprietary fonts you have included. But if you specify an uncommon typeface, you might need to provide its font file to the print shop so they can load it temporarily on their system and print the job, and if you had to purchase the font for the project, your printer will need it to complete the job. Be sure to check out the terms of use, however—some typographers disallow copying their font files, even temporarily.

DRAFTS AND MOCK-UPS

Finally, provide a good-quality printed draft of your document. Printers will use this printed version as a reference when they run across questions in the process of preparing your files for print. Make sure to print out the most recent version from the files you provide. That way you can be sure that the printer's reference (your printout) is an exact replica. If there are errors in the printer's version of the document, the printer will often refer to your hardcopy version to see if the errors are yours or theirs.

If your design involves any complex folds, die cuts, or other special features, include a *mock-up* of the design. To create a mock-up, use paper, scissors, tape, and glue to prepare a rough version of the finished document, including all of the folds and special features. This mock-up will give the printer a clear idea of what you have in mind for the design.

STORAGE MEDIA

Ask your print shop what media you can use to deliver your work to them. With our often-bewildering array of file distribution technologies, both physical and online, you need to be sure that your printer can actually access your files. There's no point in sharing the files with a link to a service the printer doesn't pay for and so can't access, or via physical technologies that might not interface with the computers they use.

RESPONDING TO PROOFS

When you deliver your document to the printer, confirm the expected production schedule, noting especially when you can expect to receive correction copies, or *proofs*, and the expected turnaround for returning those proofs to the press.

Most print shops will send you two sets of proofs: *first proofs* and *final proofs*. Make sure that you understand how you will receive the proofs (electronically or physically) and where.

More about . . . Reviewing Proofs

When reading proofs, do both linear proofing and layered proofing. For linear proofing, simply read the document all the way through with a fresh eye. Look especially for the following problems during linear proofing:

- Spelling errors (especially in proper names)
- Grammar errors
- Libelous statements
- Copyright infringement
- Errors in paragraph styles (sometimes the wrong style will get applied to a paragraph)

It's also a good idea to ask a colleague to do a linear proof. Someone who has not been intimately involved with the document might see problems that your eyes missed.

For layered proofing, go through the document again, one design element at a time:

- Titles
- Headers
- Footers
- Headings
- Art
- Captions
- Sidebars
- Pull quotes
- Front matter (table of contents, list of illustrations, publication information, etc.)
- Back matter (appendices, index)
- Covers

During layered proofing, look not only at the correctness of all of these elements but also at their visual consistency. Each first-level heading should look like all the other first-level headings, all the captions should look the same, and so on. Be especially watchful for obvious errors. Sometimes the bigger the font size, the easier it is to miss a mistake.

MAKING CORRECTIONS

If you find errors made by the printer, the print shop will typically correct those without additional costs to you (although you should verify this during early discussions about the job). But most likely, any mistakes you find on the proofs will arise from the original material you supplied to the printer. In that case, you must pay for the corrections.

What you can change in proofs is very limited. Whatever changes you specify must be performed by the print shop personnel rather than by you, which makes changes expensive. In general, stick to correcting only factual errors, misspellings or grammatical errors, and statements that might have legal consequences, such as libel or copyright issues. Anything else that just doesn't appeal to you (clumsy sentences, bad line breaks, widows [single words or lines at the top of a page or column], orphans [single words or lines at the bottom of a page or column], unattractive typefaces, etc.) will have to remain if you wish to maintain your budget and the agreed-upon delivery date. If absolutely necessary, you might use *copyfitting*, the technique of changing out words or phrases without changing the length of the line or breaking onto the next line of the paragraph. Copyfitting can be dangerous, though, as in most cases the copy you are working with has already been vetted, edited, and typeset.

Any changes that affect the page layout of the document will be especially expensive. So avoid making any changes that will change the length of paragraphs or pages or the size of graphics. Adding a single word can sometimes make a paragraph one line longer—enough that it won't fit on the page, throwing off everything in subsequent pages of the document. And changing the size of a graphic can affect text wrapping, which can have a ripple effect on subsequent paragraphs or pages. If you are working with content contributors who need to proofread the portions of the document they wrote, specify *exactly* what kind of changes they can and cannot make, as well as your deadline for receiving their corrections. Remind your contributors not to make stylistic revisions, deletions, or additions to the design or text of the document at this point. Set the contributors' deadline at least one day before the printer's deadline so you'll have time to collate their changes into a single set of instructions for the printer.

When marking proofs, use standard proofreading marks. You can find these in any style manual, such as *The Chicago Manual of Style*. In addition, follow the printer's guidelines for how to indicate corrections. Some printers prefer marked proofs, while others prefer a list itemized by page, paragraph, and line number. If you have questions about their preferences, ask if they can provide an example of the type of responses they prefer.

When you return your proofs, get a receipt or some other documentation. A simple email message is sufficient, but it's a good idea to be able to confirm your date and time submission if it becomes necessary later to retain your place in the production schedule.

CONCLUSION

Sending a job to a commercial printer might seem intimidating. But remember that print shops are as interested in producing an excellent finished product as you are. Work professionally and responsibly with commercial printers, and they'll return the favor by bringing your designs successfully into existence.

EXERCISES

In this chapter you've thought about what happens to your document as it moves from screen to page. You've considered printing technologies, and thought about inks, papers, and binding. Finally, you've read about how you might actually communicate all of that stuff to the printer you are working with, and a bit about how you'll need to proofread your work. These exercises provided here should help you further work through what you just read.

1. Go on a tour of a commercial print shop. Most large print shops will be happy to provide a short tour of their operations to potential customers and especially to students who might be customers for a long time to come. If you don't have a print shop near you, search for virtual tours online—there are quite a few out there where shop owners take viewers on a virtual tour through their workspace. Once you've seen everything in person (or virtually), talk about how what you've seen relates to what you've read. Does CMYK make more sense after seeing printers with huge tanks of those inks? What about paper types, or even offset lithography? What does seeing the physical space help with?

2. Talk to a commercial print shop operator (either in conjunction with a print shop tour or at a separate time). Consider asking these questions:

   o What are the three most essential things I should do as a designer to make the production process run smoothly?

        o What three things should I most avoid doing?

        o What are your preferences for receiving print jobs?

        o How would you describe a successful client–print shop relationship?

        Write a short report describing what you learned from this conversation.

3. Find a fairly complex print document. Imagine that you are the designer of this document and are just about to send the design for commercial printing. Using the printing terminology discussed in this chapter and others, write a specifications list that describes this document to a commercial printer and communicates what the finished product should be like.

REFERENCES AND FURTHER READING

Gaskell, P. (1995). *A new introduction to bibliography*. St. Paul's Bibliographies.

Gatter, M. (2005). *Getting it right in print: Digital prepress for graphic designers*. Harry N. Abrams.

Griffiths, A. (1996). *Prints and printmaking: An introduction to the history and techniques*. University of California Press.

Hewlett-Packard, Inc. (2016). *HP Indigo digital offset color technology*. https://objects.icecat.biz/objects/mmo_50774275_1540214855_0438_21400.pdf

International Paper. (2003) *Pocket pal: The handy little book of graphic arts production* (19th edition). International Paper.

Pines, C. C. (1931). The story of ink. *American Journal of Police Science*, 2(4), 290–301.

Robinson, S. (2014). *The book in society: An introduction to print culture*. Broadview Press.

Sanders, Linda. (1991). *47 printing headaches (and how to avoid them): How to prevent costly printing mistakes—a solution book for designers, production artists, and desktop publishers*. North Light Books.

Schriver, K. (1997). *Dynamics in document design*. John Wiley and Sons.

University of Chicago Press Staff. (2017). *The Chicago Manual of Style* (17th edition). University of Chicago Press.

Wiborg, F. B. (1926). *Printing ink: A history, with a treatise on modern methods of manufacture and use*. Harper & Brothers.

INDEX